THE CHEMISTRY OF GOLD

TOPICS IN INORGANIC AND GENERAL CHEMISTRY

A COLLECTION OF MONOGRAPHS EDITED BY

R. J. H. CLARK

Department of Chemistry, University College London, 20 Gordon Street, London WC1H 0AJ, England

MONOGRAPH 16

Other titles in the collection:

1 The Chemistry of Beryllium, by D. A. Everest (out of print)

2 Inorganic Ion Exchangers, by C. B. Amphlett (out of print)

3 Introduction to the Atomic Nucleus, by J. G. Cuninghame (out of print)

4 The Chemistry of the Rare-Earth Elements, by N. E. Topp (out of print)

5 The Chemistry of Gallium, by I. A. Sheka, I. S. Chaus and T. T. Mityureva (out of print)

6 The Chemistry of Technetium and Rhenium, by R. D. Peacock (out of print)

7 The Chemistry of Selenium, Tellurium and Polonium, by K. W. Bagnall (out of print)

8 Interactions in Electrolyte Solutions, by G. H. Nancollas

9 Plasma Chemistry in Electrical Discharges, by F. K. McTaggart

10 The Chemistry of Niobium and Tantalum, by F. Fairbrother (out of print)

11 The Chemistry of Titanium and Vanadium, by R. J. H. Clark

12 Active Carbon, by M. Smíšek and S. Černý (out of print)

13 The Chemistry of Uranium (Including Its Applications in Nuclear Technology), by E. H. P. Cordfunke (out of print)

14 The Chemistry of Thallium, by A. G. Lee (available from Applied Science Publishers, Barking, England)

15 The Actinide Elements, by K. W. Bagnall (out of print)

(Volumes 1—15 were edited by P. L. Robinson)

THE CHEMISTRY OF GOLD

RICHARD J. PUDDEPHATT

Senior Lecturer,
Department of Inorganic, Physical and Industrial Chemistry,
The University of Liverpool,
Grove St., P.O. Box 147,
Liverpool L69 3BX, England

ELSEVIER SCIENTIFIC PUBLISHING COMPANY

Amsterdam — Oxford — New York
1978

ELSEVIER SCIENTIFIC PUBLISHING COMPANY
335 Jan van Galenstraat
P.O. Box 211, 1000 AE Amsterdam, The Netherlands

Distributors for the United States and Canada:

ELSEVIER/NORTH-HOLLAND INC.
52, Vanderbilt Avenue
New York, N.Y. 10017

First edition 1978
Second impression 1980

Library of Congress Cataloging in Publication Data

Puddephatt, R. J.
The chemistry of gold.

(Topics in inorganic and general chemistry ; 16)
Includes bibliographical references and index.
1. Gold. I. Title. II. Series.
QD181.A9P8 546'.656 78-1627
ISBN 0-444-41624-2

ISBN 0-444-41624-2 (Vol. 16)
ISBN 0-444-41626-9 (Series)

Printed in The Netherlands

PREFACE

Metallic gold has been most highly valued by man since the earliest times but, since the demise of the alchemists, the chemical compounds of gold have not been studied as closely as those of other rare metals. This is perhaps partly due to the ease with which many gold compounds decompose back to the noble metal, but also due to the limited use of gold other than as a monetary standard and for jewellery and other decorative uses. Gold and its compounds have been regarded as the least useful of the noble metals as catalysts, but recent progress suggests that this situation may change significantly in the near future. Applications of gold based on its physical and chemical rather than its decorative properties account for a larger proportion of the total gold used each year. For example, the use of gold in micro-electronic circuits is increasing rapidly, and there is increasing interest in the development of new gold complexes for use in the treatment of arthritis.

Along with this expanding use of gold, there has been an impressive series of advances in the chemistry of gold compounds. For many years, the chemistry was dominated by the study of linear two-co-ordinate gold(I) complexes and square-planar four-co-ordinate gold(III) complexes. Recently, however, compounds of gold(I) and gold(III) in which the metal has higher co-ordination numbers have been studied and a new chemistry of gold(II) and gold(V) has been developed. Advances have also been made in understanding the nature of compounds with gold—metal bonds, and, in addition, exciting cluster compounds of gold have been prepared and characterised. Organogold chemistry has seen many advances, ranging from the synthesis of unstabilised alkylgold species like $[AuMe_2]^-$ and $[AuMe_4]^-$, to the preparation of carbene, alkene and alkyne complexes of gold. Many reactions are now better understood as a result of mechanistic studies, and structure determinations have been aided by the routine application of sophisticated spectroscopic techniques.

Some of the aspects discussed above have been the subjects

of excellent review articles, but, at a time when gold chemistry is undergoing an impressive renaissance, there is a clear need for a monograph covering all aspects of gold chemistry. It is hoped that this book will fill this gap and that it may stimulate further research, for there is still a great deal to be learned about gold and its compounds.

R. J. Puddephatt

CONTENTS

Chapter 3. *Gold(I) Complexes*

Chapter 4. *Gold(II) Complexes*

Chapter 5. *Gold(III) Complexes*

Chapter 6. *Gold(V) Complexes*

Chapter 7. *Organogold Complexes*

CHAPTER 1

INTRODUCTION TO THE CHEMISTRY OF GOLD

In this chapter we shall consider first the occurrence of gold and the ways used to extract and purify the element from its ores. After this, the properties of elemental gold will be covered, followed by a preliminary survey of the oxidation states and stereochemistries adopted by gold in its complexes. Finally, some general concepts which are useful in interpreting and rationalising the properties of gold and its complexes will be introduced. It is hoped that, using this general framework, the reader will be able to place the detailed properties of individual gold complexes, which are treated in subsequent chapters, into their proper perspective.

OCCURRENCE OF GOLD

From studies of the solar spectrum the abundance of gold in the sun is found to be 0.04 parts per million (ppm), but in the Earth's crust it is on average about 0.004 ppm. Thus, gold is probably more abundant in the core than in the crust, where concentration by a factor of about 10^3 is necessary before economic extraction is feasible.

Concentration of gold, which usually occurs in nature in the metallic form, has occurred in two ways. The first gives rise to alluvial or placer gold, and consists of the weathering of auriferous rocks which have been washed into river beds or similar situations. The greater specific gravity of gold (19.3) compared with residual rock (~ 2.7) leads to gold settling out while lighter rocks are washed away. Alluvial gold, once discovered, is often easy to extract as nuggets or grains by simple gravity concentration and the sources which gave rise to the great gold-rushes of California in 1848, Australia in 1850 and the Yukon in

1896 were of this form. Much of the gold found in the Ural mountains of the U.S.S.R. is also alluvial gold.

The greatest gold fields are those in South Africa, where the gold occurs as thin veins in quartz rocks. The gold, known as vein or reef gold, is present as microscopic particles so that extraction is considerably more difficult than for alluvial gold. In general reef gold is found in quartz or albite rocks, often along with iron pyrites. Deposits with silver and other rare metals often occur in volcanic regions controlled by major fault zones. Here, concentration of the gold takes place by hydrothermal metamorphism from basic rocks and deposition in sedimentary rocks [1, 2]. Gold is considered to be an inert metal but it can be dissolved quite readily by halide or sulphide ions in the presence of oxidising agents to give gold(III) or gold(I) complexes. It is probably in this way that gold is dissolved when hot volcanic rock is buried or when a hot granite intrusion rises near to the surface of the Earth's crust. Studies of model systems suggest that if the oxygen concentration is governed by the equilibrium

$$3Fe_2O_3 \rightleftharpoons 2Fe_3O_4 + \tfrac{1}{2}O_2$$

and the chloride concentration and pH by the mica—feldspar—quartz buffer system, then at the high pressures and temperatures of 400–600 °C which are likely, gold would dissolve *via* reactions of the type:

$$2Au + \tfrac{3}{2}O_2 + 12HCl \rightleftharpoons 2H_3AuCl_6 + 3H_2O^*$$

As the solution cools to 300—400 °C the concentrations of oxygen and of HCl are expected to decrease sharply, and gold is redeposited. The nature of the deposit will obviously depend on the precise conditions, and on the nature of the convection patterns or percolation routes of the cooling solutions. Hydrothermal transfer of gold as the complex ion $[Au(SH)_2]^-$ may occur in some cases [2b].

---

* There is no direct evidence for the existence of H_3AuCl_6. $H[AuCl_4]$ is perhaps more likely.

Dissolution and redeposition of gold in stream beds may also be responsible for the formation of large crystals of alluvial gold. Particularly interesting is the observation that some micro-organisms and their metabolites isolated from gold-bearing deposits are capable of dissolving gold. The maximum dissolution of gold by isolated bacterial metabolites studied so far is 10—35 g m^{-3} in 20 days. Dissolution is aided by the presence of aspartic acid, histidine, serine, alanine and glycine and by metal oxidants [3].

The only compounds of gold which form important minerals are the tellurides. Examples include calaverite and krennerite, which are different crystalline forms of $AuTe_2$, montbrayite, Au_2Te_3, and mixed gold—silver tellurides such as sylvanite, $AuAgTe_4$. Native gold also contains silver in varying proportions, one well-known example being electrum with about 20% silver. Other valuable elements obtained as minor constituents from gold ores are iridium, osmium, platinum, rhodium, ruthenium and palladium as well as copper and nickel.

The gold fields of South Africa have so far yielded over 31 million kg of gold, which is more than half the gold reserves of the world. The second largest producer is the U.S.S.R. but the precise amounts mined there are not published. No other country produces comparable quantities of gold, but Canada, the U.S.A., Australia and other African and South American countries still produce significant amounts.

GOLD IN SEA WATER

The presence of dissolved gold in sea water was discovered in 1872, and many uniformly unsuccessful attempts have been made to recover the gold commercially. The most famous attempt was made by Fritz Haber, with the intention of paying off the German war debt, in the years 1920—1927. The method was based on reduction to metallic gold using sodium polysulphide, but the available estimates of the gold concentration in sea water were too high by a factor of about 10^3 and the results were all disappointing. Most other methods rely on adsorption of gold on various media, including recently polymers containing thiol groups which complex strongly to gold. Some of these methods are useful in concentrating gold, so that accurate analyses can be performed, and it is now thought that the concentration of gold in sea water is about 2×10^{-11} M, though the concentration fluctuates depending on the geographic region and depth from which samples are collected [5, 6]. At these low concentrations, it is not possible to determine directly the nature of the gold species present, but calculations based on redox potentials of gold compounds and the composition of sea water suggest that the gold(I) complex ion $[AuCl_2]^-$ predominates, with smaller amounts of $[AuClBr]^-$ and other bromo, iodo and hydroxo complex ions of gold(I) [7] also being present.

RECOVERY AND PURIFICATION OF GOLD [8, 9]

Recovery of gold from alluvial deposits by gravity concentration dates back at least 6,000 years. The gold-containing material was conveyed by water down a sluice equipped with transverse riffles which trapped the heavier gold particles. It is thought that the use of animal hides in this process gave rise to the legend of the Golden Fleece of Jason and the Argonauts. An improvement to the method was developed which involved the use of mercury in the riffles to amalgamate the settling gold particles, the gold being recovered by distilling off the mercury in retorts and the gold finally smelted.

These methods are unsuitable for high yield recovery of reef gold in which the gold concentration may be only 10 ppm or less, and the following is a typical modern procedure. The ore is first crushed and then milled to give the fine particles needed for efficient extraction. Part of the gold is still recovered by the gravity concentration, amalgamation, retorting and smelting procedures which have been much improved by modern technology. The tailings from this process are not discarded however, but are treated with alkali cyanide and compressed air to dissolve the gold according to the equation:

$$4Au + 8CN^- + O_2 + 2H_2O \rightarrow 4[Au(CN)_2]^- + 4OH^-$$

The solution is filtered to remove the gangue and the gold is reprecipitated with zinc:

$$2[Au(CN)_2]^- + Zn \rightarrow [Zn(CN)_4]^{2-} + 2Au$$

Recovery of gold with an efficiency of around 98% can be achieved using this combined method, with the product from the final smelting process containing 88—90% gold. The residual constituents are silver (~10%) and small amounts of copper, lead, zinc, iron and platinum group metals.

Most gold is purified by the chlorination process. Chlorine gas is bubbled through the molten gold and this converts silver and base metals to the chlorides which float to the surface and can be removed easily. Gold of purity 99.5% is produced in this way and, if necessary, is further refined by electrolysis. The anode is the crude gold, the electrolyte is $H[AuCl_4]$ in hydrochloric acid, and the cathode is pure gold. Pure gold is deposited at the cathode, while platinum and palladium can be recovered from the anode mud.

A convenient method for the recovery of gold from gold residues in the laboratory consists of pyrolysis of the residues, followed by dissolution of the gold in *aqua regia.* The pH is adjusted to pH 2—4 by adding solid sodium hydroxide and the gold is reprecipitated using sodium sulphite as reducing agent. A similar procedure can be used to obtain gold powder from massive gold.

PROPERTIES OF THE ELEMENT

Gold is perhaps the most beautiful of the chemical elements and has been known and treasured by man since the earliest times. In the massive form it is a soft, yellow metal with the highest ductility and malleability of any element; thus one ounce of gold can be beaten out to 300 square feet. Gold has an atomic weight of 196.967 and a density at 20 °C of 19.32 g cm^{-3}; the pure metal melts at 1063 °C and boils at 2966 °C. The beauty and rarity of gold has led to its use in jewellery and in coinage and as a standard for monetary systems throughout the world. Since gold is such a soft metal, it is usually alloyed to give it more strength. In these alloys the term carat is used to express the amount of gold present, 24 carats being pure gold. Alloys of gold with the other elements of group IB are most frequently used. Thus, coinage gold typically contains 10% copper, while gold used in jewellery and dentistry may contain varying proportions of silver and copper. The colours of the alloys are frequently controlled by the other metal. Thus, an alloy containing 50% copper is dark red, and one containing 65% silver is white and hardly distinguishable to the eye from pure silver. The white gold commonly used in jewellery contains 75—85% gold, with 8—10% nickel and 2—9% zinc, while more expensive white alloys are palladium—gold (90% Au, 10% Pd) and platinum—gold (60% Au, 40% Pt).

Gold also has high thermal and electrical conductivity; hence its use in electronics. The high cost of gold necessitates the use of very thin films, formed by electroplating on a base metal support, for gold switching devices.

Like the other group IB elements, gold crystallises in the face-centred cubic lattice and the closest inter-nuclear distance is 288.4 pm.

At high temperature in the gas phase, gold exists as a mixture of atoms and diatomic molecules. The dissociation energy (D_0°) of Au_2 is 221 ± 2 kJ mol^{-1} and the Au—Au bond length is 250 pm [10].

COLLOIDAL GOLD

Solutions containing gold complexes such as $[AuCl_4]^-$ are easily reduced to gold, and under controlled conditions colloidal gold may be formed. Colloids of gold may be red, blue or violet depending on the mean particle size and shape and have been known for a long time, certainly since the 18th century. Various reducing agents can be used for preparing colloidal gold, some of the earliest being tannin and phosphorus, while the "potable gold" of the alchemists, to which fabulous medicinal powers were attributed, was obtained by reducing a solution of gold in *aqua regia* with ethereal oils. Formaldehyde and hydrazine hydrate are now commonly used as reducing agents. Particularly important is the "purple of Cassius" which has long been used as a colouring agent for enamel and glass. This is thought to be a mixed colloid of hydrated tin(IV) oxide and gold formed by reducing $[AuCl_4]^-$ with tin(II) chloride. A purple or ruby-red precipitate is formed on heating the solution. This forms the basis of a sensitive test for gold; thus, the purple colour is formed if a 10^{-8} M solution of $[AuCl_4]^-$ is added to a saturated solution of tin(II) chloride.

ISOTOPES OF GOLD

Gold has an atomic number of 79 and an atomic weight of 196.967. Thus, the nucleus of the only stable isotope ^{197}Au contains 79 protons and 118 neutrons. Details of the other known isotopes of gold are given in Table 1.1.

The neutron-deficient isotopes are prepared by bombarding lighter atoms with positively charged particles in a cyclotron or linear accelerator. The isotopes of mass numbers from 177 to 183 are all alpha emitters and all have half-lives of less than one minute, while those with mass numbers of 185—196 decay by electron capture accompanied by gamma radiation and in some cases by positron emission. Of these isotopes the only long-lived one is ^{195}Au with a half-life of 183 days; this isotope is prepared by the reaction:

$$^{195}_{78}Pt + ^{1}_{1}p \rightarrow ^{195}_{79}Au + ^{1}_{0}n$$

A target of platinum (which in natural abundance contains 33.8% ^{195}Pt) is bombarded with 22 MeV protons in a cyclotron. This is an expensive process made yet costlier by the need for subsequent separation of gold from platinum; hence ^{195}Au is not widely used.

The neutron-heavy isotopes are prepared by neutron capture reactions, and all decay by β-emission accompanied by gamma radiation. The most important isotope in this group is ^{198}Au, which is prepared in a nuclear reactor by the reaction:

$$^{197}_{79}Au + ^{1}_{0}n \rightarrow ^{198}_{79}Au$$

TABLE 1.1[a]

The Isotopes of Gold

Mass No.[b]	Half life	Mode of decay[c]
177	1.35 s	α
178	2.65 s	α
179	7.25 s	α
181	11.55 s	α, EC
183	45.5 s	α
185	4.3 min	α, EC
186	12 min	EC, γ
187	8 min	α, EC
188	8 min	EC, γ
189	29.7 min	EC, γ
189m	4.7 min	EC, γ
190	39 min	EC, γ
191	3.2 h	EC, γ
192	5.0 h	EC, β+, γ
193m	3.9 s	γ
193	17.5 h	EC, γ
194	39.5 h	EC, β+, γ
195m	31.0 s	β+, γ
195	183 day	EC, γ
196m	9.7 h	γ
196	6.2 day	EC, β, γ
197m	7.5 s	γ
197	stable	
198	2.70 day	β, γ
199	3.15 day	β, γ
200	48.4 min	β, γ
201	26 min	β, γ
202	30 s	β, γ
203	55 s	β, γ
204	4.05 s	β, γ

[a] From: L. Myerscough, *Gold Bull.*, 6 (1973) 62. [b] m = metastable.
[c] α = alpha emission, EC = electron capture, β+ = positron emission, β = beta particle emission, γ = gamma radiation.

This is a particularly favourable reaction since ^{197}Au, the only naturally occurring isotope, has a high neutron capture cross-section of 98.8 barns. Thus, with a moderate neutron flux of 2×10^{13} n cm^{-2} s^{-1}, a specific activity of 1 curie per gram is obtained in 6.5 hours. The isotope ^{198}Au is widely used in radiotherapy, in medical diagnosis and for tracer studies.

CHEMICAL REACTIVITY OF GOLD

Gold is the most noble of the metals, being the only one which is not attacked by either oxygen or sulphur at any temperature. Gold does however react with tellurium at high temperatures, to give $AuTe_2$, and reacts with all the halogens. Of these, bromine is the most reactive and reacts exothermally with gold powder at room temperature to give Au_2Br_6. The reaction with chlorine has been studied in more detail due to its application in the purification of gold. At temperatures below 130 °C, chlorine is adsorbed onto the surface and then forms surface compounds; at temperatures between 130—200 °C, further reaction can take place but the rate is limited by the rate of diffusion of chlorine through the surface layer of gold chlorides. Only above 200 °C is a high reaction rate achieved as the gold chlorides then sublime, thus continually exposing a clean gold surface [11].

Gold will dissolve in aqueous solutions containing a good ligand for gold and an oxidising agent, but neither condition alone is sufficient. Thus, gold will not dissolve appreciably in hydrochloric or nitric acid, but dissolves readily in *aqua regia* to give tetrachloroauric(III) acid, $H[AuCl_4]$. Similarly, gold will dissolve in hydrochloric acid in the presence of hypochlorite or iron(III) as oxidant. The dissolution of gold in cyanide solutions with air or hydrogen peroxide as oxidant is another example of this effect. The reaction with oxygen as oxidising agent apparently takes place by adsorption of oxygen onto the gold surface, followed by reaction of this surface layer with cyanide to give first AuCN and then the complex $[Au(CN)_2]^-$ which passes into solution.

Atomic gold is considerably more reactive than the massive metal. Reactions of atomic gold can be carried out by evaporation of the metal at high temperature under vacuum and then co-condensing the gold vapour with a suitable reagent onto an inert noble-gas matrix at liquid helium temperature. In this way the complexes mono-(dioxygen)gold, $Au(O_2)$, and mono(ethylene)gold, $Au(C_2H_4)$, and the carbonyls Au(CO) and $Au(CO)_2$ can be prepared and identified spectroscopically. Co-condensation of atomic gold with carbon monoxide and dioxygen gives the complex

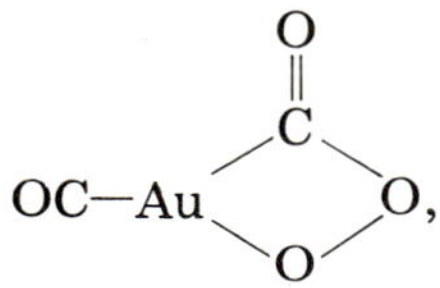

which is thought to be involved in the gold-catalysed reaction of CO and O_2 to give CO_2. All these gold compounds decompose on warming the matrix [11b].

OXIDATION STATES OF GOLD AND THE STEREOCHEMISTRIES OF GOLD COMPLEXES

In complexes containing a single gold atom, the oxidation states I, II, III and V are well established and gold also forms many complexes with metal—metal bonds in which it is difficult to assign formal oxidation states to the metal atom. Gold has the electronic configuration [Xe] $4f^{14}5d^{10}6s^1$ and belongs to group IB of the periodic table along with copper and silver. In this section, we shall study briefly the nature of typical complexes of gold in each oxidation state so that comparisons with the other group IB elements can be made, and a detailed treatment will be found in subsequent chapters.

(a) Oxidation state I

Of the group IB elements, silver alone forms a simple aquo ion in oxidation state I. Gold(I) and copper(I) compounds disproportionate in aqueous solution unless they are either of very low solubility (e.g., AuI) or strongly complexed (e.g. $[Au(CN)_2]^-$) according to the equations:

$$2Cu(I) \rightarrow Cu + Cu(II)$$

$$3Au(I) \rightarrow 2Au + Au(III)$$

Since Au^+ has the closed shell electronic configuration [Xe] $4f^{14}5d^{10}$, gold(I) complexes are diamagnetic and adopt regular structures. The most common stereochemistry is linear, the co-ordination number of the gold atom being two [12—14], and such complexes may be cationic such as $[Au(PPh_3)_2]^+$, neutral such as $[Au(CN)(PPh_3)]$ or anionic such as $[AuCl_2]^-$. In this respect gold(I) differs from silver(I) and copper(I) which have a greater tendency to form complexes with higher co-ordination numbers. Thus, the complex $MI(PEt_3)$ is monomeric when M = Au, but exists as a tetramer $[\{MI(PEt_3)\}_4]$ with bridging iodide ligands when M = Cu or Ag, so that the metal atoms have co-ordination number four. Higher co-ordination numbers are also known for gold(I). Thus, trigonal planar complexes [12, 15] with co-ordination number three such as $[Au(PPh_3)_3]^+$ and $[AuCl(PPh_3)_2]$, and tetrahedral complexes [16] with co-ordination number four such as $[Au(PMe_3)_4]^+$ are known but these readily lose phosphine ligands and revert to the more favoured linear two co-ordinate form. We shall consider later in this chapter possible explanations for this relative preference of gold(I) for co-ordination number two when compared with copper(I) and silver(I), and the similar effect observed in group IIB, where mercury(II) forms many linear two co-ordinate compounds but zinc(II) and cadmium(II) do not.

(b) Oxidation state II

The oxidation state two is dominant in the chemistry of copper, but considerably less common for silver and very rare for gold. Thus this oxidation state is established for gold only in the following situations:

(1) There is strong evidence that gold(II) species are formed as transient intermediates in redox reactions between the stable oxidation states I and III [17].

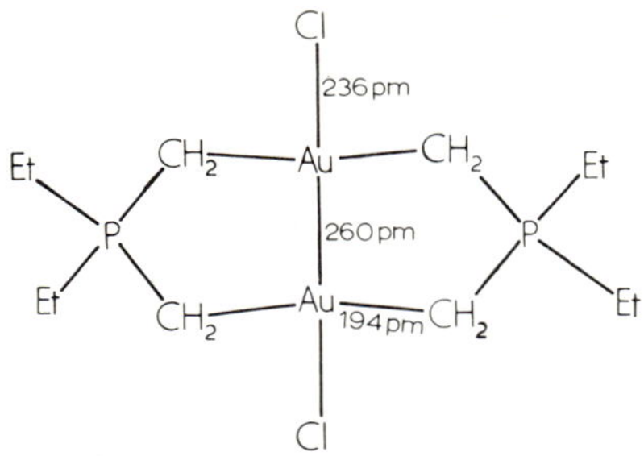

Fig. 1.1. Structure of a dimeric gold(II) complex with a gold—gold bond.

(2) Stable gold(II) species may be formed in which there is a gold—gold bond. An example is given in Fig. 1.1 [18].

(3) With certain unsaturated sulphur-donor ligands such as maleonitriledithiolate ion, mnt^{2-}, isolable complexes of gold(II) such as $[Au(mnt)_2]^{2-}$ with a square planar four co-ordinate gold(II) centre as shown below may be formed.

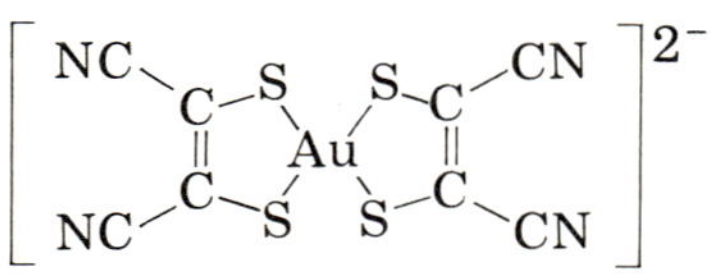

In gold(II) complexes the metal has the $5d^9$ configuration and such complexes are expected to be paramagnetic unless there is a gold—gold bond. Monomeric gold(II) complexes with dithiolate and dithiocarbamate ligands appear to be stabilised by delocalisation of the unpaired electron over the unsaturated ligand [19].

(c) Oxidation state III

This is an important oxidation state for gold but is rare for copper and silver. All known gold(III) complexes are diamagnetic with the low-spin $5d^8$ electron configuration, and the vast majority of these compounds have the four co-ordinate square-planar stereochemistry. Examples include the complex ions $[AuCl_4]^-$ and $[Au(NH_3)_4]^{3+}$, but the simple aquo ion

$[Au(OH_2)_4]^{3+}$ is never formed even on dissolution of gold(III) hydroxide in nitric acid or sulphuric acid [20]. Five co-ordinate gold(III) complexes are rare, but when formed they are square pyramidal or distorted square pyramidal, as in $[AuCl_3(2,2'$-biquinolyl)] and $[AuCl_3$(2,9-dimethyl-1,10-phenanthroline)] with rigid bidentate nitrogen—donor ligands [21]. There is evidence that the complex ion $[Au(diars)_2I]^{2+}$, where diars = 1,2-bis-(dimethylarsino)benzene, is five co-ordinate but its structure is unknown. Trigonal bipyramidal gold(III) complexes are generally invoked as intermediates during ligand substitution reactions in square planar gold(III) complexes. In co-ordination number six, a tetragonal stereochemistry is always assumed [22] and has been proved for $[Au(diars)_2I_2]^+$. The situation may be compared with that for $[CuF_6]^{3-}$ and $[AgF_6]^{3-}$ which are paramagnetic with μ_{eff} 2.8 B.M. and 2.6 B.M., respectively, at room temperature; these complex ions presumably have regular octahedral structures. However, all other Cu(III) and Ag(III) compounds are diamagnetic with square planar co-ordination.

(d) Oxidation state V

The first gold(V) complex containing the ion $[AuF_6]^-$ was reported in 1972 by Bartlett and co-workers. It is diamagnetic and has a regular octahedral structure with the low spin $5d^6$ electronic configuration [23]. The compound AuF_5 can also be prepared. These compounds are very powerful oxidising agents, and no analogous compounds of copper and silver are known.

From the foregoing account, it will be clear that there are more differences than similarities between the properties of gold compounds on the one hand, and of either copper or silver compounds on the other hand. In many cases, it is more valuable to make horizontal comparisons of gold with the neighbouring elements platinum and mercury in the periodic table than with the other group IB elements. Thus, isoelectronic species such as $[AuCl_2]^-$ and $[HgCl_2]$, $[Au(PPh_3)_3]^+$ and $[Pt(PPh_3)_3]$, and $[AuCl_4]^-$ and $[PtCl_4]^{2-}$ exist. In the following sections we shall attempt to rationalise some of the similarities and differences between gold and the other group IB elements, and to study the factors which affect the relative stability of the common oxidation states (I) and (III) in gold complexes.

IONISATION ENERGIES AND ELECTRON AFFINITIES OF THE GROUP IB ELEMENTS

The factors governing the stability of oxidation states of an element are of course complex, and consideration of complete energy cycles is desirable. However, the required data are not available for many compounds, so that we are driven to use criteria such as ionisation energies and electron affinities as a general guide. These are given in Table 1.2 and are particularly useful for the group IB elements.

TABLE 1.2

Ionisation energies / kJ mol^{-1} of Cu, Ag, Au

Element	1st	2nd	3rd	4th
Cu	745	1958	3554	5330
Ag	731	2074	3361	5020
Au	890	1980	2943	4200

It is worth noting that the first ionisation energies are considerably higher than for the alkali metals (e.g., K, 419 kJ mol^{-1}). The ionisation process for potassium is K $\{[Ar]\,4s^1\} \rightarrow$ K^+ $\{[Ar]\}$, while for copper it is Cu $\{[Ar]\,3d^{10}4s^1\} \rightarrow$ Cu^+ $\{[Ar]\,3d^{10}\}$. The outer 4s electron feels a higher effective nuclear charge in copper than in potassium because the increased nuclear charge in copper is not balanced due to the poor shielding by the electrons in the diffuse $3d^{10}$ subshell. The first ionisation energy for silver is lower than that of copper due primarily to its greater atomic radius, but there is a large increase for that of gold whose electronic configuration is $[Xe]\,4f^{14}5d^{10}6s^1$. Here, the outer 6s-electron is poorly shielded from the nuclear charge by both the 4f and 5d electrons, so that it feels a high effective nuclear charge. It is a combination of this high ionisation energy and the high binding energy which gives gold its noble character.

Variations in the second and third ionisation energies are also noteworthy. Thus, the relatively high second ionisation energy for silver accounts partly for the dominance of the +I oxidation state for this element, and the relatively low third ionisation energy for gold leads to the +III oxidation state being preferred. Subsequent ionisation energies are also lowest for gold, no doubt explaining the existence of the +V oxidation state for gold but not for silver or copper.

Accurate values of the electron affinities of the group IB elements are difficult to obtain, but there is general agreement that gold has the highest value. Values of 226 and 270 kJ mol^{-1} have been proposed by different workers [23, 24]. These relatively high values have prompted the suggestion that gold may be considered as a pseudohalogen [cf., E.A. for iodine is 296 kJ mol^{-1}] and there is indeed evidence that the compound CsAu may be considered to be $Cs^{+}Au^{-}$. Certainly gold is expected to have a high electronegativity on the basis of its high ionisation energy and electron affinity.

CO-ORDINATION NUMBER IN GOLD(I) COMPLEXES

As mentioned earlier, gold(I) complexes have linear two co-ordinate stereochemistry much more commonly than copper(I) or silver(I) complexes. There has been some debate about whether copper(I) or silver(I) has the greater tendency to linear co-ordination, and this may be dependent on the ligand to some extent. The tendency to form linear complexes can be measured by the relative magnitudes of the second and third stability constants. Thus, for the cyanide complexes the ratio of $\log \beta_3/\log \beta_2$ is for Cu(I) 1.25, for Ag(I) 1.11 and for Au(I) 1.00, so that in this case the tendency to linear complex formation is Au(I) > Ag(I) > Cu(I). The factors responsible for this sequence are not clearly understood and several theories have been put forward.

Firstly, it can be said that steric factors are unlikely to be important when considering two- and three-co-ordinate complexes, so that electronic factors must be responsible. Considering only simple electrostatic effects, a metal ion with a high polarising power should have a high co-ordination number. Thus, for the group IB metals in oxidation state I, the order should be dependent on the ionic radii which are Cu^+ 81 pm, Ag^+ 126 pm and Au^+ 137 pm. This gives the correct series, but would wrongly predict a closer similarity of silver to gold rather than to copper.

A second theory considers the nature of the orbitals used in bonding. Thus *sp* hybrid orbitals will be used in linear

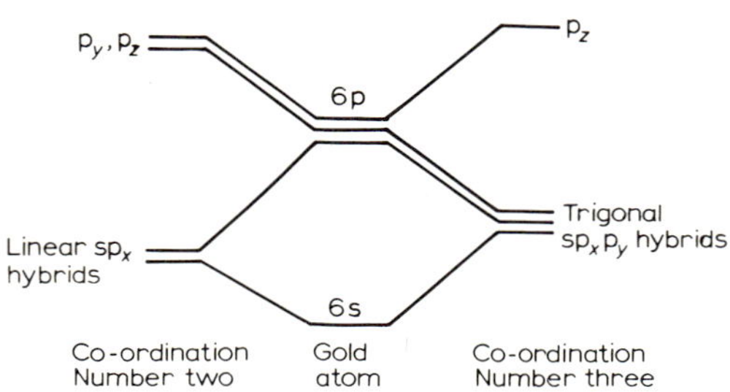

Fig. 1.2. Hybridisation scheme for gold(I).

complexes, sp^2 hybrid orbitals in trigonal planar complexes and sp^3 hybrid orbitals in tetrahedral complexes. The ease with which hybridisation of s and p orbitals takes place will depend on the energy separation between these levels. This is illustrated in Fig. 1.2.

The *n*s—*n*p separations for Cu, Ag and Au atoms are 3.79, 3.66 and 4.63 eV, respectively. Thus sp hybridisation rather than sp^2 or sp^3 hybridisation is expected to be more favourable for gold than for copper or silver, since the higher s—p separation will make hybridisation of s and p orbitals more difficult [26]. The same conclusion is reached if the bonding is considered in terms of Molecular Orbital rather than Valence Bond theory since the p orbitals of gold are at relatively high energy, and therefore less likely to take part in bonding than the p-orbitals of copper and silver, which are at relatively low energies.

A third theory is based on the Ligand Field theory of bonding [27, 28]. Thus, in a linear complex with the ligands lying along the z-axis, it may be favourable to hybridise the filled nd_{z^2} and vacant $(n + 1)$s orbitals. This will give one orbital

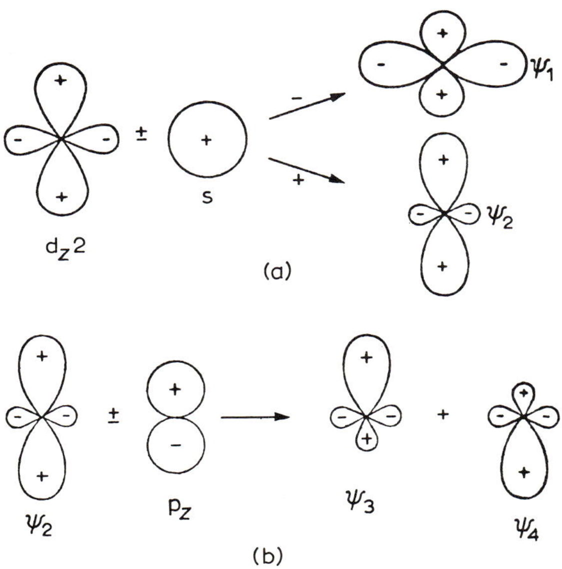

Fig. 1.3. Hybridisation of $5d_{z^2}$, 6s and 6p orbitals. (a) formation of hybrid orbitals ψ_1 and ψ_2 removing electron density into ψ_1 away from the z-axis; (b) formation of linear bonding hybrid orbitals from ψ_2 and $6p_z$ orbitals. (J. E. Huheey, *Inorganic Chemistry*, Harper and Row, London, 1975, p. 372).

with maximum electronic density in the xy plane and one with maximum electronic density along the z-axis as illustrated in Fig. 1.3.

Now, if the electron pair is in ψ_1 rather than the d_{z^2} orbital, there will be less repulsion between the electron pair and the ligands, and the complex will be stabilised. This is a second-order ligand field stabilisation energy. The orbital ψ_2 will hybridise with the $(n + 1)$p orbital to form two linear hybrid orbitals to overlap with the ligand orbitals and so give a linear complex. This d—s mixing will again become less favourable as the d—s separation increases. The energies of the lowest d^9s state above the d^{10} ground state are for Cu^+ 21,928 cm^{-1}, Ag^+ 39,164 cm^{-1} and Au^+ 15,039 cm^{-1}. Thus, the second-order ligand-field stabilisation energy on formation of linear complexes is expected to follow the series $Au^+ > Cu^+ > Ag^+$. Thus, again gold(I) is correctly predicted to have the greatest preference for linear co-ordination, but the prediction that copper(I) should resemble gold(I) more than silver(I) is not upheld.

The predictions of the three theories can be summarised in terms of the tendency to form linear complexes:

(1) Polarising power of M^+ predicts Au(I) > Ag(I) ≫ Cu(I)
(2) s—p energy separation predicts Au(I) ≫ Cu(I) > Ag(I)
(3) d—s energy separation predicts Au(I) > Cu(I) ≫ Ag(I)

Thus, if all three effects are appreciable, a combined prediction Au(I) ≫ Ag(I) ~ Cu(I) might be deduced, in agreement with experiment. It is, however, difficult to estimate the relative importance of the three contributions.

OXIDATION—REDUCTION POTENTIALS

(a) The aqueous chemistry of gold

To understand the nobility of gold, it is useful to consider the redox potentials of gold in the absence of co-ordinating ligands. Thus, for gold to react it must be oxidised, e.g.,

$$Au \rightarrow Au^+(aq) + e^-$$

The tendency for this reaction to take place is given by the Nernst equation:

$$\pi = \pi^{\ominus} + 2.303 \frac{RT}{F} \log_{10} [Au^+]/[Au]$$

where $\pi^{\ominus}$ is the standard electrode potential (estimated [25] to be +1.7 V though a value of 1.83 V has also been suggested [29]). This then reduces to (π in units of volts):

$$\pi = 1.7 + 0.059 \log_{10} [Au^+]$$

It is possible to derive similar expressions for alternative oxidation processes [30]:

$$Au \rightarrow Au^{3+} + 3e^-, \pi = 1.50 + 0.0197 \log_{10} [Au^{3+}]$$

$$Au + 3H_2O \rightleftharpoons Au(OH)_3 + 3H^+ + 3e^-, \pi = 1.46 - 0.059 \text{ pH}$$

$$Au(OH)_3 \rightleftharpoons AuO_2 + H_2O + H^+ + e^-, \pi = 2.63 - 0.059 \text{ pH}$$

These reactions can only take place if sufficiently strong oxidants are present, and reactions in aqueous solution are further limited by the limits of stability of water to oxidation to oxygen, or reduction to hydrogen.

The system can be expressed graphically in the Pourbaix diagram in which the equilibrium potential of each couple is

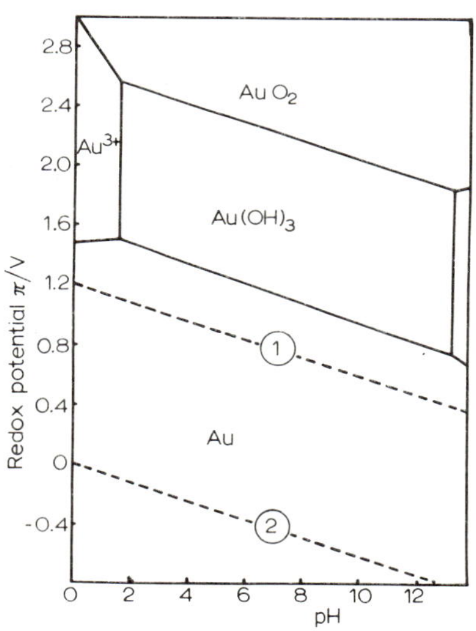

Fig. 1.4. Pourbaix diagram for the system Au—H_2O at 25 °C. The concentrations of all gold species are 10^{-4} M.

plotted against pH for particular concentrations. A simplified Pourbaix diagram for gold (using a concentration of 10^{-4} M) is shown in Fig. 1.4 and indicates the conditions under which particular species are expected to be formed [31]. The meaning of the diagram becomes clear, once it is realised that the dotted lines (1) and (2) represent the limits of stability of water. Now the oxidised forms of gold exist only at potentials greater than for line (1). Under these conditions, water will be oxidised to oxygen and the various oxidised gold species will be reduced to gold metal. Thus, in the absence of co-ordinating ligands, gold cannot be oxidised by dissolved oxygen in the presence of either strong acids or strong alkalis. Gold is the only metal for which this is the case, and is thus truly the most noble of the metallic elements.

(b) The complex chemistry of gold

The reason for the high electrode potentials for oxidation of gold in the absence of complexing ligands lies in the very low stability of the aquo ions $[Au(OH_2)_2]^+$ and $[Au(OH_2)_4]^{3+}$. The potentials are lower in the presence of complexing ligands. Thus consider the equilibrium in the presence of a ligand L.

$$[Au(OH_2)_2]^+ + 2L \overset{K}{\rightleftharpoons} [AuL_2]^+ + 2H_2O$$

Substituting in the Nernst equation used previously the equilibrium concentration for $[Au(OH_2)_2]^+$ in the presence of chloride, we find:

$$\pi = \pi^{\ominus} - \frac{2.303\,RT}{F} \log_{10}[Au(OH_2)_2]^+$$

$$= \pi^{\ominus} - \frac{2.303\,RT}{F} \log_{10}K + \frac{2.303\,RT}{F} \log_{10}[AuL_2{}^+]/[L]^2$$

$$= (\pi^{\ominus})^1 + \frac{2.303\,RT}{F} \log_{10}[AuL_2{}^+]/[L]^2$$

where $(\pi^{\ominus})^1$ is the standard electrode potential for the couple:

$$Au + 2L \rightarrow [AuL_2]^+ + e^- \qquad (1)$$

Thus, in the presence of a good ligand L for gold(I), a high value of K and a low value of $(\pi^{\ominus})^1$ will be obtained. Some typical standard potentials are given in Table 1.3 [25, 29—34]. Values corresponding to equation (1) are termed $\pi^{\ominus}_{1,0}$ to indicate that they refer to oxidation of Au(0) to Au(I). The order of decreasing $\pi^{\ominus}_{1,0}$ is then the order of increasing stability of the complexes $[AuL_2]^+$. Note in particular, the low value of $\pi^{\ominus}_{1,0}$ for oxidation of gold in the presence of cyanide. It is this ease of oxidation of gold in the presence of cyanide which makes the cyanide process for the extraction of gold feasible.

TABLE 1.3

Standard redox potentials(V) for gold in aqueous solution

Ligand	$\pi^{\ominus}_{1,0}$	$\pi^{\ominus}_{3,0}$	$\pi^{\ominus}_{3,1}$
H_2O	+1.7 (+2.12)	+1.50	+1.4
Cl^-	+1.15	+1.00	+0.93
Br^-	+0.96	+0.86	+0.81
I^-	+0.56	+0.57	+0.57
SCN^-	+0.67	+0.64	+0.62
NH_3	+0.563	+0.325	+0.206
thiourea	+0.38		
selenourea	+0.20		
CN^-	−0.61		

It should be realised, however, that the standard potentials only tell us that the reaction is feasible thermodynamically. Whether it takes place readily or not, will depend on kinetic factors.

It is, of course, possible to oxidise gold to gold(III) in the presence of suitable ligands:

$$Au + 4L \rightleftharpoons [AuL_4]^{3+} + 3e^-$$

Again, the standard potentials $\pi^{\ominus}_{3,0}$ will depend on the stability constant for formation of the complex $[AuL_4]^{3+}$. Typical values are given in Table 1.3.

A rather more complex case is the oxidation of gold(I) to gold(III) in the presence of complexing ligands, L.

$[AuL_2]^+ + 2L \rightleftharpoons [AuL_4]^{3+} + 2e^-$

It can easily be shown that the standard potentials $\pi^{\ominus}_{3,1}$ for this process are given by the expression:

$3\pi^{\ominus}_{3,0} = 2\pi^{\ominus}_{3,1} + \pi^{\ominus}_{1,0}$

Thus, whether oxidation of gold to the complex $[AuL_2]^+$ or $[AuL_4]^{3+}$ occurs in the presence of a particular ligand, will depend on the strength of the oxidising agent and on the relative magnitudes of $\pi^{\ominus}_{1,0}$ and $\pi^{\ominus}_{3,0}$. A related problem is to consider whether disproportionation of a gold(I) complex according to the equation below will occur or not.

$$3[AuL_2]^+ \overset{K}{\rightleftharpoons} [AuL_4]^+ + 2Au + 2L$$

The equilibrium constant for this reaction is given by $\log_{10} K = (2\,F/2.303\,RT)\,(\pi^{\ominus}_{1,0} - \pi^{\ominus}_{3,1})$. Thus, in general, disproportionation of gold(I) to gold and gold(III) will be favourable if $\pi^{\ominus}_{1,0} > \pi^{\ominus}_{3,1}$. The calculated values of K are for $L = H_2O$, $K = 10^{10}$; $L = Cl^-$, $K = 5 \times 10^7$; $L = Br^-$, $K = 10^5$; $L = SCN^-$, $K = 2.1$, while for $L = CN^-$ the value of K is unknown but can be estimated to be about 10^{-25}. It is apparent that disproportionation of the complex $[AuL_2]^+$ becomes less favourable as the ligand L becomes more polarisable. This effect is probably in part due to gold(I) having a relatively greater affinity than gold(III) for soft polarisable ligands.

STABILITY OF GOLD COMPLEXES

From the foregoing discussion, it will be clear that both gold(I) and gold(III) form more stable complexes with I^- than with Br^-, and with Br^- than with Cl^-. Thus, in both common oxidation states gold is a "class b" metal or a "soft acid," preferring to bond to large polarisable ligands. We have also seen that gold(I) discriminates more than gold(III) between hard and soft ligands so that gold(I) may be considered to be the softer metal ion*. Copper(I) and silver(I) are also soft acids and, indeed, there is a linear correlation between any two of $\log\beta_2$ for $[AuX_2]^-$, $[AgX_2]^-$ and $[CuX_2]^-$, where β_2 are the stability constants for different ligands X. Again, gold(I) discriminates more than copper(I) or silver(I) between the ligands. This is, of course, in agreement with the general prediction that large metal ions in low oxidation states will have the greatest "class b" or soft character.

Stability constants for gold(I) in aqueous solution are difficult to obtain due to the tendency to disproportionate, and more extensive data have been obtained [36, 37] for acetonitrile solutions by studying the equilibria:

$$[Au(MeCN)_2]^+ + 2X^- \rightleftharpoons [AuX_2]^- + 2MeCN$$

$$[Au(MeCN)_2]^+ + 2L \rightleftharpoons [AuL_2]^+ + 2MeCN$$

$$[AuX(MeCN)] + L \rightleftharpoons [AuXL] + MeCN$$

The order of stability constants is given below:

(a) Anionic complexes $\underline{[AuX_2]^-}$

$$NCO^- < NCS^- \sim Cl^- < Br^- < I^- \ll CN^-$$

(b) Cationic complexes $[AuL_2]^+$

$$Ph_3PO < Me_2S < py < AsPh_3 < NH_3 \ll PPh_3 \sim MeNC < PMePh_2 < PMe_2Ph$$

(c) Neutral complexes [AuXL]

$$Me_2O \sim NPh_3 \ll Me_2S < Me_2Se < SbPh_3 < AsPh_3 < Me_2Te < PPh_3 < PMePh_2 < PMe_2Ph$$

Again, the data show quite clearly that gold(I) is a "class b" metal ion. This is confirmed by studying complexes of gold with ambidentate ligands, in which both gold(I) and gold(III) almost invariably bond to the softer end of the ligand [38].

---

* However, evidence that gold(III) is a softer acid than gold(I) has been presented recently [29].

Thus, complexes with NCO^- ligand are N-bonded rather than O-bonded, with NCS^- they are S- rather than N-bonded, and with thiosulphate or sulphite they are S- rather than O-bonded. In certain cases, thiocyanate complexes of gold(I) have been shown to contain a small proportion of the N-bonded isomer in equilibrium with the S-bonded form [39]. The amount of this N-bonded form in the complexes LAuNCS increases with the *trans*-effect of L, decreasing in the order L = $P(OPh)_3 > PMe_3 > PPh_3 > AsPh_3 > S(CH_2Ph)_2$. This is an example of anti-symbiosis, in which a soft ligand bound to a "class b" metal destabilises the bond between the metal and another soft ligand *trans* to itself. In all complexes studied the selenocyanate ligand is Se-bonded in gold(I) complexes.

It should be realised that the stability constant for a complex is not determined only by the metal—ligand bond strength. Thus, the enthalpy of substitution of chloride by bromide in $[AuCl_4]^-$ in aqueous solution is -46 kJ mol^{-1}, but, after consideration of hydration effects, the estimated bond energy is greater for the AuCl bond than for the AuBr bond [40].

THEORETICAL STUDIES OF BONDING IN GOLD COMPLEXES

Molecular orbital calculations using the SCCC—LCAO—MO method have recently been published for some halide and pseudohalide complexes of gold(I) and gold(III), and the results are of considerable interest [41, 42].

For the halide complexes $[AuX_4]^-$ the calculated 5d-orbital energies for gold fall in the order $d_{z^2} < d_{xz}, d_{yz} < d_{xy} \ll d_{x^2-y^2}$ in agreement with the predictions of simple crystal field theory and with similar calculations on square-planar platinum(II) and palladium(II) complexes. For the corresponding gold(I) halides $[AuX_2]^-$ the 5d-orbital energies for gold follow the sequence $d_{xy}, d_{x^2-y^2} < d_{xz}, d_{yz} < d_{z^2}$ again as expected from crystal field theory. The electron configurations of Au^{3+} and Au^+ are of course $5d^8 6s^0 6p^0$ and $5d^{10} 6s^0 6p^0$, respectively, but the calculated orbital populations and charges on gold and halogens in various complexes are given in Table 1.4.

TABLE 1.4

Orbital population and atomic charges in the complexes $[AuX_4]^-$ and $[AuX_2]^-$

	$[AuX_4]^-$				$[AuX_2]^-$			
	F	Cl	Br	I	F	Cl	Br	I
Au 5d	9.008	9.188	9.258	9.439	9.723	9.817	9.854	9.955
6s	0.643	0.775	0.797	0.747	0.582	0.722	0.735	0.637
6p	0.696	0.837	0.861	0.949	0.197	0.348	0.380	0.499
q(Au)[a]	+0.653	+0.200	+0.084	−0.135	+0.498	+0.113	+0.031	−0.092
q(X)[a]	−0.413	−0.300	−0.271	−0.216	−0.749	−0.556	−0.515	−0.454

[a] q = electronic charge on the Au or X ion.

The data show very clearly that there is considerable co-valency in the metal—ligand bonds for both gold(I) and gold(III) complexes. Thus, the orbital populations indicate that the 6s and 6p orbitals of gold are engaged in bonding, and that the charge on gold in the complexes is always less than +1. The extent of the covalency, which is largely due to σ-bonding, increases as expected on passing from fluoride through to iodide complexes.

For linear gold(I) complexes the simple bonding schemes suggest that the filled 5d orbitals will play no part in bonding, and that σ-bonding will use 6sp-hybrid orbitals on gold. In agreement, the calculations show that 6s and 6p orbitals on gold are strongly involved in bonding in $[AuX_2]^-$, but also that the 5d-orbitals contribute significantly to the gold—halide σ-bonds. This is in accord with the ideas of Orgel on d—s mixing in linear gold(I) complexes, discussed earlier.

For square-planar gold(III) complexes Valence Bond theory predicts that gold(III) will have the $5d^8$ configuration and that σ-bonds will be formed using the four $5d_{x^2-y^2}6s6p_x6p_y$ hybrid orbitals. Again, the calculations support this, so that the 6p orbital population is considerably greater in the gold(III) than in the gold(I) complexes, while the reverse is true for the 5d orbital populations. Similar trends were observed in the thio-cyanate and cyanide complexes.

THE *TRANS*-EFFECT AND *TRANS*-INFLUENCE

These two concepts are widely used in interpreting the chemical properties of square planar platinum(II) complexes, but have not been exploited similarly for iso-electronic gold(III) complexes. As we shall see below, however, the concepts are potentially very useful and, as more structural and mechanistic data are accumulated, are likely to be used more frequently in rationalising the chemical properties of both gold(I) and gold(III) complexes.

The *trans*-effect is defined as the ability of a ligand to labilise the ligand *trans* to itself, and hence to direct incoming ligands to this position. For example, in the complex [$AuCl_3$-(PPh_3)] the *trans*-effect of Ph_3P is greater than that of Cl^-, so that we expect the chloride *trans* to triphenylphosphine to be more labile than the other two chloride ligands. Hence, this chloride is expected to be displaced preferentially and the incoming ligand will be directed to the stereochemical position *trans* to triphenylphosphine. The *trans*-effect is thus a kinetic concept, and we shall return to discuss it more fully in Chapter 9.

The *trans*-influence is defined as the ability of a ligand to weaken the bond between the metal and the ligand *trans* to itself, in the ground state of the molecule. Thus, in contrast to the *trans*-effect, it is a thermodynamic concept. The *trans*-influence of a ligand is deduced by measuring such parameters as the metal—ligand bond lengths by X-ray diffraction or the metal—ligand stretching force constants by analysis of the vibrational spectra. Gold—chlorine and gold—bromine bond lengths for some selected complexes of gold(I) and gold(III) are presented in Table 1.5.

From the gold—chlorine bond lengths for the square planar gold(III) complexes, it is possible to deduce the *trans*-influence series: Ph $\sim$ C_6F_5 $>$ PPh_3 $>$ SPr_2 $\sim$ Cl^-. From the gold—bromine bond lengths for the linear gold(I) complexes, it follows that the *trans*-influence of $AsPh_3$ $>$ Br^-.

It is thought that the *trans*-influence of a ligand is largely determined by the electronegativity of the donor atom or by the σ-donor ability of the ligand. Thus, in linear gold(I) complexes, for example, the gold atom is sp-hybridised so that a bromide ligand in $[AuBr_2]^-$ will overlap with a gold orbital having 50% s-character. Now, a strong σ-donor ligand like $AsPh_3$ will have a particularly high demand for the s-orbital of gold in order to give the strongest possible covalent bond. Thus, the gold orbital involved in bonding to $AsPh_3$ in $[AuBr(AsPh_3)]$

TABLE 1.5

Gold—chlorine and gold—bromine bond lengths in some complexes

Complex	*trans*-ligand	*r*(AuCl) or *r*(AuBr)/pm	Ref.
Gold(III) complexes			
$[Ph_4As][AuCl_4]$	Cl^-	227.1	43
$[AuCl_3(PPh_3)]$	Cl^-	227.3, 228.2	44
	PPh_3	234.7	
cis-$[AuCl(C_6F_5)_2(PPh_3)]$	C_6F_5	238	45
cis-$[AuCl_2(Ph)(SPr_2)]$	SPr_2	227	46
	Ph	238	
Gold(I) complexes			
$[AuCl_2]^-$	Cl^-	228.1	14
$[AuCl(PPh_3)]$	PPh_3	227.9	47
[AuCl(piperidine)]	piperidine	225.6	48
$[AuCl(PCl_3)]$	PCl_3	233	49
$[ClAuS(Ph)CH_2CH_2S(Ph)AuCl]$	SR_2	232.9, 229.3	50
$[AuBr_2]^-$	Br	234.9	51
$[AuBr(AsPh_3)]$	$AsPh_3$	237.7	52

will have more than 50% s-character, leaving the donor orbital of the *trans*-ligand, Br^-, to overlap with an orbital having more than 50% p-character. Since the p-orbitals are more diffuse than the s-orbital and, hence, overlap is less effective, the resultant Au—Br bond is weaker and more ionic in $[AuBr(AsPh_3)]$ than in $[AuBr_2]^-$. The spectroscopic properties of gold(I) and gold(III) complexes can often be rationalised using the *trans*-influence concept and we shall return to the topic in Chapter 10.

REFERENCES

1 N. A. Shilo, *Earth Sci. Rev.*, 7 (1971) 215.
2 W. S. Fyfe and R. W. Henley, *Miner. Sci. Eng.*, 5 (1973) 295.
2 (b) F. A. Letnikov, N. V. Vilov, T. A. Gantimurova, I. V. Konovalov, B. O. Shkandrii and T. A. Shkarupa, *Geokhimya*, (1975) 1827.
3 E. D. Korobushkina, A. S. Chernyak and G. G. Mineev, *Mikrobiologiya*, 43 (1974) 49; E. D. Korobushkina, G. G. Mineev and G. P. Praded, *Mikrobiologiya*, 45 (1976) 535.
4 T. Green, *The World of Gold Today*, White Lion Publishers, London, 1974.
5 J. P. Riley, in J. P. Riley and G. Skirrow (Eds.), *Chemical Oceanography*, Vol. 3, Academic Press, London, 2nd ed., 1975, Ch. 19.
6 A. I. Ryabinin, A. S. Romanov, S. Khatamov and R. Khamidova, *Geokhimya*, (1974) 158.
7 B. I. Peshchevitskii, G. N. Anoshin and A. M. Erenburg, *Dokl. Akad. Nauk. SSSR*, 162 (1965) 915; *Khim. Resur. Morei Okeavov*, (1970) 141, *Chem. Abstr.*, 75 (1971) 67332.
8 M. I. Brittan, *Am. Sci.*, 62 (1974) 402.
9 R. J. Adamson (Ed.), *Gold Metallurgy in South Africa*, Chamber of Mines of South Africa, Johannesburg, 1973.
10 J. Kordis, K. A. Gingerich and R. J. Seyse, *J. Chem. Phys.*, 61 (1974) 5114; J. Donohue, *The Structures of the Elements*, Wiley, New York, 1974.
11 (a) M. N. Zyryanov, G. A. Khlebnikova and V. A. Krenev, *Zh. Neorg. Khim.*, 18 (1973) 918; *Chem. Abstr.*, 79 (1973) 10299.
(b) D. McIntosh and G. A. Ozin, *Inorg. Chem.*, 15 (1976) 2869; 16 (1977) 51; H. Huber, D. McIntosh and G. A. Ozin, *Inorg. Chem.*, 16 (1977) 975.
12 N. C. Baenziger, K. M. Dittemore and J. R. Doyle, *Inorg. Chem.*, 13 (1974) 805.
13 P. L. Bellon, M. Manassero and M. Sansoni, *Ric. Sci.*, 39 (1969) 173.
14 J. C. M. T. Eijndhoven and G. C. Verschoor, *Mater. Res. Bull.*, 9 (1974) 1667.
15 F. Klanberg, E. L. Muetterties and L. J. Guggenberger, *Inorg. Chem.*, 7 (1968) 2272.
16 H. Schmidbauer and R. Franke, *Chem. Rev.*, 105 (1972) 2985.
17 R. L. Rich and H. Taube, *J. Phys. Chem.*, 58 (1954) 6.
18 H. Schmidbauer, J. R. Mandl, A. Frank and G. Huttner, *Chem. Ber.*, 109 (1976) 466.
19 J. G. M. van Rens, M. P. A. Viegers and E. de Boer, *Chem. Phys. Lett.*, 28 (1974) 104.
20 A. A. Bezzubenko and B. I. Peshchevitskii, *Chem. Abstr.*, 56 (1962) 20396.
21 W. T. Robinson and E. Sinn, *J. Chem. Soc., Dalton Trans.*, (1975) 726.
22 V. F. Duckworth and N. C. Stephenson, *Inorg. Chem.*, 8 (1969) 1661.
23 I. N. Bakulina and N. I. Ionov, *Dokl. Akad. Nauk SSSR*, 155 (1964) 309.
24 E. Y. Zandberg, A. G. Kamenev and V. I. Paleev, *Zh. Tekh. Fiz.*, 44 (1974) 617.
25 R. D. Hancock and N. P. Finkelstein, *Inorg. Nucl. Chem. Lett.*, 7 (1971) 477.
26 R. S. Nyholm, *Proc. Chem. Soc.* (1961) 273.
27 L. E. Orgel, *J. Chem. Soc.* (1958) 4186.
28 C. K. Jørgensen and J. Pouradier, *J. Chim. Phys.*, 67 (1970) 124.
29 C. J. Hawkins, O. Moensted and J. Bjerrum, *Acta Chem. Scand.*, 24 (1970) 1059; L. H. Skibsted and J. Bjerrum, *Acta Chem. Scand., Ser. A*, 31 (1977) 155.
30 L. M. Gedansky and L. G. Hepler, *Engelhard Ind. Tech. Bull.*, 10 (1969) 5.
31 N. P. Finkelstein and R. D. Hancock, *Gold Bull.*, 7 (1974) 72.
32 A. Hakansson and L. Johansson, *Chem. Scr.*, 7 (1975) 201.
33 V. P. Kazakov, A. I. Lapshin,and B. I. Peshchevitskii, *Zh. Neorg. Khim.*, 9 (1964) 1299.
34 L. H. Skibsted and J. Bjerrum, *Acta Chem. Scand., Ser. A.*, 28 (1974) 764.

35 A. D. Goolsby and D. T. Sawyer, *Anal. Chem.*, 40 (1968) 1978.
36 R. Roulet, N. Q. Lan, W. R. Mason and G. P. Fenske, *Helv. Chim. Acta*, 56 (1973) 2405.
37 R. Roulet and R. Favez, *Chimia*, 29 (1975) 346.
38 J. L. Burmeister and N. J. DeStefano, *Inorg. Chem.*, 10 (1971) 998.
39 J. L. Burmeister and J. B. Melpolder, *J. Chem. Soc., Chem. Commun.*, (1973) 613.
40 B. I. Peshchevitskii and V. I. Belevantsev, *Zh. Neorg. Khim.*, 14 (1969) 2393.
41 J. Reinhold, H. Zwanziger, E. Hoyer and C. Zwanziger, *Z. Chem.*, 14 (1974) 314.
42 H. Zwanziger, J. Reinhold and E. Hoyer, *Z. Chem.*, 14 (1974) 489; 15 (1974) 69.
43 P. G. Jones, J. J. Guy and G. M. Sheldrick, *Acta Crystallogr., Sect. B.*, 31 (1975) 2687.
44 G. Bandoli, D. A. Clemente, G. Marangoni and L. Cattalini, *J. Chem. Soc., Dalton Trans.* (1973) 886.
45 R. W. Baker and P. J. Pauling, *J. Chem. Soc., Chem. Commun.* (1969) 745.
46 M. McPartlin and A. J. Markwell, *J. Organomet. Chem.*, 57 (1973) C25.
47 N. C. Baenziger, W. E. Bennett and D. M. Soboroff, *Acta Crystallogr., Sect. B.*, B32 (1976) 962.
48 J. J. Guy, P. G. Jones, M. J. Mays and G. M. Sheldrick, *J. Chem. Soc., Dalton Trans.*, (1977) 8.
49 G. J. Arai, *Rec. Trav. Chim. Pays-Bas*, 81 (1962) 307.
50 M. G. B. Drew and M. J. Riedl, *J. Chem. Soc., Dalton Trans.*, (1973) 52.
51 P. T. Beurskens, H. J. A. Blaauw, J. A. Cras and J. J. Steggerda, *Inorg. Chem.*, 7 (1968) 805.
52 F. W. B. Einstein and R. Restivo, *Acta Crystallogr., Sect. B.*, B31 (1975) 624.

CHAPTER 2

BINARY COMPOUNDS OF GOLD

The chemistry of gold is dominated by the complexes of gold(I) and gold(III), and the simple binary compounds, other than alloys, are often unstable in air. Nevertheless the binary compounds, particularly the halides of gold, are frequently used as starting materials for preparing complexes of gold and have interesting properties which are often quite different from those of gold complexes.

HYDRIDES OF GOLD

There have been reports of the preparation of solid gold hydrides at room temperature but these have all been disproved. More recent reports suggest that AuH and AuH_3 may be prepared at low temperatures in ether solution, but that they decompose very rapidly to gold and hydrogen on warming [1]. For example, the reaction of Au_2Cl_6 with LiH, $Li[AlH_4]$ or $Li[BH_4]$ in ether at -120 °C apparently gave AuH_3, $Au[AlH_4]_3$ or $Au[BH_4]_3$, respectively, but the products decomposed on warming. Thus $Au[BH_4]_3$ decomposed according to equation (1).

$$Au[BH_4]_3 \rightarrow Au + 1.5H_2 + 3BH_3 \qquad (1)$$

The monomeric compounds AuH and AuD certainly exist in the gas phase when gold is heated to about 1400 °C in an atmosphere of hydrogen or deuterium, respectively. Detailed spectroscopic studies have been carried out on these compounds, and as a result the bond dissociation energy and Au—H bond distance are calculated to be 310 kJ mol^{-1} and 152.5 pm, respectively [2].

HALIDES OF GOLD

The known halides of gold, together with some thermodynamic data [3], are listed in Table 2.1. It can be seen that only fluorine gives a gold(V) derivative, while iodine gives only

TABLE 2.1

Halides of gold and thermodynamic data at 298 K for crystalline compounds

Compound	Colour	ρ_{20}^{pyk} / g cm^{-3}	$\Delta H_f^\ominus$ / kJ mol^{-1}	$\Delta G_f^\ominus$ / kJ mol^{-1}
Oxidation state (I)				
(AuF)			−75 est.	
AuCl	yellow-white	7.8	−35	−16
AuBr	light yellow	7.9	−19 [20]	−15 [20]
AuI	lemon yellow	8.25	1.7 [22]	−3.3 [22]
Oxidation state (III)				
AuF_3	golden yellow	6.7	−360 [15]	
$AuCl_3$	red	4.3	−121	−54
$AuBr_3$	dark brown	—	−67.3 [20]	−36 [20]
Oxidation state (V)				
AuF_5	dark red	—		

a gold(I) compound. This stabilisation of higher oxidation states by fluorine is of course a common phenomenon with elements other than gold.

(a) *Fluorides of gold*

There is no evidence for the existence of AuF and estimated thermodynamic data (Table 2.1) suggest that it will be unstable with respect to disproportionation to gold and AuF_3.

AuF_3 was first prepared [4] by heating $AuF_3 \cdot BrF_3$, formed by dissolution of gold in BrF_3, at 300 °C. The product formed in this way is contaminated with bromine, and the pure compound is best prepared by reaction of either gold(III) chloride or gold powder with fluorine at high temperature [5, 6].

Gold(III) fluoride can be sublimed at 300 °C under vacuum to give bright golden-yellow needles. It decomposes at 500 °C to gold and fluorine, and is a powerful fluorinating agent, causing benzene to burn. It forms the adducts $AuF_3 \cdot BrF_3$, $AuF_3 \cdot SeF_4$ and $NO[AuF_4]$.

AuF_3 is diamagnetic showing that gold(III) has the usual low spin d^8 configuration, and the compound has a unique structure which is illustrated in Fig. 2.1. It can be described as a fluorine-bridged helical chain polymer, the chains being cross-linked with weaker fluorine bridges. Each gold atom forms strong bonds to two terminal and two *cis*-bridging

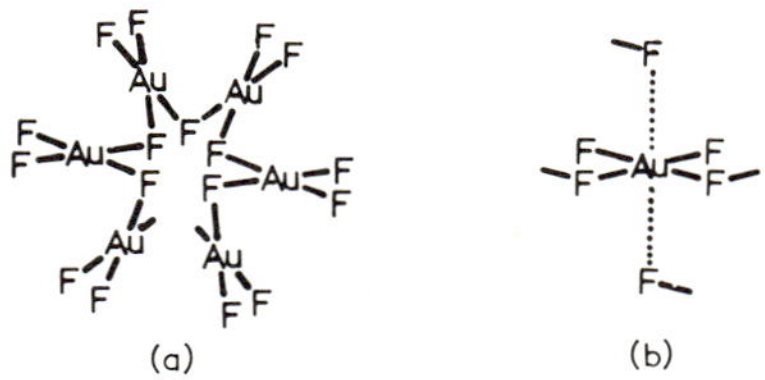

Fig. 2.1. The structure of AuF_3. (a) Part of a helical chain; (b) the co-ordination about gold(III).

fluorides to give overall a square-planar arrangement of fluorine atoms about each gold centre. These square-planar gold units are linked together by the bridging fluorines to give the helical-chain structure. There is then additional weak cross-linking in which the terminal fluorines occupy positions above and below the square-planar gold(III) centre of another chain, to give an approximately tetragonally distorted octahedral environment about each gold atom.

Gold(V) fluoride [7, 8] can be prepared by thermal decomposition of either $[KrF]^+[AuF_6]^-$ or of $[O_2]^+[AuF_6]^-$.

$$[KrF][AuF_6] \rightarrow AuF_5 + Kr + F_2$$

$$[O_2][AuF_6] \rightarrow AuF_5 + O_2 + \tfrac{1}{2}F_2$$

Gold(V) fluoride is diamagnetic so that gold(V) has the low spin $5d^6$ electron configuration. It is certainly not a monomer, but is associated by the presence of fluorine bridges. The structure is not known with certainty but, from powder X-ray photographs, a structure similar to that of RuF_5 is suggested. This is tetrameric with bent RuFRu bridges, each ruthenium having approximately octahedral stereochemistry. AuF_5 is a dark red solid which melts over the range 75—78 °C, and can be sublimed under vacuum. It combines with oxygen and fluorine to reform $[O_2][AuF_6]$, with XeF_2 to give $[Xe_2F_3][AuF_6]$ and with NOF to give $[NO][AuF_6]$.

Attempts have been made to prepare gold(VI) fluoride, which is expected to be the most powerful oxidising agent of the third transition series metal hexafluorides, but all attempts have failed.

(b) Chlorides of gold

Reaction of chlorine with gold may give either gold(I) chloride or gold(III) chloride depending on the reaction conditions. A claim that $AuCl_2$ could be formed has been disproved.

The thermodynamics of the gold—chlorine system have been thoroughly studied [3, 9, 10], and some of the more important conclusions are discussed here. At room temperature gold(I) chloride is metastable, and very slowly disproportionates to gold and gold(III) chloride, Au_2Cl_6. At a chlorine pressure of one atmosphere, Au_2Cl_6 exists as the only species in both solid (s) and gaseous (g) phases at temperatures below 254 °C. At temperatures between 254 °C and 282 °C, AuCl(s) exists together with Au_2Cl_6(g), while above 282 °C Au(s), Au_2Cl_6(g) and Au_2Cl_2(g) coexist. Thus Au_2Cl_6 can be prepared by reaction of gold with chlorine at temperatures below 254 °C. Typically a rapid flow of chlorine gas is passed over gold at 240 °C, when Au_2Cl_6 is formed and sublimes to give red crystals on the cooler parts of the tube. Alternatively, the reaction can be carried out at 250 °C in a salt bath [11]. Large crystals of the yellowish-white AuCl can be prepared by vapour transport of Au_2Cl_6(g) in an atmosphere of chlorine. Thus Au_2Cl_6(s) contained at one end of a tube is heated to 247 °C, and is deposited as AuCl(s) at the other end which is heated to 272 °C. Alternatively, AuCl can be prepared by heating Au_2Cl_6 in a stream of nitrogen at a temperature of 95 °C, or by decomposition of the carbonyl [AuCl(CO)]. Au_2Cl_6 can also be prepared conveniently by reaction of $HAuCl_4 \cdot 3H_2O$ with thionyl chloride [12].

The structure of AuCl is polymeric with zig-zag chains of Au and Cl atoms, with each chlorine bridging between two gold atoms [10, 13] (Fig. 2.2).

Thus, each gold atom has the linear co-ordination which is typical for gold(I). The bonding is evidently predominantly covalent, and the structure is quite different from that of AgCl which adopts the NaCl ionic lattice structure.

AuCl disproportionates to gold and gold(III) chloride slowly at room temperature, and decomposes to gold and chlorine on being heated above 200 °C. The disproportionation to gold

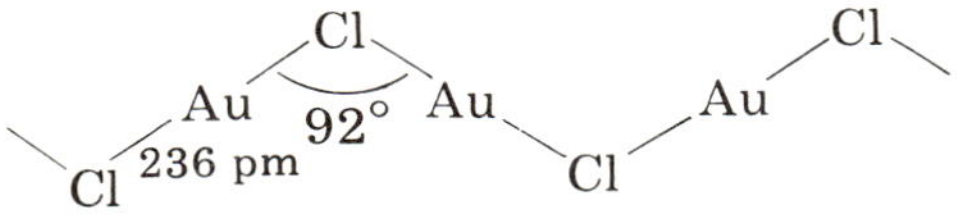

Fig. 2.2. The structure of AuCl.

and gold(III) species including $[AuCl_3(OH_2)]$ is very rapid in water, and AuCl must be stored in a desiccator to minimise this mode of decomposition. In the gas phase gold(I) chloride exists as the dimer Au_2Cl_2, which may have a cyclic structure, and for which the enthalpy of dissociation to monomeric AuCl is 68 kJ mol^{-1} [14].

The most typical reaction of AuCl is to form complexes of the type LAuCl, for which the ligand L may be ammonia, pyridine, a trialkylphosphine, a dialkylsulphide, carbon monoxide, an alkene and many others. These, and complexes derived by addition of more than one equivalent of the ligand L, will be discussed in Chapter 3. AuCl dissolves in concentrated sodium chloride solution to give $Na[AuCl_2]$ but the $[AuCl_2]^-$ ion is itself decomposed by water to give Au and $[AuCl_4]^-$.

Gold(III) chloride has the dimeric chlorine bridged structure shown in Fig. 2.3, in which each gold atom has square-planar stereochemistry.

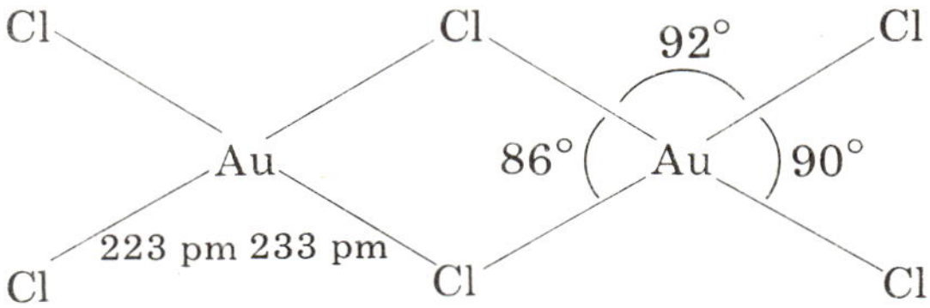

Fig. 2.3. The structure of Au_2Cl_6.

It dissolves in chloride solutions to give the $[AuCl_4]^-$ ion, and in water to give various hydrolysed species such as $[AuCl_3(OH)]^-$. The chlorine bridges of $[Au_2Cl_6]$ are easily broken by many neutral ligands, L, to give complexes $[LAuCl_3]$ where L may be pyridine for example. With easily oxidised ligands $[Au_2Cl_6]$ may act as a chlorinating agent, being itself reduced to gold(I). For example, tertiary phosphines typically give $[R_3PAuCl]$ and R_3PCl_2. Alkenes and alkynes, and organo-tin and -lead compounds are also chlorinated by $[Au_2Cl_6]$ [15, 16]. A particularly interesting reaction is the auration of aromatic compounds to give arylgold(III) compounds [17, 18]

$$Au_2Cl_6 + 2C_6H_6 \rightarrow [(AuCl_2Ph)_2] + 2HCl$$

These reactions do not take place, however, if the aromatic compound contains a substituent which can co-ordinate to gold. Thus benzonitrile and azobenzene form simple nitrogen—donor complexes of the type $[LAuCl_3]$ [19].

A particularly stable graphite insertion compound of composition $C_{12.6}AuCl_3$ is formed by direct reaction of graphite with $[Au_2Cl_6]$. The stability is due to the short inter-planar distance of 680 pm, no doubt made possible by the planar structure of $[Au_2Cl_6]$. This dimeric structure of $[Au_2Cl_6]$ persists in the gas phase [9, 14], and the enthalpy of dissociation to monomeric $[AuCl_3]$ is estimated to be 77 kJ mol^{-1}.

Many claims for the formation of the compound "$AuCl_2$" have been proved to be false, but a black compound of this empirical formula is formed by reaction of gold(III) chloride in thionyl chloride with a deficiency of carbon monoxide.

The reaction gives [Au(CO)Cl] in the first stage and this reacts with excess of $[Au_2Cl_6]$ as follows:

$$[Au(CO)Cl] + \frac{1}{2}(Au_2Cl_6) \rightarrow Au_2(CO)Cl_4$$

$$Au_2(CO)Cl_4 + Au_2Cl_6 \rightarrow 4\,AuCl_2 + COCl_2$$

The product $AuCl_2$ is diamagnetic and so is most likely a mixed gold(I)—gold(III) species; this is confirmed by the structure determination [19b] which shows the presence of cyclic $[Au_4Cl_8]$ molecules with linear gold(I) and square planar gold(III) centres as shown in Fig. 2.4.

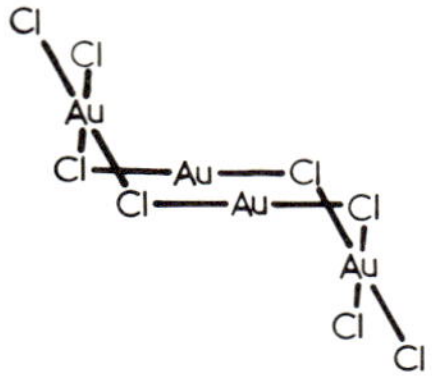

Fig. 2.4. The structure of Au_4Cl_8.

(c) Bromides and iodides of gold

Gold(III) bromide is formed by the exothermic reaction of gold powder with liquid bromine. It forms dark red-brown crystals. The structure is dimeric like that of $[Au_2Cl_6]$ though the full crystal structure has not been completed [13]. Its chemical properties also resemble those of $[Au_2Cl_6]$. Thus it reacts with nitrogen donor ligands, L, to give complexes $[LAuBr_3]$ and is reduced by tertiary phosphines and arsines and by dialkylsulphides to give complexes [LAuBr]. On heating to about 200 °C, $[Au_2Br_6]$ decomposes to bromine and AuBr. Again like the corresponding chloride, this is moisture sensitive, decomposing to gold and gold(III) species, and forms complexes [LAuBr] with many ligands L and the complex ion $[AuBr_2]^-$ with bromide. AuBr decomposes to the elements on being heated above 250 °C.

Several claims have been made that gold(III) iodide can be prepared from solutions of gold(III) complexes containing iodide ions. However, many gold(III) complexes readily oxidise iodide to iodine, and it now seems likely that the isolated compounds with empirical formula AuI_3 are actually compounds of gold(I) with I_3^- rather than gold(III) iodide [20]. The compound AuI_3 readily eliminates iodine to give gold(I) iodide, which may be obtained directly by reaction of iodide with aqueous solutions of $H[AuCl_4]$ or, slowly, by direct reaction of gold with iodine at elevated temperature. It forms yellow crystals which are much more stable to water than AuCl, probably due to its much lower solubility [21]. It dissolves in water in the presence of iodide ions to give $[AuI_2]^-$, and in iodide—iodine solutions to give mixtures of $[AuI_2]^-$, and $[AuI_4]^-$.

The structure of AuI is like that of AuCl. Each gold is linearly co-ordinated to two iodine atoms at a distance of 262 pm with four more iodines further away at 308 pm [22].

PSEUDOHALIDES OF GOLD

Gold(I) cyanide is readily formed by heating $K[Au(CN)_2]$ with hydrochloric acid. It forms yellow crystals, which have a polymeric structure with linear —AuCNAuCN— chains. It is insoluble in water but dissolves readily in the presence of cyanide as the complex ion $[Au(CN)_2]^-$. This reaction has been invoked as the final step in the dissolution of gold by cyanide in the presence of oxygen [23].

Anhydrous gold(III) cyanide is not known, probably due to its inability to dimerise like the gold(III) halides, but the complex $Au(CN)_3 \cdot 3H_2O$ can be prepared by dehydration of solutions of $H[Au(CN)_4]$. Its structure is not known, but it decomposes readily to gold(I) cyanide on being heated.

An alternative route to gold(I) cyanide is by heating dialkylgold(III) cyanides, and this method has also been used to prepare gold(I) thiocyanate [24]. Thus, heating $[(Et_2AuSCN)_2]$ in xylene at 125 °C gives AuSCN as an insoluble colourless powder, which decomposes to gold at 140 °C. AuSCN dissolves in water or alcohol in the presence of thiocyanate ions to give $[Au(SCN)_2]^-$, which decomposes rather rapidly in water. The potassium salt $K[Au(SCN)_2]$ can be obtained from ethanol, however.

OXIDES AND HYDROXIDES OF GOLD

Gold(III) hydroxide $Au(OH)_3$ is formed by addition of sodium hydroxide to solutions containing $[AuCl_4]^-$. This hydroxide can be dehydrated to give first Au(O)OH and then Au_2O_3. Au_2O_3 decomposes to gold and oxygen above 160 °C. The most recent determination gives $\Delta H_f^{\ominus}(Au_2O_3) = -13.0 \pm 2.4$ kJ mol^{-1}, and Au/Au_2O_3 acts as a true metal/metal oxide electrode system with $\pi^{\ominus} = 1.36$ V [25, 26].

Gold(III) hydroxide reacts with perchloric acid on being heated in a sealed tube at 150 °C for 2 weeks to give red transparent crystals of AuOCl. The gold(III) oxide chloride decomposes to the elements at 290 °C [27]. $Au(OH)_3$ is generally known as auric acid since it behaves as a weak acid. Thus it dissolves readily in KOH solution to give after evaporation $KAuO_2 \cdot 3H_2O$, which probably contains the $[Au(OH)_4]^-$ ion. Neutralisation of such solutions leads to reprecipitation of gold(III) hydroxide. Although the hydroxide is soluble to some extent in sulphuric or nitric acid, there is no evidence for formation of the aquogold(III) ion in the solutions. There is evidence that the species $[H_2AuO_3]^-$, $[HAuO_3]^{2-}$ and $[AuO_3]^{3-}$ are formed on treating $Au(OH)_3$ with increasingly concentrated hydroxide solutions, but the structures of these ions are not known [28].

Reports of the existence of gold(I) oxide and gold(I) hydroxide are now regarded with scepticism, but the oxidation of CsAu with oxygen at 400 °C has been shown to give the aurate(I) derivative CsAuO [29].

SULPHIDES, SELENIDES AND TELLURIDES OF GOLD

Gold does not react directly with sulphur, but the compounds Au_2S and Au_2S_3 can be prepared. Gold(I) sulphide is formed by reaction of $[Au(CN)_2]^-$ with H_2S. It is a black-brown powder which is only very slightly soluble in water [30] ($K_{sp} = [Au^+]^2[S^{2-}] = 4 \times 10^{-69}$ at 25 °C) but which dissolves in cyanide or polysulphide solutions. The standard Gibbs free energy of formation $\Delta G_f^{\ominus}$ is +29 kJ mol^{-1}. In the gas phase at high temperature Au_2S molecules apparently exist [31].

Gold(III) sulphide can be prepared by reaction of H_2S with $[Au_2Cl_6]$ in anhydrous ether at low temperature. Attempts to prepare it in aqueous solution lead to reduction to metallic gold. The black powder decomposes to the elements on being heated to 200 °C. It dissolves with decomposition to $[Au(CN)_2]^-$ in cyanide solution.

Two selenides of gold are known, AuSe and Au_2Se_3. There are two crystalline forms of AuSe, both of which form monoclinic crystals. The β-form of AuSe contains gold in two forms, one with linear co-ordination to two selenium atoms and the other with square-planar co-ordination to four selenium atoms. It is, therefore, best represented as Au(I)Au(III)Se_2.

Gold reacts directly with tellurium to give $AuTe_2$, which also occurs in the minerals calaverite and sylvanite. The crystals contain linear TeAuTe units. The standard enthalpy of formation $\Delta H_f^{\ominus}$ is (− 18.6 ± 3) kJ mol^{-1} [33]. In addition Au_2Te_3 occurs in the mineral montbrayite and a compound of composition $AuTe_{1.75}$ is formed by reduction of solutions containing gold(III) and tellurium(IV) with hydrazine [34].

In addition to the binary selenides and tellurides, the mixed compounds AuBrSe, AuITe, $AuClTe_2$, $AuBrTe_2$ and $AuITe_2$ are known. Typically, $AuITe_2$ is prepared by reaction of elemental gold and tellurium with hydriodic acid in a sealed tube at a temperature of 450 °C, the mixture then being cooled slowly to 150 °C over a period of ten days [35, 36]. The compounds $AuXTe_2$ all exhibit metallic conductivity, and the structures have gold co-ordinated to four tellurium atoms [37]. The structure of AuBrSe contains gold in square-planar co-ordination with some gold atoms co-ordinated to four selenium atoms and some to two bromine and two selenium atoms [38].

GOLD(III) FLUOROSULPHATE AND NITRATE

Reaction of gold with bromine(I) fluorosulphate at 65 °C gives a complex $Au(SO_3F)_3 \cdot 2BrSO_3F$ which loses $BrSO_3F$ on being heated under vacuum to give $Au(SO_3F)_3$. This is a bright orange-yellow solid which melts at 94 °C. Its structure is not known but it appears to be polymeric with bridging fluorosulphate groups [39].

Dissolution of Au_2O_3 in concentrated nitric acid does not give anhydrous gold(III) nitrate but the complex acid $H[Au(NO_3)_4] \cdot 3H_2O$. It has been claimed that reaction of this product with N_2O_5 followed by heating of the product under vacuum gives a golden-yellow sublimate of $Au(NO_3)_3$ at 80 °C/4 x 10^{-3} Torr. The yield was very low and the characterisation relied entirely on analysis for gold [40]. The claim must therefore be considered doubtful though complex nitratoaurates(III), $M^+[Au(NO_3)_4]^-$, are well characterised [41].

NITRIDES AND OTHER GROUP V DERIVATIVES

No simple gold nitrides are known but there was great interest in the Middle Ages in an explosive substance called fulminating gold. This is formed by reaction of Au_2O_3 with aqueous ammonia or ammonium salts, by reaction of a gold(III) salt and an ammonium salt with base and by several other routes. It has been variously formulated as $AuN \cdot NH_3 \cdot 1\frac{1}{2}H_2O$, $Au(NH)NH_2$ or $Au_2O_3 \cdot 4NH_3$ and probably several different mixtures of compounds may be formed depending on the preparative route. All detonate to give gold, nitrogen, ammonia and water.

The compounds Au_2P_3 ($\Delta H_f^{\ominus}$ −97.5 or −48.5 kJ mol^{-1}) and $AuSb_2$ ($\Delta H_f^{\ominus}$ −13.1 kJ mol^{-1}) can be prepared by direct combination of the elements at high temperature [42]. $AuSb_2$ has the electrical properties of a metal [42], and has the cubic fluorite structure, the shortest Au—Sb and Sb—Sb distances being 276 and 286 pm, respectively [43].

REFERENCES

1 E. Wiberg and H. Neumaier, *Inorg. Nucl. Chem. Lett.*, 1 (1965) 35.
2 U. Ringstrom, *Nature*, 198 (1963) 981; *Ark. Fys.*, 27 (1964) 227; J. P. Desclaux and P. Pyykko, *Chem. Phys. Lett.*, 39 (1976) 300.
3 L. M. Gedansky and L. G. Hepler, *Engelhard Ind. Tech. Bull.*, 10 (1969) 5.

4 A. G. Sharpe, *J. Chem. Soc.*, (1949) 2901.
5 L. B. Asprey, F. H. Kruse, K. H. Jack and R. Maitland, *Inorg. Chem.*, 3 (1964) 602.
6 F. W. B. Einstein, P. R. Rao, J. Trotter and N. Bartlett, *J. Chem. Soc. A*, (1967) 478.
7 J. H. Holloway and G. J. Schrobilgen, *J. Chem. Soc. Chem. Commun.*, (1975) 623; V. B. Sokolov, V. N. Prusakov, A. V. Ryzhkov and Yu. V. Drobyshevskii, *Dokl. Akad. Nauk SSSR*, 229 (1976) 884.
8 M. J. Vasile, T. J. Richardson, F. A. Stevie and W. E. Falconer, *J. Chem. Soc., Dalton Trans.*, (1976) 351.
9 E. M. W. Janssen, F. Pohlmann and G. A. Wiegers, *J. Less-Common Met.*, 45 (1976) 261.
10 E. M. W. Janssen, J. C. W. Folmer and G. A. Wiegers, *J. Less-Common Met.*, 38 (1974) 71.
11 T. Mundorf and K. Dehnicke, *Z. Naturforsch., Teil B*, 28 (1973) 506.
12 D. B. Dell'Amico and F. Calderazzo, *Gazz. Chim. Ital.*, 103 (1973) 1099.
13 J. Straehle and K. P. Loercher, *Z. Naturforsch., Teil B*, 29 (1974) 266.
14 M. N. Zyryanov and G. A. Khlebnikova, *Izv. Vyssh. Uchebn. Zaved., Tsvetn. Metal.*, 16 (1973) 112.
15 G. Tagliavini, U. Belluco and G. Pilloni, *Ric. Sci. Parte 2, Sez. A*, 3 (1963) 889.
16 R. Hüttel, H. Reinheimer and K. Nowak, *Chem. Ber.*, 101 (1968) 3761.
17 K. S. Liddle and C. Parkin, *J. Chem. Soc., Chem. Comm.*, (1972) 26.
18 P. W. J. de Graaf, J. Boersma and G. J. M. van der Kerk, *J. Organomet. Chem.*, 105 (1976) 399.
19 (a) F. Calderazzo and D. B. Dell'Amico, *J. Organomet. Chem.*, 1974, 76, C59. (b) D. B. Dell'Amico, F. Calderazzo and F. Marchetti, *J. Chem. Soc., Dalton Trans.*, (1976) 1829; D. B. Dell'Amico, F. Calderazzo, F. Marchetti, S. Merlino and G. Perego, *J. Chem. Soc., Chem. Commun.*, (1977) 31.
20 J. L. Ryan, *Inorg. Chem.*, 8 (1969) 2058.
21 A. Hakansson and L. Johansson, *Chem. Scr.*, 7 (1975) 201.
22 R. W. G. Wyckoff, *Crystal Structures*, Vol. 1, Interscience, New York, 2nd ed., 1963, p. 159.
23 I. A. Kakovskii and G. F. Cherkasov, *Chem. Abstr.*, 81 (1974) 173471.
24 W. L. Gent and C. S. Gibson, *J. Chem. Soc.*, (1949) 1835.
25 S. J. Ashcroft and E. Schwarzmann, *J. Chem. Soc., Faraday Trans. 1*, 68 (1972) 1360.
26 J. P. Hoare, *Electrochim. Acta*, 11 (1966) 203.
27 E. Schwarzmann, E. Schulze and J. Mohn, *Z. Naturforsch., Teil B*, 29 (1974) 561.
28 H. L. Johnston and H. L. Leland, *J. Am. Chem. Soc.*, 60 (1938) 1439.
29 K. Hestermann and R. Hoppe, *Z. Anorg. Allg. Chem.*, 360 (1968) 113.
30 A. de Cugnac-Pailliotet and J. Pouradier, *C. R. Acad. Sci., Ser. C*, 275 (1972) 551.
31 S. Smoes, F. Mandy, A. V. Auwera-Mahieu and J. Drowart, *Bull. Soc. Chim. Belg.*, 81 (1972) 45.
32 J. E. Cretier and G. A. Wiegers, *Mater. Res. Bull.*, 8 (1973) 1427.
33 R. F. L. Andon, J. F. Martin and K. C. Mills, *J. Chem. Soc. A*, (1971) 1788.
34 F. H. Howie and C. R. Veale, *J. Inorg. Nucl. Chem.*, 28 (1966) 1149.
35 A. Rabenau and H. Rau, *Inorg. Synth.*, 14 (1973) 160.
36 A. Rabenau, H. Rau and G. Rosenstein, *Angew. Chem. Int. Ed. Eng.*, 8 (1969) 145.
37 H. M. Haendler, D. Mootz, A. Rabenau and G. Rosenstein, *J. Solid State Chem.*, 10 (1974) 175.
38 D. Mootz, A. Rabenau, H. Wunderlich and G. Rosenstein, *J. Solid State Chem.*, 6 (1973) 583.
39 W. M. Johnson, R. Dev and G. H. Cady, *Inorg. Chem.*, 11 (1972) 2260.

40 B. O. Field and C. J. Hardy, *J. Chem. Soc.*, (1964) 4428.
41 C. C. Addison, G. S. Brownlee and N. Logan, *J. Chem. Soc., Dalton Trans.*, (1972) 1440.
42 W. D. Johnston, R. C. Miller and D. H. Damon, *J. Less-Common Met.*, 8 (1965) 272; C. E. Myers, T. J. Conti and N. F. Marley, *J. Less-Common Met.*, 48 (1976) 213.
43 S. Furuseth, K. Selte and A. Kjekshus, *Acta Chem. Scand.*, 19 (1965) 735.

CHAPTER 3

GOLD(I) COMPLEXES

In this chapter the complexes of gold(I) are discussed according to the nature of the donor atom(s) of the ligand(s) complexed to gold. We shall examine complexes with group IV, V, VI and VII donors in that sequence. Many of the group IV donor complexes will be discussed in subsequent chapters on organogold complexes and on compounds with gold—metal bonds, and hence only the carbonyl, cyanide and alkyl and aryl isocyanide complexes are considered here.

CARBONYL COMPLEXES [1]

The only well-characterised compound is [AuCl(CO)]. This was originally prepared by treatment of CO with solid AuCl at 90 °C. The carbonyl complex is a white solid which sublimes in a stream of CO, but which dissociates back to AuCl and CO on being heated under vacuum. It can be prepared more easily by bubbling CO through a suspension of AuCl in benzene, in which solvent the product [AuCl(CO)] is soluble, or by bubbling CO through a solution of $[Au_2Cl_6]$ in tetrachloroethylene at 120 °C. In the latter case phosgene is formed [2].

$$[Au_2Cl_6] + 4CO \rightarrow 2[AuCl(CO)] + 2COCl_2$$

A particularly convenient preparative route involves dehydration of the readily available tetrachloroauric(III) acid, $H[AuCl_4] \cdot 3H_2O$, with thionyl chloride to give $[Au_2Cl_6]$, which is treated *in situ* with carbon monoxide to give [AuCl(CO)] and phosgene [3]. Anhydrous conditions are necessary for the preparation since decomposition to gold and CO occurs very readily on reaction with water. Even moist air or alcohol decomposes [AuCl(CO)] in this way.

Chloro(carbonyl)gold(I) is monomeric in solution, and hence has the usual linear two-co-ordinate gold(I) centre. The carbonyl stretching frequency of 2080 cm^{-1} is not much lower than that for free carbon monoxide and indicates that back-bonding is weak. The gold—carbonyl bond is certainly weak and the most typical reaction of [AuCl(CO)] involves displacement of CO by other ligands [2, 4].

$$[AuCl(CO)] + C_5H_5N \longrightarrow [C_5H_5N \rightarrow AuCl] + CO$$

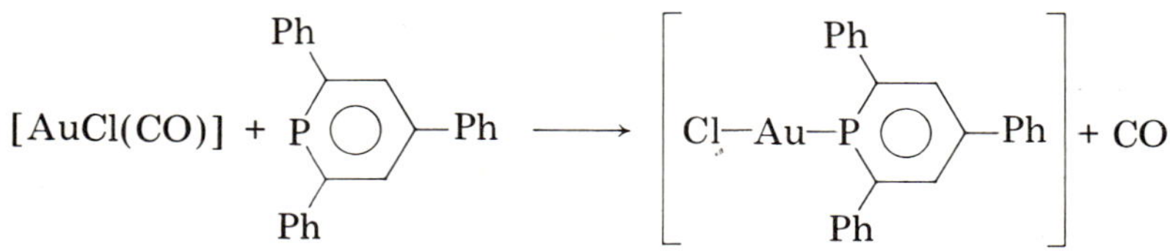

Treatment of [AuCl(CO)] with Grignard reagents RMgX gives carbon monoxide and products arising from decomposition of the corresponding organogold species $[(RAu)_n]$ [2].

ISOCYANIDE COMPLEXES

Alkyl and aryl isocyanides are better σ-donors than carbon monoxide, and hence are able to form a wide range of complexes with gold. The complexes may be prepared by reaction of tetrachloroauric(III) acid with isocyanides [5—7]. The first step involves formation of $[Au(RNC)Cl_3]$ and excess of isocyanide causes reduction to give [Au(RNC)Cl] [5]. However, the preferred route to isocyanide complexes involves displacement of dimethylsulphide from $[Au(SMe_2)Cl]$ [8, 9].

Addition of excess of isocyanide gives the linear complex ions $[Au(RNC)_2]^+$ which can be isolated as the hexafluorophosphate salts, and there is also evidence for formation of the ions $[Au(RNC)_4]^+$ which are probably tetrahedral.

The nature of the bonding in isocyanide complexes of gold has been investigated using i.r. and n.m.r. spectroscopy. The N≡C stretching frequency is always higher in the gold complex than in the parent isocyanide. Thus, $\nu(NC)$ is 2123 cm^{-1} in 4-fluorophenylisocyanide but 2227 cm^{-1} and 2200 cm^{-1} in $[Au(4\text{-}FC_6H_4NC)Cl]$ and $[Au(4\text{-}FC_6H_4NC)(4\text{-}FC_6H_4)]$, respectively. This effect is attributed to polarisation in the sense $Ar{-}\overset{\delta+}{N}{\equiv}C{-}\overset{\delta-}{Au}Cl$, and indicates that backbonding from d-orbitals on gold to π^* orbitals of the isocyanide, which should decrease $\nu(NC)$, is very low. This is confirmed by comparing the ^{19}F chemical shifts in complexes $[Au(4\text{-}FC_6H_4NC)X]$ and $[Au(3\text{-}FC_6H_4NC)X]$, where X = halide, with those for the parent isocyanides. The fluorine atoms in both complexes are deshielded compared with those in the free isocyanides, this effect being greater for the 4-fluorophenylisocyanide complex. Since the ^{19}F chemical shift will be affected by both metal—ligand σ- and π-bonding in the 4-fluorophenyl derivative but only by σ-bonding effects in the 3-fluorophenylisocyanide complex, it seems that the aryl isocyanides act as both σ- and π-donors to gold(I). The polarisation $R{-}\overset{\delta+}{N}{\equiv}C{-}\overset{\delta-}{Au}$ appears greater[6] in the cation $[Au(^iPrNC)_2]^+$ with $\nu(NC)$ = 2265 cm^{-1} than in the neutral $[AuCl(^iPrNC)]$ molecule with $\nu(NC)$ = 2240 cm^{-1}.

It is the polarisation of the isocyanide which makes the co-ordinated ligand susceptible to attack by nucleophilic reagents, such as amines and alcohols with the formation of carbene complexes of gold. These, and related reactions will be considered further in Chapter 7.

$$[Au(CNC_6H_4Me)Cl] + MeOH \rightarrow [Au\{C(OMe)NHC_6H_4Me\}Cl]$$

$$[Au(CN^tBu)_2]^+ + {}^tBuNH_2 \rightarrow [Au\{C(NH^tBu)_2\}]^+$$

CYANIDE COMPLEXES OF GOLD(I) [10]

The linear complex ion $[Au(CN)_2]^-$ is perhaps the most stable formed by gold(I) and the stability constant has been estimated in aqueous solution as:

$$Au^+ + 2CN^- \rightleftharpoons [Au(CN)_2]^- \quad K \sim 10^{39}$$

There is no evidence for formation of species with higher co-ordination number analogous to the copper(I) complex ion $[Cu(CN)_4]^{3-}$. The potassium salt $K[Au(CN)_2]$ can be formed by dissolving gold powder in KCN solution in the presence of air. Alternatively, an electrolytic method based on dissolution of a pure gold anode in aqueous potassium cyanide solution can be used, or AuCN can be dissolved in KCN solution.

On boiling an aqueous solution of colourless $K[Au(CN)_2]$ with 2 M HCl, decomposition occurs and lemon-yellow AuCN crystallises from solution. The acid $H[Au(CN)_2]$ can be prepared by ion exchange from $K[Au(CN)_2]$, but decomposes to HCN and AuCN at 100 °C [11].

Oxidation of $[Au(CN)_2]^-$ by halogens occurs readily to give the corresponding ions *trans*-$[Au(CN)_2X_2]^-$ where X = Cl, Br, or I. A study of the kinetics of the oxidative addition of iodine to $[Au(CN)_2]^-$ leads to the conclusion that a concerted, one-step *trans* addition occurs. Oxidation by I_3^- takes place over 100 times faster than by free I_2, since the bond separation needed to give the *trans* adduct is the greater as illustrated below [12].

Other cyanide complexes include $Tl[Au(CN)_2]$, in which there is considerable Tl—Au interaction as evidenced by the infrared spectrum [13], $[Au(PPh_3)_3]$ $[Au(CN)_2]$ in which both cationic and anionic gold(I) species are present [14], and the linear neutral molecule $[Au(CN)(PPh_3)]$ [14—16].

There is good evidence that the gold—cyanide bond has significant contributions from σ-bonding and from π-back-bonding involving donation of electrons from filled 5d orbitals on gold to vacant π*-orbitals of the cyanide ligand. This has been made clear in X-ray structural studies [15] and especially in the analysis of the vibrational spectra [17, 18] of salts of $[Au(CN)_2]^-$. Some force constants for the

TABLE 3.1

Force constants (F) and intensity (k_6) for vibrations of complexes $[M(CN)_2]^-$

Complex	F(MC)/ N m^{-1}	F(CN)/ N m^{-1}	F(MC, MC)/ N m^{-1}	k_6/ 10 m mol^{-1}
$[Ag(CN)_2]^-$	183	1704	22	3100
$[Au(CN)_2]^-$	281	1675	51	4980
$[Hg(CN)_2]$	261	1762	14	71

$[Au(CN)_2]^-$ ion are compared with those for $[Ag(CN)_2]^-$ and $[Hg(CN)_2]$ are given in Table 3.1.

The magnitudes of the absolute intensity, k_6, of the infra-red-active C—N stretching band and of the interaction force constant F(MC,MC) are considered good measures of the extent of π-bonding. These follow the order $[Au(CN)_2]^- > [Ag(CN)_2]^- > [Hg(CN)_2]$, and hence Au—CN π-bonding is the strongest for $[Au(CN)_2]^-$. It will be seen that the metal—carbon stretching force constants are similar for $[Au(CN)_2]^-$ and for $[Hg(CN)_2]$. The M—C σ-bond is therefore stronger in $[Hg(CN)_2]$ than in $[Au(CN)_2]^-$. Confirmation comes from the magnitude of F(CN). Thus increased M$\leftarrow$C σ-bonding leads to an increase in the C$\equiv$N bond energy but increased M$\rightarrow$C π-bonding leads to a decrease. Thus, as expected, the force constant F(CN) is lowest for $[Au(CN)_2]^-$. The higher metal—carbon bond strength in $[Au(CN)_2]^-$ compared with $[Ag(CN)_2]^-$ is no doubt partially responsible for the high overall stability constant for formation of $[Au(CN)_2]^-$ (K ca. 10^{39}) compared with $[Ag(CN)_2]^-$ (K ca. 10^{20}).

NITROGEN DONOR COMPLEXES

(a) Preparative methods

Complexes of gold(I) with neutral nitrogen—donor ligands are generally prepared in the following ways.

1. By direct reaction with gold(I) halides

This is the method originally used to prepare ammine complexes of gold(I). Thus, dissolution of the gold(I) halides in liquid ammonia followed by evaporation of the solvent gives many gold—ammine complexes. The species stable at room temperature however are $AuCl \cdot 3NH_3$, $AuBr \cdot 2NH_3$ and $AuI \cdot NH_3$. The complex $AuCl \cdot NH_3$ can be prepared by dissolving AuCl in aqueous ammonia followed by acidification with HCl. The compounds $AuX \cdot NH_3$ are almost certainly linear, but the other complexes have never been fully characterised.

Cationic gold complex ions can be prepared in the presence of $SbCl_5$ as below

$$AuCl + SbCl_5 \xrightarrow{MeCN} [Au(MeCN)_2]^+[SbCl_6]^-$$

2. By oxidation of gold in a nitrogen—donor solvent

The oxidation of gold by $[NO]^+[ClO_4]^-$ in acetonitrile offers an alternative route to the ion $[Au(MeCN)_2]^+$ as the perchlorate salt, but the method has not been commonly used.

3. By ligand displacement

The acetonitrile ligands in $[Au(MeCN)_2]^+$ are easily displaced, for example by pyridine derivatives to give $[Au(py)_2]^+$. Derivatives of 2-, 3- and 4-cyanopyridine have been prepared in this way and isolated as the perchlorate salts. The 3- and 4-substituted derivatives are bound to gold by the pyridine nitrogen atom, but 2-cyanopyridine co-ordinates through the nitrile nitrogen atom [23]. Similarly, in the complex [LAuCl] where L is 2-diphenylphosphinopyridine the ligand is bound by the phosphorus rather than the nitrogen atom [24].

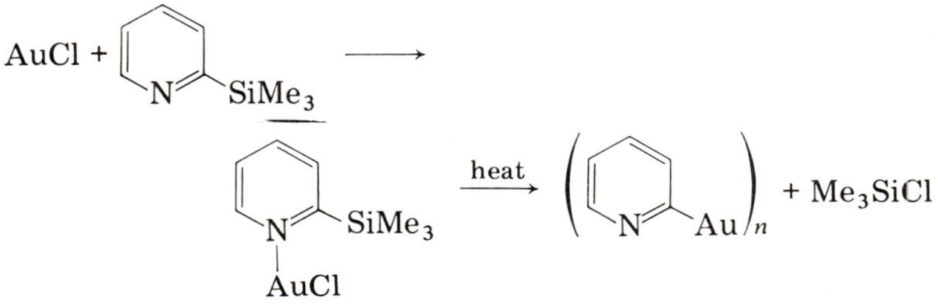

A particularly interesting rearrangement takes place on heating the 2-trimethylsilylpyridine complex of gold(I) chloride to give 2-pyridylgold [25].

Complexes are also known which contain bidentate ligands with both nitrogen and arsenic or sulphur donor atoms [26, 27]. Examples include the tetrahedral complexes $[AuL_2]^+$,

where L is 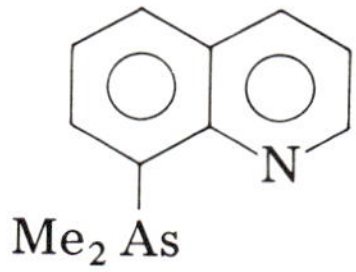 or 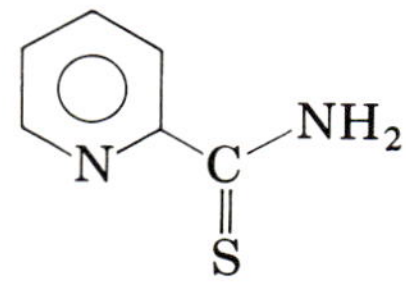

Rather surprisingly, the complexes $K[Au(CN)_2(2,2'\text{-bipyridine})]$ and $K[Au(CN)_2(1,10\text{-phenanthroline})]$, formed by direct reaction of the appropriate ligand with $K[Au(CN)_2]$, appear to have square-planar stereochemistry at the gold(I) centres. Further investigation of these complexes is merited. The complex $[Au(PPh_3)(2,2'\text{-bipyridine})]^+[PF_6]^-$ has an interesting structure in which the 2,2′-bipyridine is asymmetrically bound to gold with one Au–N bond (216.6 pm) considerably shorter than the other (240.6 pm). The distortion is such as to make the co-ordination about gold(I) intermediate between linear and trigonal planar [27b].

(b) Azide complexes of gold(I)

These compounds are usually prepared by reaction of the corresponding gold(I) chloride complex with lithium azide or sodium azide [16, 28—30].

$$[AuCl(PPh_3)] + LiN_3 \rightarrow [AuN_3(PPh_3)] + LiCl$$

Azidogold(I) complexes can also be prepared by decomposition of azidogold(III) complexes [28].

$$[Ph_4As][Au(N_3)_4] + 3PPh_3 \rightarrow [Ph_4As]N_3 + [AuN_3(PPh_3)_3] + 3N_2$$

The complex $[Ph_4As][Au(N_3)_4]$ also decomposes in daylight to give $[Ph_4As][Au(N_3)_2]$ [31].

The azido complexes undergo 1,3-dipolar cycloaddition reactions with many unsaturated reagents, but with carbon monoxide the corresponding isocyanatogold complex is formed with displacement of N_2. Some of these reactions are summarised in Fig. 3.1 [28–30].

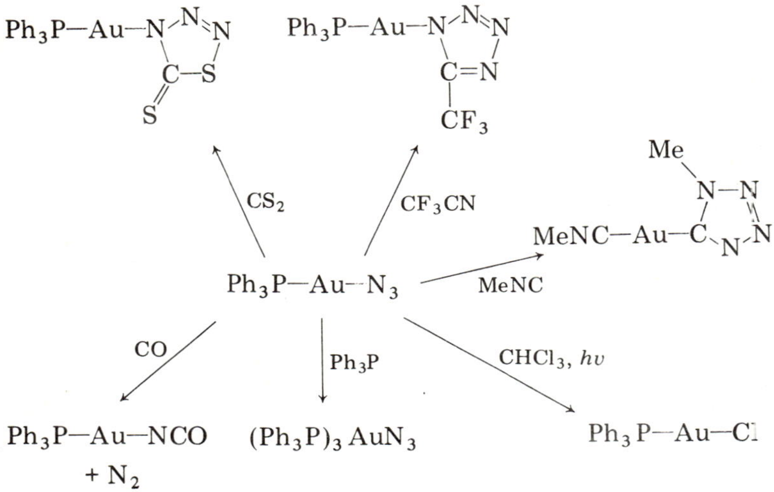

Fig. 3.1. Some reactions of azido(triphenylphosphine)gold(I).

(c) Isocyanate complexes of gold(I)

Isocyanatogold(I) complexes are prepared by metathesis from the corresponding chloro— or acetato—gold(I) complex [16, 32] or from azidogold(I) or azidogold(III) complexes with carbon monoxide [28].

$$[Ph_4As][Au(N_3)_4] \xrightarrow{CO} [Ph_4As][Au(NCO)_2]$$

The NCO^- ligand can bind through either the nitrogen or oxygen atom, but with gold only the N-bonded form is found. This is predicted in terms of the soft or "class b" character of the gold centre, which is also illustrated by the following reaction.

$$[OCN{-}Au{-}PPh_2CH_2CH_2PPh_2{-}Au{-}NCO] + (SCN)_2 \rightarrow$$
$$[Au(NCO)(SCN)_2(PPh_2CH_2CH_2PPh_2)Au(NCO)(SCN)_2]$$

The product of oxidative addition of thiocyanogen contains only N-bonded NCO^- and S-bonded SCN^- groups [32].

(d) Amide complexes of gold(I)

The simplest known amide complexes are the bis(trimethylsilyl) amides.

$$Me_3P{-}Au{-}Cl + NaN(SiMe_3)_2 \rightarrow Me_3P{-}Au{-}N(SiMe_3)_2$$

The trimethylphosphine complex is monomeric, and can be distilled under vacuum [33].

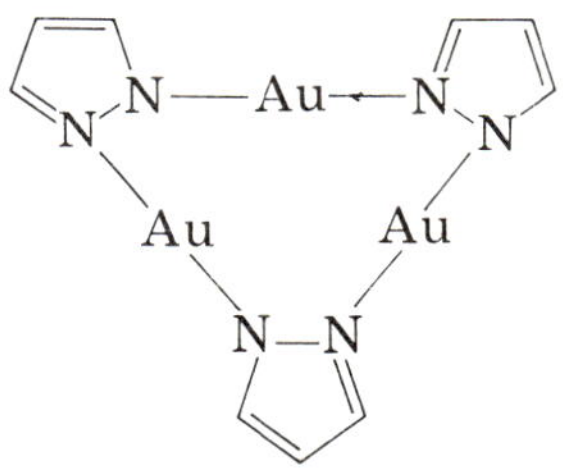

Fig. 3.2. The proposed structure of pyrazolylgold(I).

A more extensive series of compounds is derived from heterocyclic nitrogen compounds, including pyrazolyl [34], 1,2,4-triazolyl [35] and tetrazolyl [36] derivatives. The pyrazolyl derivatives $C_3H_3N_2Au$ are trimeric with a 9-membered ring, thus enabling each gold(I) centre to retain linear 2-coordination (Fig. 3.2) [34].

PHOSPHORUS, ARSENIC AND ANTIMONY DONOR COMPLEXES

A great number of tertiary phosphine complexes of gold(I) are known and it will not be possible to mention them all here [37]. A selection of complexes is made with the aim of illustrating general principles behind the synthetic methods therefore.

(a) Methods of preparation

1. Reduction of gold(III) complexes

This is the most common method of preparation and is illustrated by the preparation of $[AuCl(PPh_3)]$.

$$Na[AuCl_4] + 2Ph_3P \rightarrow [AuCl(PPh_3)] + PPh_3Cl_2 + NaCl$$

The mechanism of reaction has been studied and involves the initial formation $[AuCl_3(PPh_3)]$ which is then reduced by more triphenylphosphine [38].

$$[AuCl_4]^- + PPh_3 \rightarrow [AuCl_3(PPh_3)] + Cl^-$$

$$[AuCl_3(PPh_3)] + PPh_3 \rightarrow [AuCl(PPh_3)] + PPh_3Cl_2$$

However, the analogous reaction with triphenylarsine or triphenylstibine follows a different mechanism in which the first step involves reduction of $[AuCl_4]^-$ to $[AuCl_2]^-$.

$$[AuCl_4]^- + AsPh_3 \rightarrow [AuCl_2]^- + AsPh_3Cl_2$$

$$[AuCl_2]^- + AsPh_3 \rightarrow [AuCl(AsPh_3)] + Cl^-$$

The difference probably arises because the tertiary phosphine is a better ligand for gold(I) but a weaker reducing agent than triphenylarsine or triphenylstibine [38]. Some of the complexes which may be prepared by similar methods are given in Table 3.2.

TABLE 3.2

Some tertiary phosphine, arsine and stibine complexes of gold(I)

Complex[a]	m.p./°C	Ref.
$[AuCl(PPh_3)]$	242	39
$[AuBr(PPh_3)]$	250—251	40
$[AuI(PPh_3)]$	230—231	40
$[AuCl(AsPh_3)]$	219—220	40
$[AuCl(SbPh_3)]$	90 decomp.	40
$[AuCl\{P(CH_2SiMe_3)_3\}]$	166—168	41
$[AuCl(AsMe_3)]$	169—172	42
$[AuCl\{P(cyclohexyl)_3\}]$	222—225	43

[a] All complexes are white.

2. Preparations from other gold(I) complexes

Most frequently a more weakly bound ligand is replaced by a tertiary phosphine ligand [44, 45]

$$[AuCl(SMe_2)] + PPh_3 \rightarrow [AuCl(PPh_3)] + SMe_2$$

$$[AuCl(\text{cyclo-octene})] + P(OMe)_3 \rightarrow [AuCl\{P(OMe)_3\}] + \text{cyclo-octene}$$

It is also possible to prepare complexes directly from the gold(I) halides or by substitution at the phosphorus centre [46].

$$AuCl + PCl_3 \rightarrow [AuCl(PCl_3)]$$

$$[AuCl(PCl_3)] + 3MeOH \rightarrow [AuCl\{P(OMe)_3\}] + 3HCl$$

In some cases, one tertiary phosphine ligand may replace another [47, 48].

$$[AuMe(PMe_3)] + PEt_3 \longrightarrow [AuMe(PEt_3)] + PMe_3$$

$$2[AuCl(PEt_3)] + 2Et_2PCH_2CH_2SH \longrightarrow$$

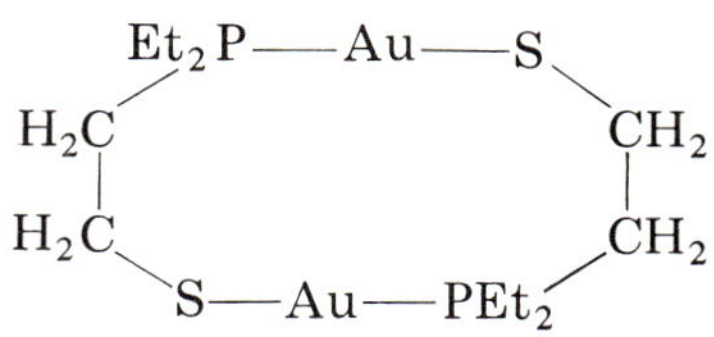

$+ 2PEt_3$

$+ 2HCl$

(b) Structure and reactions

The complexes LAuX almost invariably have linear two-co-ordination at the gold(I) centre as evidenced by structure determinations on $[AuX(PPh_3)]$ where X = Cl, Me, or CN [15, 49] and $[AuBr(AsPh_3)]$ [50]. Even when the anionic group is potentially bidentate, this structure is usually maintained. For example, $[Au(O_2CMe)(PPh_3)]$ appears to have a unidentate acetate group [16], though in the complex $[Au(S_2CNEt_2)(PPh_3)]$ there is some deviation from linearity due to weak interaction of gold with the second sulphur atom of the dithiocarbamate ligand as shown in Fig. 3.3 [51].

The strength of binding of gold(I) to tertiary phosphines is greater than to analogous arsines or stibines. Thus the equilibrium constants for the reaction below are 0.15 when $L = SbPh_3$, 13 when $L = AsPh_3$ and immeasurably large when $L = PPh_3$ [38].

$$[AuCl_2]^- + L \rightleftharpoons [AuClL] + Cl^-$$

The complexes [AuClL] do not interact with excess of ligand when $L = Ph_3As$ or Ph_3Sb, but with $L = Ph_3P$ further association can take place. The first step involves the equilibria:

$$[AuCl(PPh_3)] + Ph_3P \rightleftharpoons [AuCl(PPh_3)_2] \rightleftharpoons [Au(PPh_3)_2]^+ + Cl^-$$

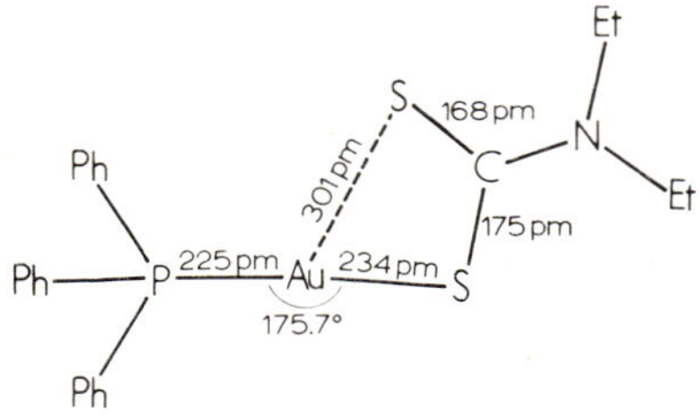

Fig. 3.3. The structure of N,N-diethyldithiocarbamato(triphenylphosphine)gold(I).

In the solid state $[AuCl(PPh_3)_2]$ has a trigonal planar structure [52], but in solution rapid reversible dissociation of triphenylphosphine or chloride takes place. Ionisation of chloride takes place to a greater extent in more polar solvents. Addition of more triphenylphosphine gives $[Au(PPh_3)_3]^+$ and finally $[Au(PPh_3)_4]^+$ which is co-ordinatively saturated. In general, the equilibrium constants decrease with increasing association, so that disproportionation reactions such as that shown below take place only to a minor extent.

$$2[AuCl(PPh_3)_2] \rightleftharpoons [AuCl(PPh_3)] + [Au(PPh_3)_3]^+ + Cl^-$$

Similar reactions have been studied with several other phosphine and phosphite ligands [45, 46, 53–56]. For example, with $[Au(MeCN)_2]^+$ the ligand

L = 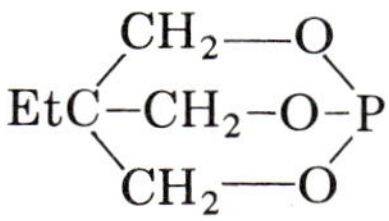

forms successively $[AuL(MeCN)]^+$, $[AuL_2]^+$, $[AuL_3]^+$ and $[AuL_4]^+$. With L = P(p-tolyl)$_3$ the complexes [AuClL], $[AuL_2]^+$, $[AuL_3]^+$ and $[AuL_4]^+$ have been identified in dichloromethane solution using ^{31}P n.m.r. spectroscopy [53], and an equilibrium constant K = 0.1 was found for the disproportionation reaction:

$$2[AuL_2]^+Cl^- \rightleftharpoons [AuL_3]^+Cl^- + [AuClL]$$

In general, the complexes $[AuL_4]^+$ are expected to have tetrahedral co-ordination at gold(I), and since the gold centre has the 18-electron configuration, co-ordination of the anionic counter-ion is not possible. Although the complexes [AuClL] have the 14-electron configuration at gold(I), there is much less tendency to expand the co-ordination shell by association than in the copper and silver analogues. The structures of the complexes $AuXL_2$ and $AuXL_3$ appear to depend critically on the nature of X and L, and in solution also on the solvent polarity. Thus, in the solid state $[AuCl(PPh_3)_2]$ exists as a trigonal planar molecule, but in $[Au(PPh_3)_2]^+(TCNQ)^-$ the radical anion of tetracyano-quinodimethane, TCNQ, is not co-ordinated to gold which therefore has the linear two-co-ordinate stereochemistry [52].

Similarly, in the complexes $AuX(PPh_3)_3$ the gold atom is four-co-ordinate when X = $SnCl_3$ in $[Au(SnCl_3)(PPh_3)_3]$ [57] but three-co-ordinate with trigonal planar stereochemistry when X = $B_9H_{12}S$ [58] in $[Au(PPh_3)_3]^+[B_9H_{12}S]^-$.

Thus, for both $AuXL_2$ and $AuXL_3$, the structures depend on the co-ordinating power of the group X for gold(I). In solution the equilibrium constants for the following reaction follow the series X = $Cl^- > Br^- > I^-$ as expected on this basis [40].

$$[AuX(PPh_3)_2] \rightleftharpoons [Au(PPh_3)_2]^+ + X^-$$

When X is a poor ligand such as nitrate, perchlorate or tetrafluoroborate for gold(I) the ionic $[AuL_2]^+$ or $[AuL_3]^+$ are probably formed exclusively.

(c) Complexes with bidentate ligands

Neutral complexes of the type [XAuL⌒LAuX] where L⌒L is a bidentate ligand such as $Ph_2PCH_2PPh_2$, $Ph_2PCH_2CH_2PPh_2$, $Ph_2AsCH_2CH_2AsPh_2$, $Ph_2PC{\equiv}CPPh_2$ or 2,2′-biphenylenebis(diethylphosphine) are well known [40, 59, 60, 60b], but these bidentate ligands also form many complexes in which gold(I) has a co-ordination number of three or four.

The ligand 2,11-bis(diphenylphosphinomethyl)benzo-[*c*]-phenanthrene, P⌒P, which can potentially span mutually *trans* co-ordination sites at a metal centre, forms three-co-ordinate complexes [MCl(P⌒P)] with all coinage metals. Interestingly, the angle PMP increases from 132° (M = Cu), 141° (M = Ag) to 176° (M = Au) and this change is accompanied by an increase in the ionic character of the metal—chlorine bond [61]. Thus, the gold complex partially ionises to $[Au(P⌒P)]^+$ in polar solvents, providing another example of the greater tendency of gold(I) over copper(I) or silver(I) to form linear two-co-ordinate complexes.

Chelating ligands also form more stable complexes of the form $[Au(L⌒L)_2]^+$ than do monodentate ligands [62]. However, attempts to resolve complexes of unsymmetrical bidentate ligands $[Au(L⌒L^1)_2]^+$, where $L⌒L^1$ may be

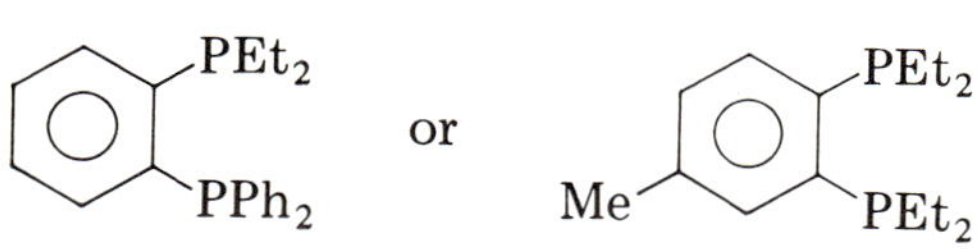

have been unsuccessful [63]. This is no doubt due to racemisation by reversible dissociation of the ligands from gold.

(d) *The nature of the gold—phosphorus bond*

The structures of several triphenylphosphine complexes of gold(I) have been determined by X-ray diffraction, and AuP bond lengths correlate with the bond angles in the complexes as shown in Table 3.3. It can be seen that as the ∠PMP decreases from 180° in the linear complexes, where gold(I) is sp-hybridised, to 120° in the trigonal planar complexes, where gold(I) is sp^2-hybridised, the Au—P bond lengths increase. Thus increasing s-character in the Au—P bond leads to a stronger σ-bond being formed [52]. Similarly, variations in

TABLE 3.3

Au—P bond lengths in some complexes

Complex	r(AuP)/pm	∠(PAuP)/°	Ref.
$[AuCl(PPh_3)]$	224.3	179.6[a]	52
$[Au(PPh_3)_2]^+$	228.6	180	52
$[AuCl(PPh_3)_2]$	232.3	132.1	52
$[Au(PPh_3)_3]^+$	238.2	120	58

[a] ∠ PAuCl.

the Au—P bond lengths with X in the compounds $[Ph_3P—Au—X]$ may be due to variations in the demand of the ligands X for the s-orbital of gold [49].

All the Au—P bond lengths in Table 3.3 are, however, shorter than the predicted Au—P single bond distance of 242 pm and this shortening has been ascribed to the effects of π-bonding. Thus filled 5d-orbitals of gold(I) of π-symmetry (d_{xz} and d_{yz}) may overlap with vacant 3d-orbitals of the phosphorus atom. There is good evidence for this effect in linear gold(I) complexes [40, 59], but π-bonding is apparently much weaker in trigonal planar and tetrahedral gold(I) complexes [58, 59].

Overall then the strength of the gold—phosphorus bond is dependent on both σ- and π-factors and the relative importance of these is difficult to assess. The extent of π-bonding is expected to increase with the electronegativity of the substituents on the phosphine ligand and this may explain the shorter AuP bond length in $[(Cl_3P)AuCl]$ of 219 pm [64] compared with that in $[(Ph_3P)AuCl]$ of 224.3 pm.

(e) Gold(I) diphenylphosphide and diphenylarsenide

These complexes are precipitated as insoluble white solids on reaction of diphenylphosphine or diphenylarsine with methylgold(I) complexes [65].

$$AuMe(PMe_2Ph) + Ph_2Ph \rightarrow \{Au(PPh_2)\}_n + CH_4 + PMe_2Ph$$

They presumably have polymeric structures with bridging Ph_2P or Ph_2As groups as in the gold(I) halides.

COMPLEXES WITH GOLD—OXYGEN BONDS

Gold(I) forms few complexes with oxygen-donor ligands, and these almost invariably contain a tertiary phosphine ligand in addition. Often the complexes have low thermal stability and decompose readily to gold, and the most common reactions involve displacement of the oxygen-donor ligand by ligands with a greater affinity for gold. Some typical reactions are shown in Fig. 3.4.

Particularly interesting are the tris(triphenylphosphine-gold)-oxonium salts, $[(Ph_3PAu)_3O]^+[BF_4]^-$ and $[(Ph_3PAu)_3O]^+$-$[MnO_4]^-$ prepared from $[AuCl(PPh_3)]$ and silver(I) oxide in the presence of $NaBF_4$ or $KMnO_4$, respectively [67], and also the complex $[Au(OSiMe_3)(PMe_3)]$ which has high thermal stability and can be sublimed under vacuum without decomposition [68].

COMPLEXES WITH SULPHUR- AND SELENIUM-DONOR LIGANDS

(a) Thiosulphate complexes

Solutions of $H[AuCl_4]$ or $[Au(NH_3)_4]^{3+}$ are reduced by thiosulphate to give the complex ion $[Au(S_2O_3)_2]^{3-}$. The reaction with $[Au(NH_3)_4]^{3+}$ has been shown to take place

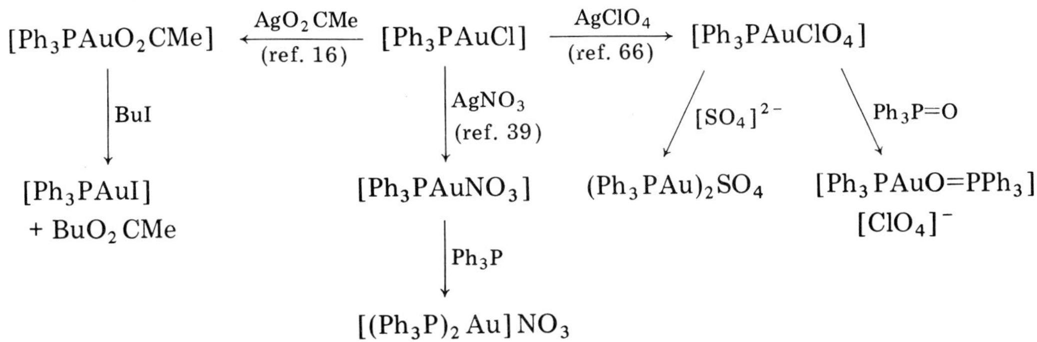

Fig. 3.4. Some complexes of gold(I) with oxygen-donor ligands.

[69] by substitution of ammonia by thiosulphate to give $[Au(S_2O_3)_3(OH_2)]^{3-}$, which is then reduced by more thiosulphate to give $[Au(S_2O_3)_2]^{3-}$ and $[S_4O_6]^{2-}$. The gold atom in $Na_3[Au(S_2O_3)_2] \cdot 2H_2O$ is linearly bonded to one sulphur atom of each thiosulphate group [70].

(b) Dialkylsulphide and dialkylselenide complexes

Dialkylsulphides reduce many gold(III) complexes in aqueous or alcoholic solution according to the equation

$$[AuCl_4]^- + 2R_2S + H_2O \rightarrow [AuCl(R_2S)] + R_2SO + 2H^+ + 3Cl^-$$

The reactions take place by co-ordination of the dialkylsulphide to gold(III) followed by reduction of the resulting complex by more dialkylsulphide. In the case of the ligand, L = thiomorpholin-3-one,

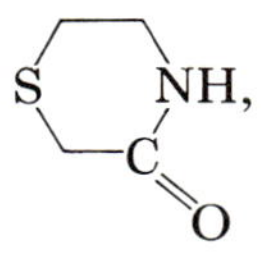

the precise sequence of reactions has been shown to be [71]:

$$[AuCl_4]^- + H_2O \rightleftharpoons [AuCl_3(OH)]^- + H^+ + Cl^-$$

$$[AuCl_3(OH)]^- + L \rightleftharpoons [AuCl_3L] + OH^-$$

$$[AuCl_3L] + H_2O \rightleftharpoons [AuCl_2(OH)L] + H^+ + Cl^-$$

$$[AuCl_2(OH)L] + L \rightleftharpoons [AuClL] + L{=}O + H^+ + Cl^-$$

Interestingly, the oxidation of S-methionine by $[AuCl_4]^-$ takes place stereospecifically to give S-methionine-S-sulphoxide [72].

Alternatively, complexes may be prepared by direct reaction with gold(I) halides.

$$AuCl + Me_2Se \rightarrow [AuCl(SeMe_2)]$$

Further addition of dialkylsulphide or dialkylselenide fails to give complexes with higher co-ordination number, however, even when chelating ligands are used [73]. Instead, only complexes such as $ClAuS(Ph)CH_2CH_2S(Ph)AuCl$ are formed, with the ligand bridging between linear gold(I) centres [73—75].

The complexes [AuClL] are thermally stable colourless solids, but the dialkylsulphide ligand L can easily be replaced, for example by tertiary phosphine ligands. Related complexes are formed by triphenylphosphine sulphide [76—78], which gives $[Ph_3P{=}S{-}AuCl]$, and by thiourea and its derivatives. Ionic species such as $[Au\{S{=}C(NH_2)_2\}_2]^+$ are also formed by thiourea [79, 80] and the structure of a derivative has been determined [80b].

Yellow solids of composition $(AuSR)_n$ are precipitated on reaction of $H[AuCl_4]$ with the corresponding thiol RSH [81].

$$H[AuCl_4] + 3RSH \rightarrow (AuSR) + RSSR + 4HCl$$

These compounds are polymeric with bridging thio groups and are used in some gold-coating procedures. For this reason, there has been some interest in alternative preparative methods [82—85]. In one procedure gold(III) chloride in aqueous solution is reduced by bis(2-hydroxyethyl)sulphide. The resulting gold(I) complex, $[AuCl\{S(CH_2CH_2OH)_2\}]$, is soluble in water, and further treatment with RSH leads to precipitation of $(AuSR)_n$.

$$[AuCl\{S(CH_2CH_2OH)_2\}] + RSH \rightarrow (AuSR) + HCl + S(CH_2CH_2OH)_2$$

Complexes are known with R = methyl, phenyl and many other alkyl and aryl groups. Derivatives of thioglucose are used in chemotherapy against arthritis.

(c) Dithiocarbamate and related derivatives of gold(I)

These complexes have particularly interesting structures, being dimeric with gold—gold bonding [81]. Thus the structure of (N,N-dipropyldithiocarbamato)gold(I) is shown in Fig. 3.5.

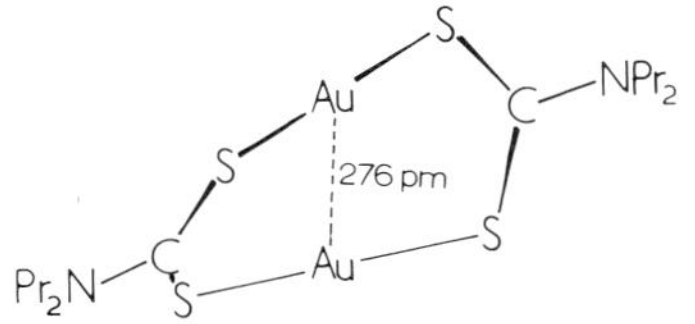

Fig. 3.5. The structure of N,N-dipropyldithiocarbamatogold(I).

The Au—Au bond distance (276 pm) is less than in metallic gold (288 pm) [86], but study by Raman spectroscopy indicates that the gold–gold bond order is only about 0.25 [87]. There are further weak interactions between gold atoms in neighbouring dimeric units, leading to chains being formed. This feature is more pronounced in the structure of O,O′-di-isopropyldithiophosphato—gold(I) which is best represented as shown in Fig. 3.6.

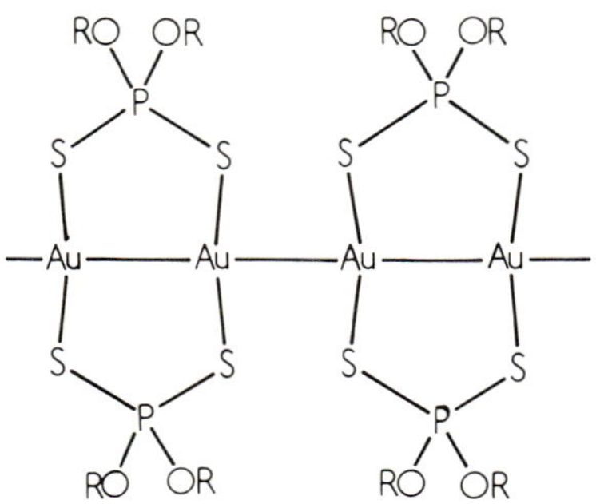

Fig. 3.6. The structure of O,O′-di-isopropyldithiophosphatogold(I). R = Me_2CH.

Here, there is no significant difference between the Au—Au bond distances within a dimeric unit and between dimeric units, the mean value being 304 pm. The co-ordination geometry about each gold centre is approximately square planar [88]. The corresponding dithiophosphinate derivatives are also dimeric and probably have similar structures [89]. They are prepared by the reaction:

$$H[AuCl_4] + 3Na[S_2PEt_2] \rightarrow [(AuS_2PEt_2)_2] + (Et_2PS_2)_2 + 3NaCl + HCl$$

(d) Complexes with tertiary phosphine ligands

A great number of complexes of gold(I) which contain both a tertiary phosphine and a sulphur or selenium donor ligand

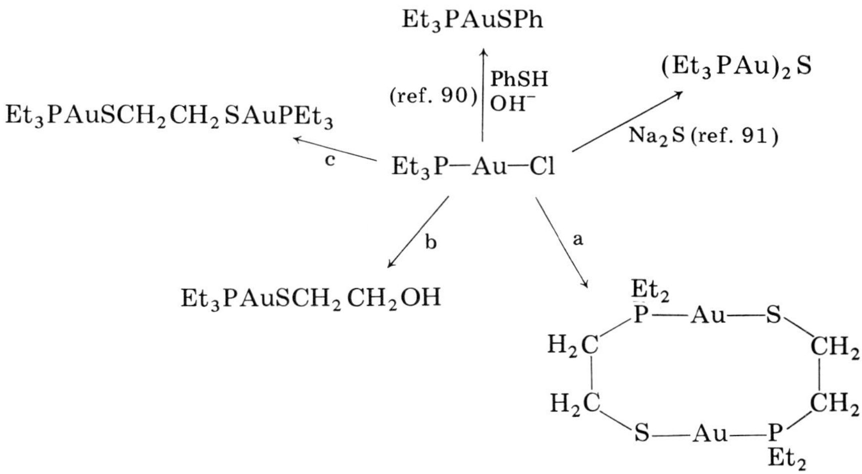

Fig. 3.7. Complexes of gold(I) with thio ligands.
a = $Et_2PCH_2CH_2SH$ (ref. 92); b = $HSCH_2CH_2OH$ (ref. 93);
c = $HSCH_2CH_2SH$ (ref. 92).

are known. The preparative routes to some of the compounds are shown in Figs. 3.7 and 3.8.

A particularly interesting compound is the tungsten sulphide derivative $WS_4(AuPMePh_2)_2$ which has a structure with trigonal-planar gold atoms, each bound to a tertiary phosphine and two sulphur atoms of the WS_4^{2-} tetrahedron [98a] as shown in Fig. 3.9.

HALIDE COMPLEXES OF GOLD(I)

We shall only discuss the complex ions $[AuX_2]^-$ here, since neutral complexes of the general type [AuXL] have been described earlier in this chapter. The stability of the ions $[AuX_2]^-$ increases with the heavier halide ions. Thus the fluoride complex is unknown, while the iodide complex is readily prepared by dissolution of gold(I) iodide in aqueous sodium iodide solution [99].

$$AuI + I^- \rightleftharpoons [AuI_2]^-$$

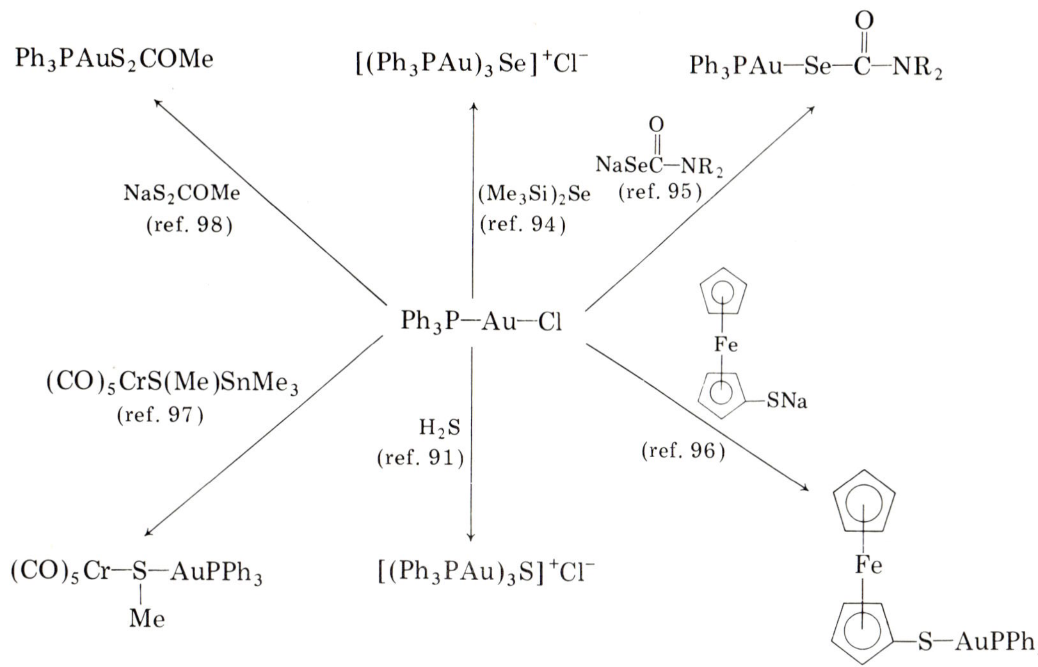

Fig. 3.8. Complexes of gold(I) with sulphur and selenium-donor ligands.

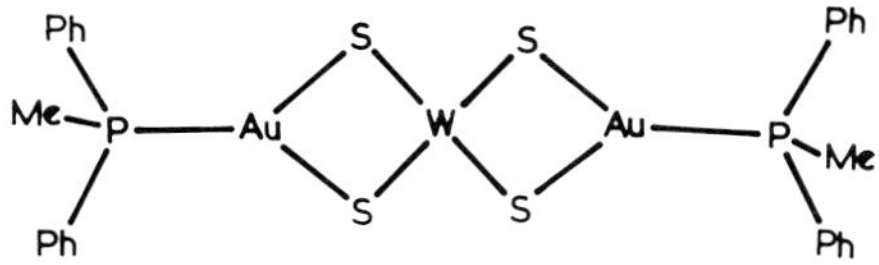

Fig. 3.9. The structure of $[WS_4(AuPMePh_2)_2]$.

The resulting $[AuI_2]^-$ can be oxidised by iodine to give $[AuI_4]^-$, the equilibrium constant for the reaction being 200 l mol^{-1}.

$[AuI_2]^- + I_2 \rightleftharpoons [AuI_4]^-$

The complex ions $[AuCl_2]^-$ and $[AuBr_2]^-$ may be formed in acetonitrile solution by the reaction [100]:

$[Au(MeCN)_2]^+ + 2Cl^- \rightleftharpoons [AuCl_2]^- + 2MeCN$

They may also be prepared by careful reduction of $[AuCl_4]^-$ or $[AuBr_4]^-$ in ethanol or acetone solutions [101].

$[AuBr_4]^- + (CH_3)_2CO \rightarrow [AuBr_2]^- + CH_2BrCOCH_3 + HBr$

The complex ions $[AuX_2]^-$ can be isolated as salts with large cations such as $[Et_4N]^+$ or $[Au(S_2CNEt_2)_2]^+$. Structure determinations by X-ray diffraction have shown that $[AuCl_2]^-$ and $[AuBr_2]^-$ have the expected linear structures with r(AuCl) 228.1 pm and r(AuBr) 234.9 pm, respectively [102, 103].

The existence of $[AuCl_2]^-$ and $[AuBr_2]^-$ in aqueous solution has been the subject of some debate. Early reports suggested that reduction of $[AuCl_4]^-$ by sulphur dioxide in aqueous solution gave $[AuCl_2]^-$ but it now seems likely that sulphito-gold complexes are really formed, and the pure compounds $Et_4N[AuCl_2]$ and $Et_4N[AuBr_2]$ have been shown to dissolve in water with disproportionation to gold and $[AuCl_4]^-$ and $[AuBr_4]^-$, respectively [101]. On the other hand calculations suggest that $[AuCl_2]^-$ should be present in sea water, and so may be stabilised in aqueous solution if an excess of chloride ions is present, as expected from the equation:

$3[AuCl_2]^- \rightleftharpoons [AuCl_4]^- + 2Au + 2Cl^-$

REFERENCES

1 M. I. Bruce, *J. Organomet. Chem.*, 44 (1972) 209.
2 M. S. Kharasch and H. S. Isbell, *J. Am. Chem. Soc.*, 52 (1930) 2919.
3 D. B. Dell'Amico and F. Calderazzo, *Gazz. Chim. Ital.*, 103 (1973) 1099.
4 K. C. Dash, J. Eberlein and H. Schmidbaur, *Synth. Inorg. Met-Org. Chem.*, 3 (1973) 375.
5 A. Sacco and M. Freni, *Gazz. Chim. Ital.*, 86 (1956) 195.
6 J. A. McCleverty and M. M. M. da Mota, *J. Chem. Soc., Dalton Trans.* (1973) 2571.
7 J. E. Parks and A. L. Balch, *J. Organomet. Chem.*, 71 (1974) 453.
8 L. G. Vaughan and W. A. Sheppard, *J. Am. Chem. Soc.*, 91 (1969) 6151.
9 F. Bonati and G. Minghetti, *Gazz. Chim. Ital.*, 103 (1973) 373.
10 B. M. Chadwick and A. G. Sharpe, *Adv. Inorg. Chem. Radiochem.*, 8 (1968); A. G. Sharpe, *The Chemistry of Cyano Complexes of the Transition Metals*, Academic, London, 1976.
11 D. F. Evans, D. Jones and G. Wilkinson, *J. Chem. Soc.* (1964) 3164.
12 M. H. Ford-Smith, J. J. Habeeb and J. H. Rawsthorne, *J. Chem. Soc., Dalton Trans.*, (1972) 2116.
13 H. Stammreich, B. M. Chadwick and S. G. Frankiss, *J. Mol. Spectrosc.*, 1 (1968) 191.
14 F. Cariati, D. Galizzioli and L. Naldini, *Chim. Ind. (Milan)*, 52 (1970) 995.
15 P. L. Bellon, M. Manassero and M. Sansoni, *Ric. Sci.*, 39 (1969) 173.
16 D. I. Nichols and A. S. Charleston, *J. Chem. Soc. A.* (1969) 2581.
17 L. H. Jones, *Inorg. Chem.*, 2 (1963) 777.
18 B. M. Chadwick and S. G. Frankiss, *J. Mol. Struct.*, 31 (1976) 1.
19 F. Meyer, *Compt. Rend.*, 143 (1906) 280.
20 M. E. Diemer, *J. Am. Chem. Soc.*, 35 (1913) 553.
21 A. P. Zuur and W. L. Groeneveld, *Rec. Trav. Chim. Pays-Bas*, 86 (1967) 1089.
22 G. Bergerhoff, *Z. Anorg. Allg. Chem.*, 327 (1964) 139.
23 F. Farha and R. T. Iwamoto, *Inorg. Chem.*, 4 (1965) 844.
24 H. G. Ang. W. E. Kow and K. F. Mok, *Inorg. Nucl. Chem. Lett.*, 8 (1972) 829.
25 P. Jutzi and H. Heusler, *J. Organomet. Chem.*, 8 (1976) 829.
26 G. A. Barclay, M. A. Collard, C. M. Harris and J. V. Kingston, *J. Inorg. Nucl. Chem.*, 31 (1969) 3509
27 G. J. Sutton, *Aust. J. Chem.*, 19 (1966) 2059; H. J. Dothie, F. J. Llewellyn, W. Wardlaw and A. J. E. Welch, *J. Chem. Soc.*, (1939) 426.
27b W. Clegg, *Acta Crystallogr., Sect. B*, B32 (1976) 2712.
28 W. Beck, W. P. Fehlhammer, P. Poellmann and H. Schaechl, *Chem. Ber.*, 102 (1969) 1976.
29 W. Beck, K. Burger and W. P. Fehlhammer, *Chem. Ber.*, 104 (1971) 1816.
30 R. F. Ziolo, J. A. Thich and Z. Dori, *Inorg. Chem.*, 11 (1972) 626.
31 W. Beck, W. P. Fehlhammer, P. Poellman, E. Schwierer and K. Feldl, *Chem Ber.*, 100 (1967) 2335.
32 J. L. Burmeister and E. T. Weleski, *Synth. Inorg. Met.-Org. Chem.*, 2 (1972) 295; J. L. Burmeister and N. J. de Stefano, *Inorg. Chem.*, 10 (1971) 998.
33 A. Shiotani and H. Schmidbaur, *J. Am. Chem. Soc.*, 92 (1970) 7003.
34 F. Bonati and G. Minghetti, *J. Chem. Soc., Chem. Commun.*, (1974) 88.
35 W. J. Chambers, *Ger. Pat.*, 2,245,447; *Chem. Abstr.*, 79 (1973) 5460.
36 R. L. Kieft, W. M. Peterson, G. L. Blundell, S. Horton, R. A. Henry and H. B. Jonassen, *Inorg. Chem.*, 15 (1976) 1721.
37 C. A. McAuliffe (Ed.), in *Transition Metal Complexes of Phosphorus, Arsenic and Antimony Ligands*, MacMillan, London, 1973.

38 R. Roulet, N. Q. Lan, W. R. Mason and G. P. Fenske, *Helv. Chim. Acta*, 56 (1973) 2405.
39 L. Malatesta, L. Naldini, G. Simonetta and F. Cariati, *Coord. Chem. Rev.*, 1 (1966) 255.
40 A. D. Westland, *Can J. Chem.*, 47 (1969) 4135.
41 A. T. T. Hsieh, J. D. Ruddick and G. Wilkinson, *J. Chem. Soc., Dalton Trans.*, (1972) 1966.
42 D. A. Duddell, P. L. Goggin, R. J. Goodfellow, M. G. Norton and J. G. Smith, *J. Chem. Soc. A* (1970) 545.
43 J. Bailey, *J. Inorg. Nucl. Chem.*, 35 (1973) 1921.
44 L. G. Vaughan, *U. S. Pat.*, 3,661,959, 1972.
45 D. A. Couch and S. D. Robinson, *Inorg. Chim. Acta*, 9 (1974) 39.
46 H. Schmidbaur and R. Franke, *Chem. Ber.*, 105 (1972) 2985.
47 H. Schmidbaur and A. Shiotani, *Chem. Ber.*, 104 (1971) 2821.
48 B. M. Sutton and J. Weinstock, *U.S. Pat.*, 3,911,119, 1973; *Chem. Abstr.*, 84 (1976) 31242; *U.S. Pat.* 3,842,108, 1974; *Chem. Abstr.*, 82 (1975) 86410.
49 N. C. Baenziger, W. E. Bennett and D. M. Soboroff, *Acta Crystallogr., Sect. B*, B32 (1976) 962; P. D. Gavens, J. J. Guy, M. J. Mays and G. M. Sheldrick, *Acta Crystallogr., Sect. B*, 33 (1977) 137.
50 F. W. B. Einstein and R. Restivo, *Acta Crystallogr., Sect. B*, B31 (1975) 624.
51 J. G. Wijnhoven, W. P. J. H. Bosman and P. T. Beurskens, *J. Cryst. Mol. Struct.*, 2 (1972) 7.
52 N. C. Baenziger, K. M. Dittemore and J. R. Doyle, *Inorg. Chem.*, 13 (1974) 805.
53 E. L. Muetterties and C. W. Allegranti, *J. Am. Chem. Soc.*, 92 (1970) 4114.
54 D. A. Couch and S. D. Robinson, *Inorg. Chem.*, 13 (1974) 456.
55 J. W. Collier, A. R. Fox, I. G. Hinton and F. G. Mann, *J. Chem. Soc.*, (1964) 1819.
56 G. P. Fenske and W. R. Mason, *Inorg. Chem.*, 13 (1974) 1783.
57 J. A. Dilts and M. P. Johnson, *Inorg. Chem.*, 5 (1966) 2079.
58 L. J. Guggenberger, *J. Organomet. Chem.*, 81 (1974) 271.
59 A. J. Carty and A. Efraty, *Inorg. Chem.*, 8 (1969) 543.
60 (a) D. W. Allen, I. T. Millar, F. G. Mann, R. M. Canadine and J. Walker, *J. Chem. Soc. A*, (1969) 1097.
60 (b) H. Schmidbaur, A. Wohlleben, F. Wagner, O. Orama and G. Huttner, *Chem. Ber.*, 110 (1977) 1748.
61 M. Barrow, H. B. Buergi, D. K. Johnson and L. M. Venanzi, *J. Am. Chem. Soc.*, 98 (1976) 2356.
62 R. S. Nyholm, *Nature (London)*, 168 (1951) 705.
63 M. Davis and F. G. Mann, *J. Chem. Soc.*, (1964) 3791.
64 G. J. Arai, *Rec. Trav. Chim. Pays-Bas*, 81 (1962) 307.
65 R. J. Puddephatt and P. J. Thompson, *J. Organomet. Chem.*, 117 (1976) 395.
66 R. Uson, P. Royo, A. Laguna and J. Garcia, *Rev. Acad. Cienc. Exactas, Fis.-Quim. Natur. Zaragoza*, 28 (1973) 67.
67 A. N. Nesmeyanov, K. I. Grandberg, V. P. Dyadchenko, D. A. Lemenovskii and E. G. Perevalova, *Izv. Akad. Nauk SSSR, Ser. Khim.*, (1974) 740.
68 H. Schmidbaur, J. Adlkofer and A. Shiotani, *Chem. Ber.*, 105 (1972) 3389.
69 G. Nord, L. H. Skibsted and A. S. Khalonin, *Acta Chem. Scand., Ser. A*, A29 (1975) 505.
70 H. Ruben, A. Zalkin, M. O. Faltens and D. H. Templeton, *Inorg. Chem.*, 13 (1974) 1836.
71 D. de Filippo, F. Devillanova and C. Preti, *Inorg. Chim. Acta*, 5 (1971) 103.
72 E. Bordignon, L. Cattalini, G. Natile and A. Scatturin, *J. Chem. Soc., Chem. Commun.*, (1973) 878.
73 K. C. Dash and H. Schmidbaur, *Chem. Ber.*, 106 (1973) 1221.
74 G. W. A. Fowles, D. A. Rice and M. J. Riedl, *J. Less-Common Met.*, 32 (1973) 379

75 M. G. B. Drew and M. J. Riedl, *J. Chem. Soc., Dalton Trans.*, (1973) 52.
76 I. M. Keen, *J. Chem. Soc.*, (1965) 5751.
77 R. A. Potts and A. L. Allred, *J. Inorg. Nucl. Chem.*, 28 (1966) 1479.
78 M. G. King and G. P. McQuillan, *J. Chem. Soc. A*, (1967) 898.
79 G. Marcotrigiano, R. Battistuzzi and P. Morini, *Inorg. Nucl. Chem. Lett.*, 10 (1974) 641.
80 E. N. Ovespyan, V. M. Tarayan and G. N. Shaposhnikova, *Arm. Khim. Zh.*, 19 (1966) 412.
80b P. G. Jones, J. J. Guy and G. M. Sheldrick, *Acta Crystallogr., Sect. B*, B32 (1976) 3321.
81 S. Åkerstrom, *Ark. Kemi*, 14 (1959) 387.
82 E. W. Abel and C. R. Jenkins, *J. Organomet. Chem.*, 14 (1968) 285.
83 *Ger. Pat.*, 1,216,296, 1966; *Chem. Abstr.*, 65 (1966) 5071.
84 L. Vegh and E. Hardegger, *Helv. Chim. Acta*, 56 (1973) 2079.
85 H. M. Fitch, *Belg. Pat.*, 621,886 (1963); *Chem. Abstr.*, 62 (1965) 9065.
86 R. Hesse and P. Jennische, *Acta Chem. Scand.*, 26 (1972) 3855.
87 T. G. Spiro and F. J. Farrell, *Inorg. Chem.*, 10 (1971) 1606.
88 S. L. Lawton, W. J. Rohrbaugh and G. T. Kokotailo, *Inorg. Chem.*, 11 (1972) 2227.
89 W. Kuchen and H. Mayatapek, *Chem. Ber.*, 101 (1968) 3454.
90 G. E. Coates, C. Kowala and J. M. Swan, *Aust. J. Chem.*, 19 (1966) 539.
91 C. Kowala and J. M. Swan, *Aust. J. Chem.*, 19 (1966) 547
92 B. M. Sutton and J. Weinstock, *Chem. Abstr.*, 84 (1976) 31242.
93 E. R. McGusty and B. M. Sutton, *Chem. Abstr.*, 80 (1974) 100212.
94 H. Schmidbaur, R. Franke and J. Eberlein, *Chem.- Ztg.*, 99 (1975) 91.
95 F. W. Pijpers, A. H. Dix and J. G. M. van der Linden, *Inorg. Chim. Acta*, 11 (1974) 41.
96 E. G. Perevalova, D. A. Lemenovskii, K. I. Grandberg and A. N. Nesmeyanov, *Dokl. Akad. Nauk SSSR*, 203 (1972) 1320.
97 W. Ehrl and H. Vahrenkamp, *Chem. Ber.*, 103 (1970) 3563.
98 C. Kowala and J. M. Swan, *Aust. J. Chem.*, 19 (1966) 999.
98a J. C. Huffman, S. R. Roth and A. R. Siedle, *J. Am. Chem. Soc.*, 98 (1976) 4340.
99 A. Hakansson and L. Johansson, *Chem. Scr.*, 7 (1975) 201.
100 R. Roulet and R. Favez, *Chimia*, 29 (1975) 346.
101 P. Braunstein and R. J. H. Clark, *J. Chem. Soc., Dalton Trans.*, (1973) 1845.
102 J. C. M. T. Eijndhoven and G. C. Verschoor, *Mater. Res. Bull.*, 9 (1974) 1667.
103 P. T. Beurskens, H. J. A. Blaauw, J. A. Cras and J. J. Steggarda, *Inorg. Chem.*, 7 (1968) 805.

CHAPTER 4

GOLD(II) COMPLEXES

This oxidation state becomes progressively less stable for the elements on descending group IB of the periodic table. Thus, from being a dominant oxidation state for copper, it is unusual for silver and very rare for gold. For square planar complexes the odd d-electron for metal ions with the d^9 electron configuration will be in the σ^*-orbital $d_{x^2-y^2}$. Since the ligand field splitting of d-orbitals is greater for the heavier transition elements, this orbital is expected to have a relatively higher energy for gold(II) complexes than for copper(II), and hence gold(II) complexes are expected to be oxidised or to disproportionate more readily than copper(II) complexes.

Many complexes of gold whose empirical formulae suggest the presence of gold(II) have been shown to be mixed gold(I)—gold(III) species. For example, the complex $CsAuCl_3$ is properly formulated $Cs_2[AuCl_2][AuCl_4]$[1]; $(PhCH_2)_2SAuCl_2$ is $[AuCl\{S(CH_2Ph)_2\}][AuCl_3\{S(CH_2Ph)_2\}]$[2]; AuCl(dmg), where dmg is dimethylglyoximate, is $[Au(dmg)_2]^+[AuCl_2]^-$[3]. In each case the complexes contain equal proportions of linear gold(I) and square planar gold(III) centres. Other examples of such complexes have been found recently. For example, oxidation of the gold(I) dithiocarbamate $[(AuS_2CNBu_2)_2]$ with bromine gives a complex $AuBr(S_2CNBu_2)$ which has been shown to have the structure [4]:

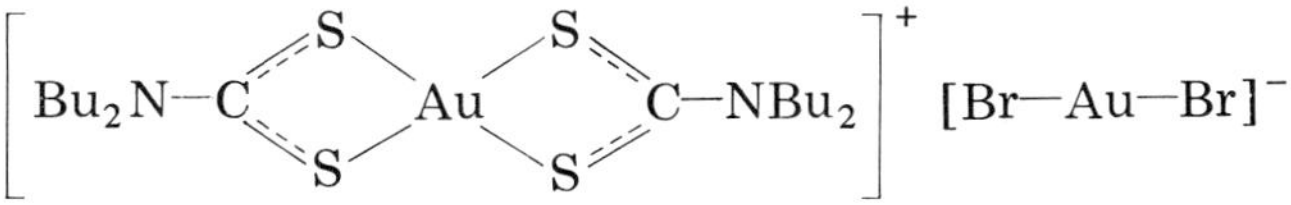

Here we have a square-planar gold(III) cation and linear gold(I) anion, which can be compared with the maleonitriledithiolate derivative which is thought to have a linear gold(I) cation and square-planar gold(III) anion [5, 6]:

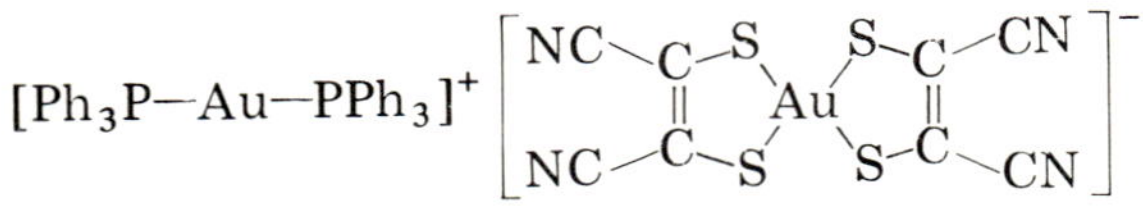

DITHIOLATE AND RELATED COMPLEXES OF GOLD(II)

Of the true gold(II) complexes, those with 1,1-dithiolate or 1,2-dithiolate ligands are the best characterised. This work began with a study of the reversible reaction [7]:

$$(AuS_2CNR_2)_2 + 2(R_2NCS_2)_2 \rightleftharpoons 2[Au(S_2CNR_2)_3]$$

The e.s.r. spectrum of the mixture showed that a gold(II) complex was present, presumably formed by the reaction

$$[Au(S_2CNR_2)_3] \rightleftharpoons [Au(S_2CNR_2)_2] + \tfrac{1}{2}(R_2NCS_2)_2$$

The gold(II) complex is present in the reaction mixture only as a fairly low equilibrium concentration and is too unstable to be isolated in pure form [8]. Complexes $[Au(S_2CNR_2)_2]$ are known with R = ethyl, propyl and butyl, and the complexes can be co-crystallised with the stable diamagnetic nickel(II) derivatives $[Ni(S_2CNR_2)_2]$. The presence of paramagnetic gold(II) species is deduced from the e.s.r. spectra. Since the only naturally occurring isotope of gold, ^{197}Au, has a nuclear spin of 3/2, the e.s.r. spectrum contains four lines of equal intensity and with equal spacing due to coupling with the ^{197}Au nucleus. The structure of the complexes $[Au(S_2CNR_2)_2]$ is thought to be square planar, and, if the axes are defined as in Fig. 4.1, the unpaired electron has

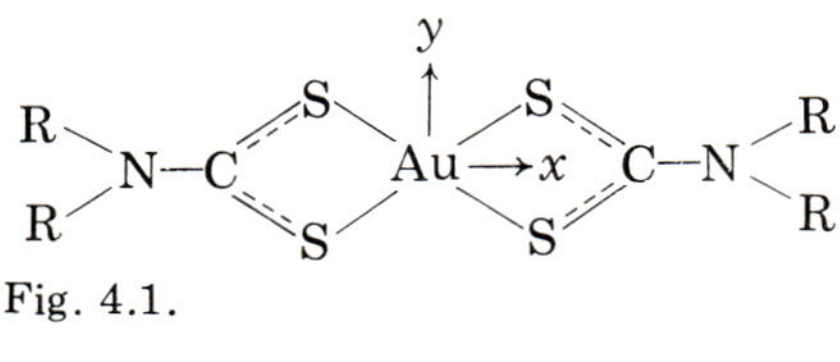

Fig. 4.1.

been shown to occupy the σ^*-orbital $5d_{xy}$ as expected for a d^9 square planar complex [9].

The corresponding N,N-diethyldiseleno-carbamate derivative $[Au(Se_2CNEt_2)_2]$ has been detected [10] by e.s.r. spectroscopy as a product of the disproportionation of $AuSe_2CNEt_2$. The e.s.r. parameters for this and for some other gold(II) complexes are given in Table 4.1.

TABLE 4.1

E.s.r. data for some gold(II) complexes

Compound	g	$A(^{197}Au)/cm^{-1}$	Ref.
$[Au(S_2CNEt_2)_2]$	2.040	28.0×10^{-4}	10
	2.037	31.9×10^{-4}	9
$[Au(Se_2CNEt_2)_2]$	2.013	28.9×10^{-4}	10
$[Au(S_2CN^iPr_2)_2]$	2.040	28×10^{-4}	7
$[Au(S_2CNBu_2)_2]$	2.039	26.2×10^{-4}	8
$(Bu_4N)_2[Au(mnt)_2]$	2.0002	40.4×10^{-4}	9
	2.009	27.6×10^{-4}	11
[Au(phthalocyanine)]	2.065	61×10^{-4}	15
Au(II) in FeAsS	2.001	—	14

The first gold(II) complex to be prepared pure was $(Bu_4N)_2[Au(mnt)_2]$, where mnt = maleonitriledithiolate [11]. The air-sensitive complex was isolated after reduction of the corresponding gold(III) complex $Bu_4N[Au(mnt)_2]$ with sodium borohydride. The complex is green, and the gold(II) centre is thought to have square-planar stereochemistry as shown below:

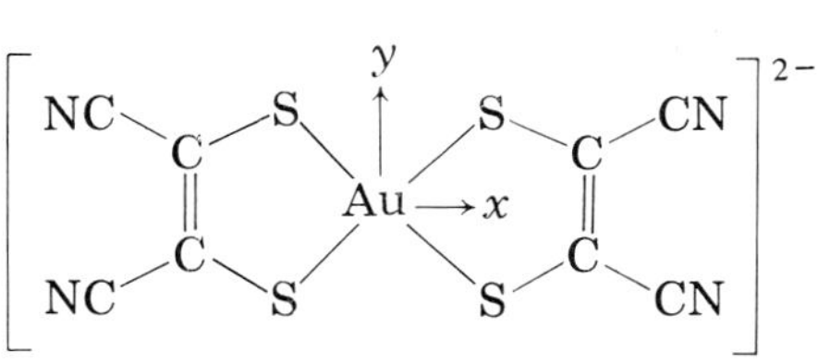

The magnetic moment is 1.85 B.M. [11] and the e.s.r. spectrum shows that the odd electron is again in the σ^* $5d_{xy}$ orbital on gold [9]. Interestingly, this complex can also be obtained from the mixed gold(I)—gold(III) compound $[AuCl\{S(CH_2Ph)_2\}]$-$[AuCl_3\{S(CH_2Ph)_2\}]$ by reaction with maleonitriledithiolate [12]. Normally, of course, gold(II) complexes are expected to undergo the reverse reaction, that is to disproportionate to gold(I) and gold(III) species.

In addition to the complexes $[Au(S_2CNEt_2)_2]$ and $[Au(mnt)_2]^-$ discussed above, there is also good evidence for formation of the mixed complex $[Au(S_2CNEt_2)(mnt)]^-$ in solution by electrochemical reduction of $[Au(S_2CNEt_2)(mnt)]$[13].

The magnetic properties of the charge-transfer complex of $[Au\{S_2C_2(CF_3)_2\}_2]$ with tetrathiafulvalene have been investigated, and the complex is shown to exhibit a spin-Peierls transition [13b].

Gold(II) certainly is stabilised by these sulphur-donor ligands and it has even been proposed that gold(II) occurs naturally in gold-containing arsenopyrite ores, FeAsS. The evidence, however, is based only on the observation of an e.s.r. signal in gold-containing ores, which was absent in samples not containing gold [14]. However, the expected coupling to the nuclear spin of ^{197}Au was not seen in the e.s.r. spectrum.

OTHER STABLE GOLD(II) COMPLEXES

Treatment of either gold or gold(I) bromide with 1,3-diiminoisoindoline at 250 °C gives the stable square planar gold(II) phthalocyanine, whose e.s.r. spectrum shows coupling to both gold ($I = \frac{3}{2}$) and to the four nitrogen donor atoms of the phthalocyanine ligand [15].

An unusual example of gold(II) is found in the carborane derivative $(Et_4N)_2[Au(B_9C_2H_{11})_2]$, μ_{eff} 1.79 B.M., which is formed on reduction of the corresponding gold(III) derivative $Et_4N[Au(B_9C_2H_{11})_2]$ with sodium amalgam [16].

So far, we have considered complexes containing a single gold(II) centre, but recently some interesting dimeric complexes of gold(II) have been prepared by the reactions [17, 18]:

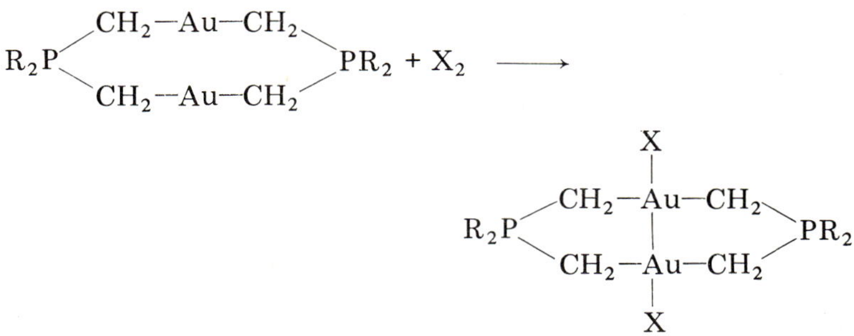

The group R may be methyl or ethyl, and X may be chlorine, bromine or iodine. Since the compounds have a gold—gold bond they are diamagnetic. When R=Et and X=Cl, the Au—Au bond distance is 259.7 pm. Further oxidation by excess of halogen causes cleavage of the Au—Au bond, with formation of gold(III) complexes.

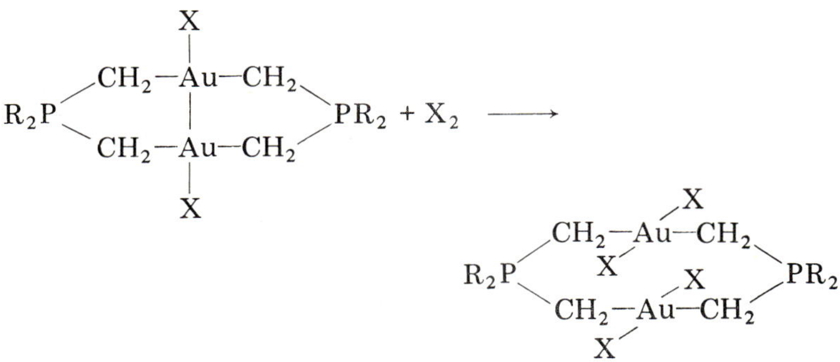

Similarly, oxidation of bis(diphenylphosphino)methane complexes of gold(I) with halogens gives gold(II) complexes with a gold—gold bond [18].

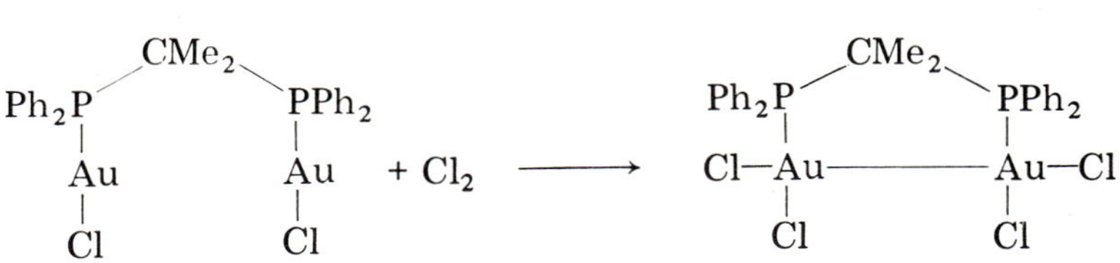

GOLD(II) COMPLEXES AS REACTION INTERMEDIATES

Besides the gold(II) compounds which are sufficiently stable to be isolated or to be detected by spectroscopic methods, there is evidence for formation of some gold(II) complexes as short-lived reaction intermediates which are not sufficiently long-lived to be detected directly. Thus a gold(II) complex, perhaps $[AuCl_4]^{2-}$, is thought [19] to be a reaction intermediate in the iron(II)-catalysed exchange of radioactive chloride with $[AuCl_4]^-$.

$$Fe(II) + [AuCl_4]^- \rightarrow Fe(III) + [AuCl_4]^{2-}$$

Some reactions involving oxidation of gold(I) to gold(III) complexes are thought to occur by free-radical mechanisms and gold(II) intermediates are proposed. Thus, oxidation of $[AuMe(PMe_3)]$ by CF_3I to give $[AuIMe(CF_3)(PMe_3)]$ is thought to involve the intermediacy of $[AuMe(CF_3)(PMe_3)]$ [20]. Similarly, reaction of $[AuMe(PMe_3)]$ with benzenethiol is thought to take place *via* the gold(II) intermediate $[AuMe(SPh)(PMe_3)]$[21]. The best example of this type of process is the oxidation of $[\{Au(S_2CNEt_2)\}_2]$ by $(Et_2NCS_2)_2$ to give $[Au(S_2CNEt_2)_3]$, in which the gold(II) intermediate $[Au(S_2CNEt_2)_2]$ is sufficiently stable to be detected directly [7].

REFERENCES

1 N. Elliot and L. Pauling, *J. Am. Chem. Soc.*, 60 (1938) 1846.
2 F. H. Brain, C. S. Gibson, J. A. J. Jarvis, R. F. Phillips, H. M. Powell and A. Tyabji, *J. Chem. Soc.* (1952) 3686. F. Coletta, R. Ettore and A. Gambaro, *Inorg. Nucl. Chem. Lett.*, 8 (1972) 667.
3 R. E. Rundle, *J. Am. Chem. Soc.*, 76 (1954) 3101.
4 P. T. Beurskens, H. J. A. Blaauw, J. A. Cras and J. J. Steggarda, *Inorg. Chem.*, 7 (1968) 805.
5 T. J. Bergendahl and J. H. Waters, *Inorg. Chem.*, 14 (1975) 2556.
6 T. J. Bergendahl, *J. Chem. Educ.*, 52 (1975) 731.
7 T. Vänngard and S. Akerström, *Nature (London)*, 184 (1959) 183.
8 T. J. Bergendahl and E. M. Bergendahl, *Inorg. Chem.*, 11 (1972) 638.
9 J. G. M. van Rens, M. P. A. Viegers and E. de Boer, *Chem. Phys. Lett.*, 28 (1974) 104.
10 R. Kirmse, B. Lorenz and W. Windsch, *Z. Anorg. Chem.*, 384 (1971) 160.
11 J. H. Waters and H. B. Gray, *J. Am. Chem. Soc.*, 87 (1965) 3534.

12 J. H. Waters, T. J. Bergendahl and S. R. Lewis, *J. Chem. Soc., Chem. Commun.*, (1971) 834.
13 J. G. M. van der Linden and H. G. J. van de Roer, *Inorg. Chim. Acta*, 5 (1971) 254.
13b I. S. Jacobs, J. W. Bray, H. R. Hart, L. V. Interrante, J. S. Kasper, G. D. Watkins, D. E. Prober and J. C. Bonner, *Phys. Rev. B*, 14 (1976) 3036; I. S. Jacobs, H. R. Hart, L. V. Interrante, J. W. Bray, J. S. Kasper, G. D. Watkins, D. E. Prober, E. P. Wolf and J. C. Bonner, *Physica B + C*, 86—88 B + C (1977) 655.
14 V. N. Voitsekhovskii, B. P. Berkovskii, L. V. Yashchurzhinskaya, L. V. Chugaev and M. V. Nikitin, *Izv. Vyssh. Uchebn. Zaved., Tsvetn. Metall.* (1975) 60; *Chem. Abstr.*, 83 (1975) 167183.
15 A. MacCragh and W. S. Koski, *J. Am. Chem. Soc.*, 87 (1965) 2496.
16 I. F. Warren and M. F. Hawthorne, *J. Am. Chem. Soc.*, 90 (1968) 4823.
17 H. Schmidbaur, *Acc. Chem. Res.*, 8 (1975) 62; H. Schmidbaur, J. R. Mandl, A. Frank and G. Huttner, *Chem. Ber.*, 109 (1976) 466.
18 H. Schmidbaur, *Angew. Chem., Int. Ed. Engl.*, 15 (1976) 728; H. Schmidbaur, A. Wohlleben, F. E. Wagner, D. F. van der Vondel and G. P. van der Kelen, *Chem. Ber.*, 110 (1977) 2758.
19 R. L. Rich and H. Taube, *J. Phys. Chem.*, 58 (1954) 6.
20 A. Johnson and R. J. Puddephatt, *J. Chem. Soc., Dalton Trans.*, (1976) 1360.
21 A. Johnson and R. J. Puddephatt, *J. Chem. Soc., Dalton, Trans.*, (1975) 115.

CHAPTER 5

GOLD(III) COMPLEXES

The chemistry of gold(III) complexes is about as extensive overall as that of gold(I). However, whereas gold(I) chemistry is dominated by complexes with soft bases such as tertiary phosphines, with fluorides and oxygen-donor complexes unknown or very scarce, this feature is less prominent in the chemistry of gold(III). Thus, as predicted from its higher oxidation state, gold(III) can be considered to be a harder Lewis acid than gold(I), and there is an extensive chemistry of gold(III) not only with soft ligands like tertiary phosphines and thiols but also with hard ligands like fluoride and nitrogen-donors.

COMPLEXES WITH CARBON-DONOR LIGANDS

The complex $K[Au(CN)_4]$ is prepared by addition of a solution of $Na[AuCl_4]$ to concentrated potassium cyanide solution [1]. The ion $[Au(CN)_4]^-$ contains square-planar gold(III), and studies of the vibrational spectra indicate that backbonding from filled 5d-orbitals on gold to vacant π^*-orbitals of the cyanide ligands is much weaker than in the corresponding gold(I) complexes [2]. This is also true for the ions *trans*-$[Au(CN)_2X_2]^-$ which are formed by oxidation of $[Au(CN)_2]^-$ with the halogens, $X_2 = Cl_2$, Br_2 or I_2 [3, 4].

From $K[Au(CN)_4]$ it is possible to prepare the acid $H[Au(CN)_4] \cdot 2H_2O$ [5], and also the complex ions $[Au(CN)_5]^{2-}$ and $[Au(CN)_6]^{3-}$ by reaction with cyanide in aqueous solution [6].

Some aryl isocyanide complexes of gold(III), such as $[AuCl_3(CNPh)]$, have been prepared but they are readily reduced to gold(I)–isocyanide complexes by treatment with excess of aryl isocyanide [7].

NITROGEN-DONOR COMPLEXES OF GOLD(III)

(a) Amine and related complexes

These complexes are usually prepared by displacement of chloride in $H[AuCl_4]$ by the nitrogen-donor ligand. For example, on bubbling gaseous ammonia into an aqueous solution of $H[AuCl_4]$ and ammonium nitrate the complex $[Au(NH_3)_4](NO_3)_3$ is obtained [8]. This contains square-planar gold(III) with $r(AuN) = 202$ pm [9]. Complexes with chelate ligands are more readily prepared [10]. Thus 1,2-diaminoethane, en, gives $[Au(en)_2]Br_3$, and the tridentate ligands $(H_2NCH_2CH_2)_2NH$, dien, and $(Et_2NCH_2CH_2)_2NH$, Et_4dien, give complexes containing the ions $[Au(dien)Cl]^{2+}$ and $[Au(Et_4dien)Cl]^{2+}$.

(b) Acidity of amine complexes of gold(III)

The high polarising power of gold(III) causes weakening of the NH bonds in amine complexes of gold(III), and the complexes may behave as weak acids due to facile loss of a proton from the co-ordinated ligand. The weakening of the NH bonds in $[Au(NH_3)_4]^{3+}$ is demonstrated [9] by the low values of ν_a(NH) 3240 cm^{-1} and ν_s(NH) 3100 cm^{-1} compared with corresponding values of 3277 and 3182 cm^{-1} for the iso-electronic complex ion $[Pt(NH_3)_4]^{2+}$.

Some pK_a values for amine complexes of gold(III) in aqueous perchlorate medium are given in Table 5.1.

TABLE 5.1

pK_a Values of some amine complexes of gold(III)

Complex	pK_a	Ref.
$[Au(NH_3)_4]^{3+}$	7.48	8
$[Au(en)_2]^{3+}$	6.3	10, 11
$[Au(dien)OH]^{2+}$	5.8	11
$[Au(dien)Br]^{2+}$	4.5	11
$[Au(dien)Cl]^{2+}$	4.0	11
$[Au(Et_4dien)Cl]^{2+}$	2.2	12

It will be seen that chelation greatly increases the acidity of the complexes, as is also found in complexes of platinum(IV). The effect is great enough to have a significant bearing on the chemical properties of the complexes. Thus, complexes such as $[Au(dien)Cl]Cl_2$ must be isolated by crystallisation from hydrochloric acid at pH 2. Crystallisation at pH 6 leads to formation of the salt of the conjugate base [Au(dien-H)Cl]Cl. The remarkably high acidity of the complex ion $[Au(Et_4dien)\text{-}Cl]^{2+}$ is attributed largely to the poorer solvation of this ion compared with $[Au(dien)Cl]^{2+}$, due to the bulky ethyl groups blocking access of solvent molecules to the gold(III) centre. Thus the loss in solvation energy in going to $[Au(Et_4dien\text{-}H)\text{-}Cl]^+$ is less than for the dien complex. The steric effects of the ethyl groups in the complex $[Au(Et_4dien\text{-}H)Cl][PF_6]$ are also felt in substitution reactions as described in Chapter 9. The acidity of $[Au(Et_4dien)Cl]^{2+}$ is so great that the ion is only formed in strong acid solution, and only salts of the conjugate base $[Au(Et_4dien\text{-}H)Cl]^+$ can be isolated.

Another example of the easy conversion of amine to amide when complexed to gold(III) can be seen in the reaction of gold(III) hydroxide with succinimide to give

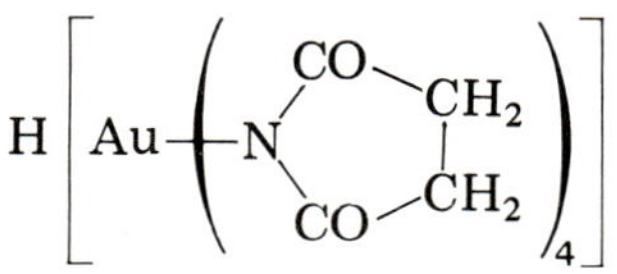

This complex is stable in acetic acid and is not readily reduced to gold [13].

As well as acid—base equilibria of the type:

$$[Au(NH_3)_4]^{3+} + H_2O \rightleftharpoons [Au(NH_3)_3(NH_2)]^{2+} + H_3O^+$$

it is also possible for the ammine ligands to be displaced from gold, though such reactions are slower [8].

$$[Au(NH_3)_4]^{3+} + H_2O \rightleftharpoons [Au(NH_3)_3(OH)] + [NH_4]^+$$

Chelate ligands can also be displaced in a similar way. Thus reaction of $[Au(Et_4dien\text{-}H)Cl]^+$ with azide ions [12] gives $[Au(N_3)_4]^-$.

Reactions of the co-ordinated ligand may also take place, for example in the reaction of $[Au(en)_2]Cl_3$ with chlorine to give $[AuCl_2(NHClCH_2CH_2NHCl)]Cl$ [14], which involves displacement of one ligand and chlorination of the other.

(c) Complexes of gold(III) with monodentate nitrogen-donor ligands

A wide variety of complexes $[AuCl_3L]$ and $[AuBr_3L]$ have been prepared by reaction of the ligand L with $[Au_2Cl_6]$ and $[Au_2Br_6]$, respectively. The ligand L may be pyridine or its derivatives [15, 16], a nitrile such as MeCN [17] or PhCN [18], or an azo derivative such as MeN=NMe [19], PhN=NPh [16, 18, 19] or 11H-dibenzo[*c*,*f*]-[1,2]diazepine [20].

Pyridine will displace a nitrile from the complex $[AuCl_3(NCC_6H_4Me)]$ to give $[AuCl_3(py)]$, and further reaction may also take place:

$$[AuCl_3(py)] + py \rightarrow [AuCl_2(py)_2]Cl$$

Such reactions are reversible and chloride will also displace pyridine from gold(III).

$$[AuCl_3(py)] + Cl^- \rightarrow [AuCl_4]^- + py$$

Thermal decomposition of $[AuCl_3(PhN{=}NPh)]$, in which only one nitrogen atom of the azobenzene ligand is thought to be co-ordinated to gold, apparently gives metallation of the phenyl ring to give HCl and the complex,

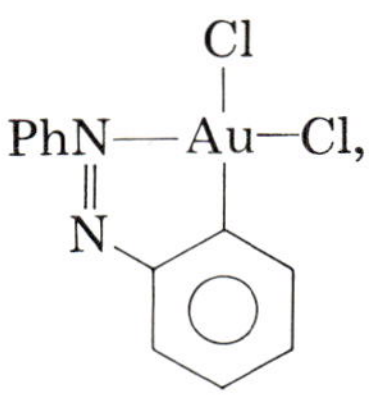

which rapidly decomposes further to gold(I) chloride and 2-chloroazobenzene [19].

(d) Chelating nitrogen-donor ligands

Complexes of gold(III) with 1,10-phenanthroline (phen) and 2,2′-bipyridine (bipy) are formed by reaction of $H[AuCl_4]$ with the appropriate ligand in aqueous solution [21, 22]. The complex ion $[AuCl_2(phen)]^+$ or $[AuCl_2(bipy)]^+$ is formed first, and can be isolated as the perchlorate or tetrachloroaurate (III) salt. These contain square planar gold(III), but the complexes $AuCl_3(phen)$ or $AuBr_3(phen)$ may contain gold(III) in a higher co-ordination number. Thus, the colours are not those of the ions $[AuCl_2(phen)]^+$ or $[AuBr_2(phen)]^+$, and it is suggested that the third halide ion in each case may be co-ordinated to gold. Perhaps surprisingly, the bidentate ligands are readily replaced by chloride or bromide in nitrobenzene solution, for example:

$$[AuCl_2(phen)]^+ + 2Cl^- \rightarrow [AuCl_4]^- + phen$$

The chloride ligands in $[AuCl_2(bipy)]^+$ are readily replaced in aqueous solution by bromide to give $[AuBr_2(bipy)]^+$ or by hydrolysis to give $[AuCl(OH)(bipy)]^+$. Comparison of the infrared and Raman spectra of the compounds $[AuCl_2$(2,2′-bipyridine)]$[AuCl_4]$ and $AuCl_3$(4,4′-bipyridine), leads to the conclusion that the latter contains the polymeric $[\{AuCl_2$(4,4′-bipyridine)$\}_n]^{n+}$ ion with the 4,4′-bipyridine as *cis*-bridging ligands between gold(III) centres [23].

An interesting change of stereochemistry is observed in complexes of gold(III) with the ligands 2,2′-biquinolyl and 2,9-dimethyl-1,10-phenanthroline. Due to steric effects, these ligands are incapable of forming square-planar complexes, and the five co-ordinate complexes $[AuCl_3$(biquinolyl)] and $[AuX_3$(2,9-dimethyl-1,10-phenanthroline)] with X = Cl or Br are formed instead [24, 25]. As seen in Fig. 5.1, these are best considered as distorted square-pyramidal structures. Steric effects are minimised in each case since there is only one halide ligand in the plane of the bidentate ligand, where steric interactions are expected to be severe.

It is suggested that the structures adopted are such as to minimise repulsion between the axial nitrogen atom

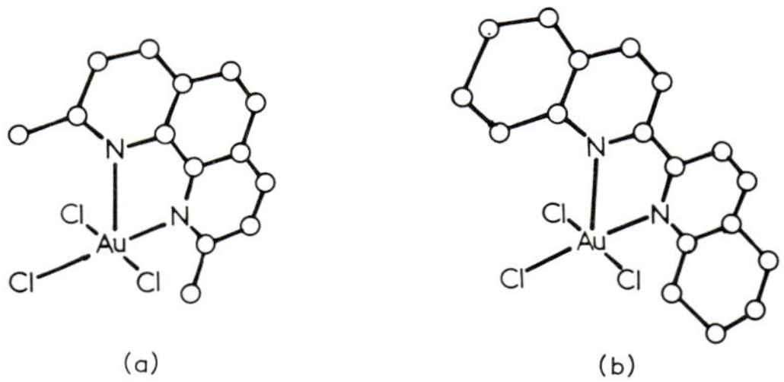

Fig. 5.1. Structures of (a) trichloro(2,9-dimethyl-1,10-phenanthroline)-gold(III), and (b) trichloro(2,2′-biquinolyl)gold(III). Circles represent carbon atoms; hydrogen atoms are omitted.

and the $5d_{z^2}$ orbital of gold(III), which in turn will maximise the energy gap between the highest occupied d-orbital (d_{z^2}) and the vacant $d_{x^2-y^2}$ orbital [25]. As expected the complexes are diamagnetic with the low spin $5d^8$ configuration.

In the above complexes, five-co-ordination is mandatory since co-ordination of one nitrogen atom of the bidentate ligand necessarily brings the second nitrogen atom within bonding distance of the gold centre. However, with more flexible chelate ligands the strong affinity of gold(III) for square-planar co-ordination is dominant. Thus, in the potentially five-co-ordinate complex $[AuCl_2(pz_3BH)]$, where $[HBpz_3]^-$ is the potentially tridentate ligand tripyrazolylborate, one of the pyrazolyl groups is not co-ordinated to gold. The structure is therefore as shown in Fig. 5.2 with square-planar gold(III).

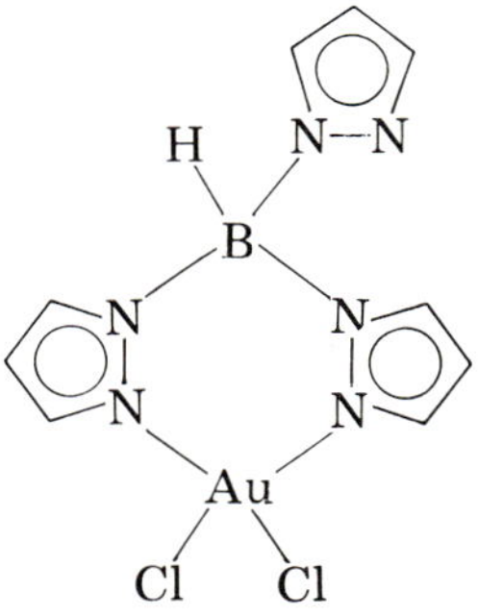

Fig. 5.2. The proposed structure of dichloro(tripyrazolylborato)gold(III).

A five-co-ordinate transition state is readily available however, since the co-ordinated and non-co-ordinated pyrazolyl groups exchange rapidly on the n.m.r. time scale at room temperature but not at −60 °C [26].

Many other square-planar gold(III) complexes with chelate ligands containing nitrogen-donor atoms are known. Some of these are given in Table 5.2. In addition complexes with the planar tetradentate ligand tetraphenylporphine^{2-}, L, characterised as $[AuL]^+ [AuCl_4]^-$ and the five-co-ordinate [AuLCl], in which the stereochemistry at gold(III) is distorted square

TABLE 5.2

Some gold(III) complexes with nitrogen-donor chelate ligands

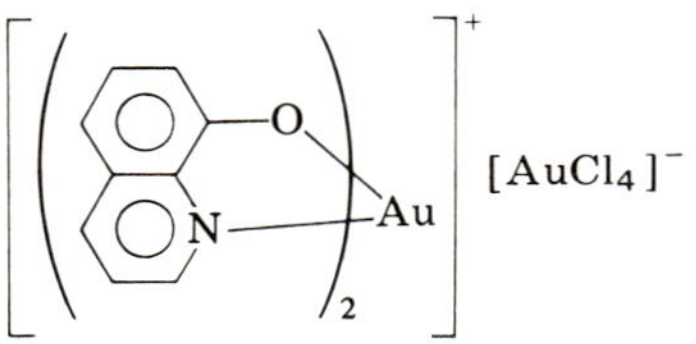

cis- and *trans*-isomers
(ref. 27)

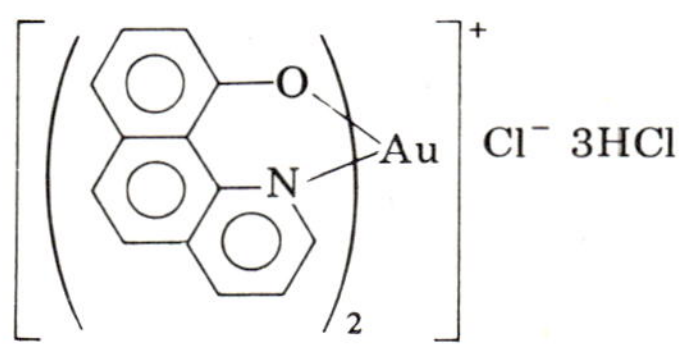

(ref. 27)

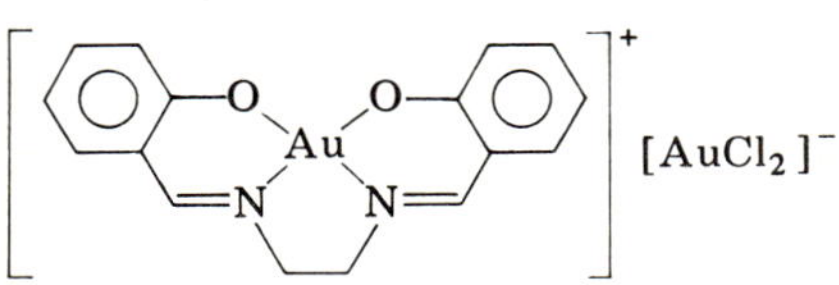

(ref. 27)

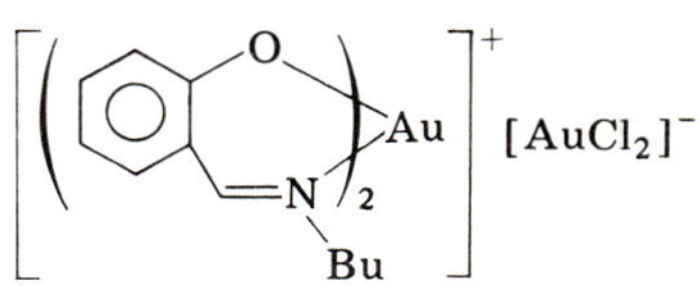

(ref. 28)

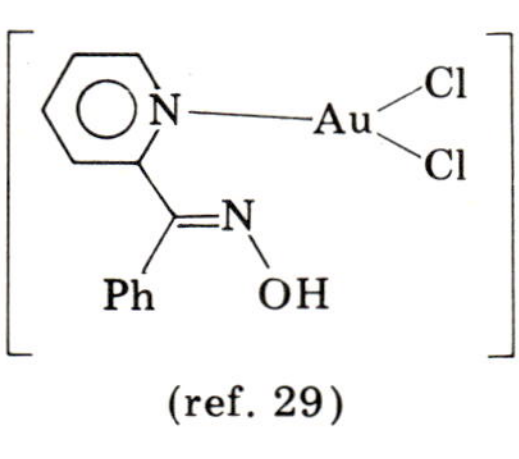

(ref. 29)

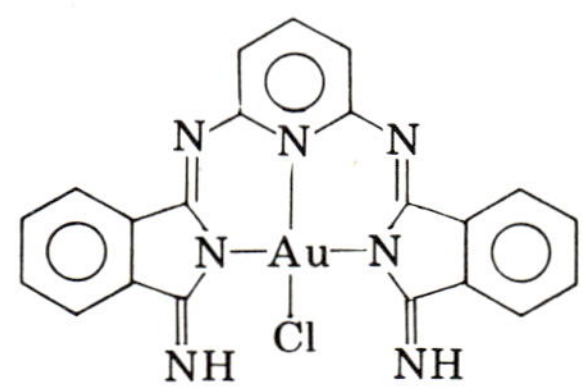

(ref. 30)

pyramidal, have been isolated [31]. Also complexes with chelating mixed N,S- and N,As-donor ligands have been studied [32, 33].

COMPLEXES WITH PHOSPHORUS-, ARSENIC- AND ANTIMONY-DONOR LIGANDS

Oxidation of complexes [$AuX(PR_3)$] with the corresponding halogen, X_2, gives [$AuX_3(PR_3)$][34]. The colours of the complexes are darker for the heavier halogens, so that [$AuCl_3(PEt_3)$] is yellow, [$AuBr_3(PEt_3)$] is red and [$AuI_3(PEt_3)$] is black. In addition to these complexes, mixed halogen compounds can be prepared:

$$[AuCl(PEt_3)] + I_2 \rightarrow [AuClI_2(PEt_3)] \leftarrow [AuI(PEt_3)] + ICl$$

These mixed halogen compounds were for many years thought to be formed as pure complexes, but recently they have been shown to be mixtures of all possible species [$AuX_nY_{3-n}(PR_3)$] [35, 36]. These are formed by rapid redox reactions between gold(I) and gold(III) species, as shown in the following equations:

$$[AuCl(PEt_3)] + Br_2 \rightarrow [AuClBr_2(PEt_3)]$$

$$[AuClBr_2(PEt_3)] + [AuCl(PEt_3)] \rightleftharpoons [AuBr(PEt_3)] + [AuCl_2Br(PEt_3)]$$

$$[AuCl_2Br(PEt_3)] + [AuCl(PEt_3)] \rightleftharpoons [AuBr(PEt_3)] + [AuCl_3(PEt_3)]$$

Since both *cis* and *trans* isomers of [$AuCl_2Br(PEt_3)$] and [$AuClBr_2(PEt_3)$] are formed, a total of six species are present in this reaction mixture. The same type of redox reaction can take place with platinum(II) complexes.

$$[AuBr_3(PMe_2Ph)] + cis\text{-}[PtMe_2(PMe_2Ph)_2] \rightarrow [AuBr(PMe_2Ph)] + trans,\ cis,\ cis\text{-}[PtBr_2Me_2(PMe_2Ph)_2]$$

The complex [$AuBr_3(PMe_3)$] contains a square-planar gold(III) centre [37], but [$AuCl_3(PPh_3)$] has a considerable distortion from square planar towards tetrahedral stereochemistry [38]. The AuCl bond *trans* to phosphorus (234.7 pm) is longer than the other two (mean 227.8 pm), and the ∠PAuCl is only 170° compared with 180° expected for a regular square-planar complex. The distortion is probably necessary in order to minimise steric interactions between the phosphine ligand and the *cis*-chlorine atoms.

The complex *trans*-$[AuI_2(diars)_2]^+I^-$ formed by reaction of iodine with $[Au(diars)_2]^+I^-$, contains a tetragonally distorted octahedral gold(III) centre as shown in Fig. 5.3. The AuI bond lengths (335 pm) are longer than the sum of the covalent radii (273 pm) indicating that these bonds are weak [39]. The large distortion from regular octahedral stereochemistry allows the gold(III) centre to keep the low spin configuration with the $5d_{z^2}$ orbital doubly occupied, and the complex is diamagnetic [40].

Similarly, with the ligand 1,2-bis(dimethylphosphino)-benzene, diphos, the complex ions $[Au(diphos)_2]^{3+}$, $[AuCl(diphos)_2]^{2+}$ and $[AuCl_2(diphos)_2]^+$ with four, five- and six-co-ordinate gold(III) centres can all be isolated as the perchlorate salts [40].

Another complex which is thought to contain six-co-ordinate gold(III) is $[Ph_4As]_3[Au(SC_6F_5)_4\{Sb(SC_6F_5)_4\}_2]$. In solution one of the antimony-donor ligands dissociates to give the presumably five-co-ordinate $[Au(SC_6F_5)_4\{Sb(SC_6F_5)_4\}]^{2-}$

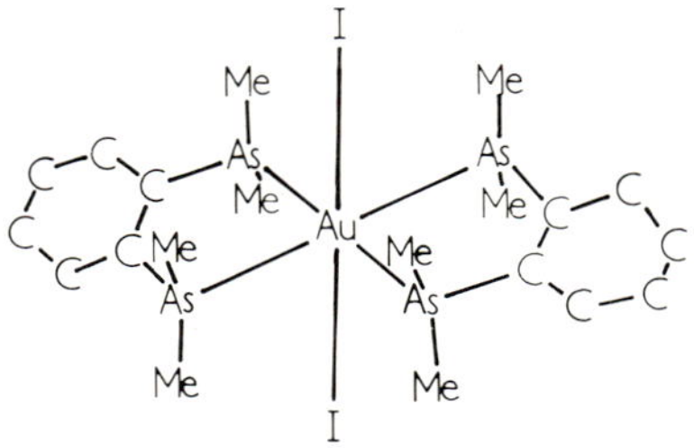

Fig. 5.3. The structure of di-iodobis[1,2-bis(dimethylarsino)benzene]-gold(III) cation.

ion [41]. The complexes catalyse the hydrogenation of alkenes in ethanol solution.

NITRATOGOLD(III) COMPLEXES

Gold(III) complexes with oxygen-donor ligands have not been studied extensively with the exception of the nitrate complexes [42]. Reaction of gold with excess of N_2O_5 gives the very air-sensitive $[NO_2]^+[Au(NO_3)_4]^-$, which decomposes readily with evolution of nitrogen oxides. More stable salts can be obtained by recrystallisation from pure nitric acid in the presence of an alkali metal nitrate, when the corresponding salts $M[Au(NO_3)_4]$ are obtained. The potassium salt contains a square-planar gold(III) centre with monodentate nitrate ligands [43]. Reaction of $[NO_2][Au(NO_3)_4]$ with N_2O_4 gives $[NO]^+[Au(NO_3)_4]^-$ which decomposes slowly under vacuum to give $Au_2O(NO_3)_4$. All of these complexes are decomposed instantly in water.

COMPLEXES WITH SULPHUR- AND SELENIUM-DONOR LIGANDS

As expected for a class b metal like gold, sulphur-donor ligands form more stable complexes than oxygen-donor ligands, and when a ligand has both oxygen- and sulphur-donor sites, the sulphur atom is always preferred by the gold(III) centre. Thus sulphite [44] in $[Au(SO_3)_2(en)]^{2-}$, dithio-oxalate [45] in $K[Au(C_2O_2S_2)_2]$, and dimethylsulphoxide [46] in $[AuCl_3(Me_2SO)]$ all bind to gold through the sulphur atoms.

In ligands containing both nitrogen and sulphur donor sites the situation is less clearcut [47—49]. Thus N,N,N′,N′-tetramethyldithio-oxamide forms the sulphur-donor complex,

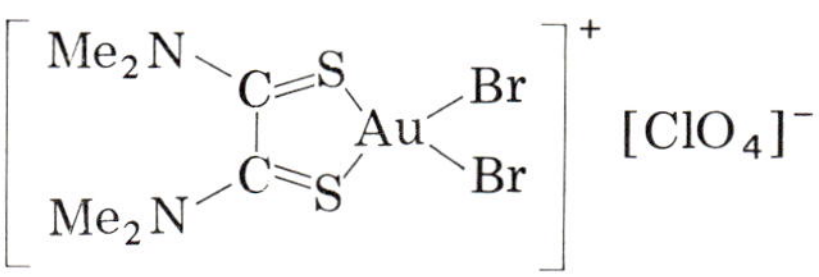

but N,N′-dimethyldithio-oxamide forms the N,S-donor complex, in which one of the nitrogen atoms is deprotonated.

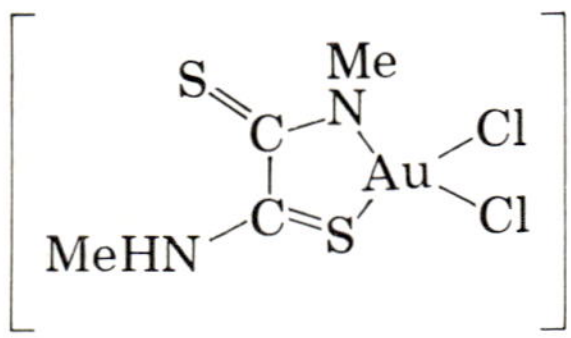

(a) Dialkylsulphide and dialkylselenide complexes

Complexes of the general type [$AuCl_3(SR_2)$] can be prepared by oxidation of the corresponding gold(I) complex [$AuCl(SR_2)$] with chlorine. They are also formed by reaction of dialkylsulphides with H[$AuCl_4$], it is difficult to prevent reduction to [$AuCl(SR_2)$] by excess of sulphide. Bidentate thiol ligands act as bridging groups in gold(I) complexes, but oxidation to gold(III) is accompanied by rearrangement to the chelate form [50].

$ClAuS(Me)CH_2CH_2S(Me)AuCl + 2Cl_2 \longrightarrow$

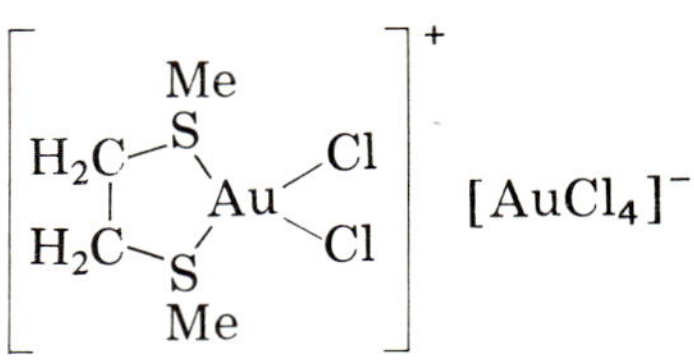

The chelating selenium-donor ligands L = $MeSe(CH_2)_nSeMe$, where n = 2, or 3, or

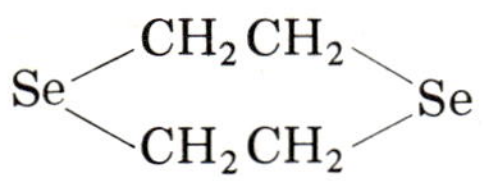

form complexes of empirical formula $AuCl_3L$ but their structures are not known [51, 52].

In the complexes $[AuCl\{S(CH_2Ph)_2\}]$ and $[AuCl_3\text{-}\{S(CH_2Ph)_2\}]$, it has been shown that inversion at the sulphur atom takes place much more rapidly for the gold(I) complex than for the gold(III) complex. The process probably occurs by concerted cleavage of the gold—sulphur bond with formation of a new gold—sulphur bond using the free lone-pair of electrons on the sulphur atom [53].

(b) Dithiolate complexes of gold(III)

Dithiocarbamate complexes of gold(III) can be prepared by oxidation of the dimeric gold(I) complexes $[\{Au(S_2CNR_2)\}_2]$ with chlorine or bromine [54—57].

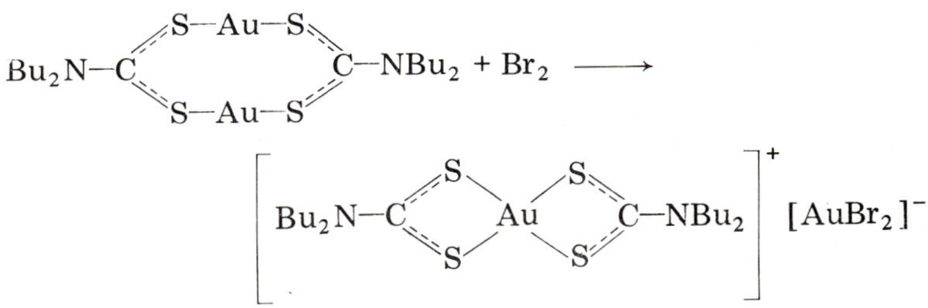

The resulting gold(I)—gold(III) complexes can be oxidised further.

$[Au(S_2CNBu_2)_2][AuBr_2] + Br_2 \longrightarrow$

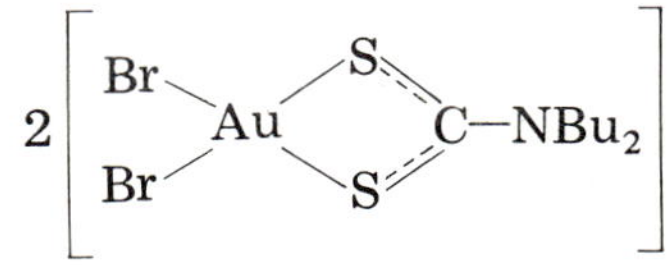

The exchange of dithiocarbamate groups between gold centres, which is illustrated by this reaction, also occurs in the isomerisation of $[Au(S_2CNBu_2)_2][AuBr_4]$ to the neutral $[AuBr_2\text{-}(S_2CNBu_2)]$ and in the following reaction [55, 56]:

$$2[AuBr_2(S_2CNBu_2)] + [\{Au(S_2CNBu_2)_2\}] \rightarrow 2[Au(S_2CNBu_2)_2][AuBr_2]$$

In addition, tris(dithiocarbamato)gold(III) complexes can be prepared by reaction of the thiuram disulphides $(R_2NCS_2)_2$ with the corresponding complex $[\{Au(S_2CNR_2)\}_2]$ or with $Na[AuCl_4]$. These complexes are soluble in organic solvents, in which they readily revert to the gold(I) precursor [58, 59].

There are thus three types of dithiocarbamatogold(III) complex, namely complexes $[AuX_2(S_2CNR_2)]$, $[Au(S_2CNR_2)_2]^+$ and $[Au(S_2CNR_2)_3]$. Analogous diselenocarbamates and thioselenocarbamates are prepared similarly, typical examples being $[AuBr_2(Se_2CNR_2)]$ [60], $[Au(Se_2CNEt_2)_2]^+Br^-$, $[Au(Se_2CNBu_2)_3]$ [61] and $[AuCl_2(SSeCNMe_2)]$ [62]. Probably related structurally to the tris(dithiocarbamato)gold(III) complexes is the complex $[Au\{SP(S)Ph_2\}_3]$ which is prepared by reaction of $H[AuCl_4]$ with the corresponding acid $Ph_2P(S)SH$ [63, 64].

1,2-Dithiolate complexes of gold(III) can also be prepared by oxidation of gold(I) complexes with S—S bonded species [65, 66].

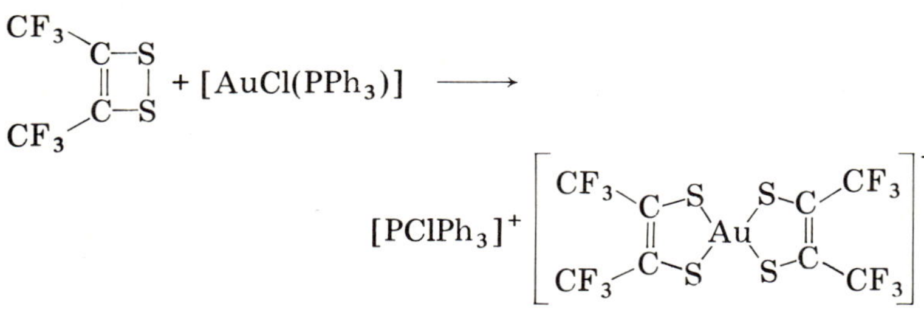

However, more usually they are prepared by substitution reactions [67, 68].

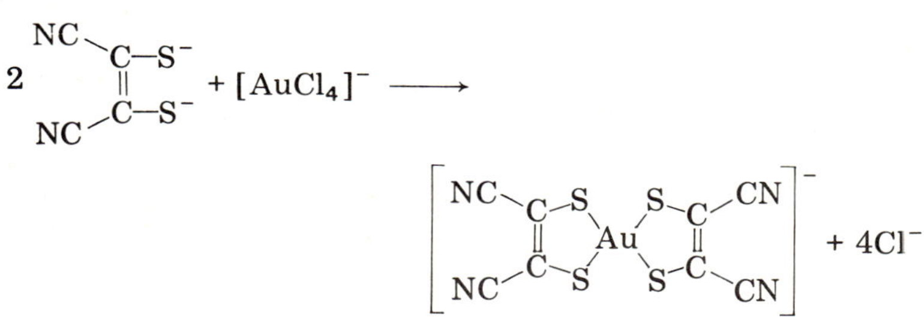

In a similar way the mixed 1,1-dithiolate-1,2-dithiolate complex can be prepared [69, 70].

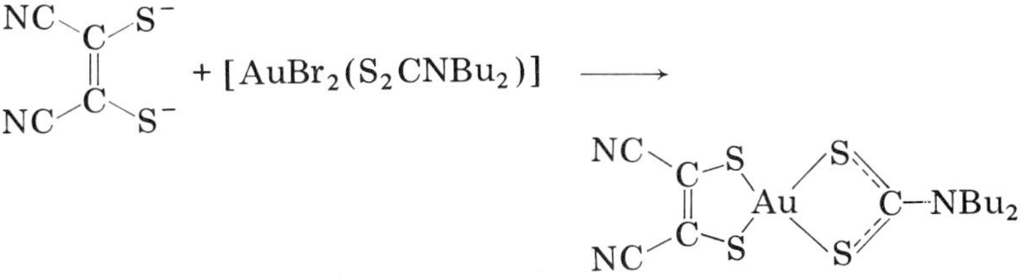

Interestingly, the isomeric ionic complex $[Au(S_2CNBu_2)_2]^+$ $[Au\{S_2C_2(CN)_2\}_2]^-$ is also known [71] and the structures of both isomers have been determined by X-ray crystallography [71, 72].

All dithiocarbamate and maleonitriledithiolate complexes of gold(III) have square-planar co-ordination about the gold(III) centre, and the dithiolate ligands act usually as bidentate ligands. Some bond lengths and angles in typical complexes are shown in Table 5.3. They show that the Au—S bonds are shorter and the SAuS bond angles greater for the 1,2-dithiolates than for the dithiocarbamates. The complexes $[Au(S_2CNR_2)_3]$ must contain one bidentate and two mono-

TABLE 5.3

Bond lengths and angles in gold(III)-dithiolate complexes

Complex	r(AuS)/pm	∠(SAuS)/°	Ref.
$[Au(S_2CNBu_2)_2]^+$	233.3	75.1	55, 56, 71, 73
$[Au\{S_2C_2(CN)_2\}_2]^-$	230.9	90.7	71
$[Au\{S_2C_2(CF_3)_2\}_2]^-$	229.1	—	66
$[Au(S_2CNBu_2)\{S_2C_2(CN)_2\}]$	232.7[a]	75.0[a]	72
	229.4[b]	90.9[b]	72
$[Au(\eta^2\text{—}S_2CNEt_2)(\eta^1\text{—}S_2CNEt_2)_2]$	234.3[a]	75.0[a]	74
	234.1[c]	—	74

[a] for η^2—S_2CNR_2.
[b] for $S_2C_2(CN)_2$.
[c] for η^1—S_2CNEt_2.

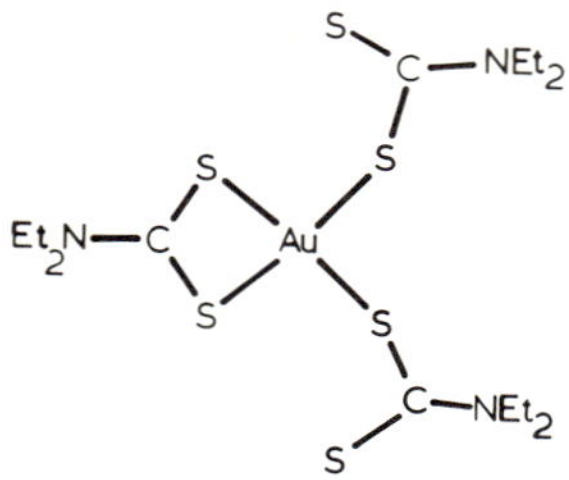

Fig. 5.4. The structure of tris(N,N-diethyldithiocarbamato)gold(III).

dentate dithiocarbamate ligands in order for gold(III) to be square planar, and the expected structure has been confirmed for the case where R = Et [74], as shown in Fig. 5.4.

COMPLEX HALIDES OF GOLD(III)

The complex halides of gold of general formula $M[AuX_4]$ are perhaps the mostly commonly used gold complexes in the laboratory. There are, of course, many halogenogold(III) complexes containing other ligands also, but these have mostly been discussed earlier in this chapter.

(a) Fluoride complexes of gold(III)

Reaction of bromine trifluoride with a mixture of gold and potassium chloride gives $K[AuF_4]$, and with a mixture of gold and silver it gives $Ag[AuF_4]$ [75, 76]. Tetrafluoroaurates-(III) can also be prepared by reaction of the corresponding tetrachloroaurates(III) with fluorine or with bromine trifluoride [75, 77].

$K[AuF_4]$ has a structure with four fluorine atoms in a square-planar arrangement about gold with $r(AuF)$ = 195 pm [78]. There are also contacts with neighbouring $[AuF_4]^-$ ions, with the closest interionic $r(AuF)$ = 312 pm. Tetrafluoroaurates are very reactive and, like all other fluorides of gold, are immediately hydrolysed by water.

The only other important fluorides of gold(III) are the complexes $AuF_3 \cdot BrF_3$ and $AuF_3 \cdot SeF_4$, in which the gold atom is thought to bind to a fluorine atom of the BrF_3 or SeF_4 group rather than to bromine or selenium [79]. Thermal decomposition of $AuF_3 \cdot BrF_3$ at 300 °C gives gold(III) fluoride contaminated with a little bromine, while $AuF_3 \cdot SeF_4$ decomposes at 350 °C according to the equation

$$3AuF_3 \cdot SeF_4 \rightarrow AuF_3 + 2Au + 3SeF_6$$

(b) Chloride and bromide complexes of gold(III)

Gold dissolves easily in aqua regia or in a mixture of hydrobromic and nitric acids to give tetrachloroauric(III) acid, $H[AuCl_4]$, or tetrabromoauric(III) acid, $H[AuBr_4]$, respectively. The acids can be isolated in hydrated forms such as $H[AuCl_4] \cdot 3H_2O$ and $H[AuBr_4] \cdot 4H_2O$ by evaporation of the solutions, with repeated addition of hydrochloric or hydrobromic acid so as to remove all nitric acid and nitrogen oxides. The acids are very soluble in organic solvents such as diethyl ether and ethyl acetate, which can therefore be used to extract $H[AuCl_4]$ from acidic aqueous solutions with very high efficiency. This extraction procedure is commonly used in analysis for gold.

$H[AuCl_4]$ and $H[AuBr_4]$ are strong acids, which can be neutralised with bases to give such salts as $Na[AuCl_4] \cdot 2H_2O$, $K[AuCl_4]$, and $K[AuBr_4] \cdot 2H_2O$. In addition, many mixed complexes such as $Cs_2[AuCl_2][AuCl_4]$, $Cs_2[AgCl_2][AuCl_4]$, $(NH_4)_6[Ag_2Cl_5][AuCl_4]_3$ and $Cs_4[PdBr_4][AuBr_4]$ are known [80, 81]. Some of these have interesting electrical conductivity properties [81]. In all these complexes the $[AuX_4]^-$ ions have square-planar stereochemistry, though there may also be weak interactions with halide atoms of neighbouring complex ions to give overall tetragonal stereochemistry about gold(III). This feature is seen in the structure of $(NH_4)_6[Ag_2Cl_5][AuCl_4]_3$ for example [80]. The gold—chlorine bond distance in the $[AuCl_4]^-$ ion is somewhat dependent on the cation, but falls in the range 227.1—229.5 pm [82—84]. The $[AuCl_4]^-$ and

$[AuBr_4]^-$ ions are also square planar in aqueous solution, with gold—halogen bond lengths of 229 and 243 pm, respectively [84b].

It has been claimed [85] that association of bromide with $[AuBr_4]^-$ takes place in nitromethane solution to give $[AuBr_6]^{3-}$ but this is now thought to be incorrect. Reduction to gold(I) takes place instead. However, there is evidence for formation of the five-co-ordinate species $[AuCl_4(SCN)]^{2-}$ as an intermediate in the substitution of chloride in $[AuCl_4]^-$ by SCN^- to give finally $[Au(SCN)_4]^-$ [86].

In aqueous solution both the $[AuCl_4]^-$ and $[AuBr_4]^-$ ions are hydrolysed to some extent. Thus, the equilibrium constants for the reaction below are $K = 4.6 \times 10^{-5}$ when X = Cl, and $K = 3.4 \times 10^{-6}$ when X = Br in 0.4 M perchloric acid solution [87].

$$[AuX_4]^- + H_2O \rightleftharpoons [AuX_3(OH_2)] + X^-$$

The product acts as a weak acid [87—89], owing its acidity to the reaction below for which the most recent value of the acid dissociation constant is given (at 20 °C and infinite dilution) [89].

$$[AuCl_3(OH_2)] \rightleftharpoons [AuCl_3(OH)]^- + H^+ \quad \log_{10} K = -1.0 \pm 0.3$$

In alkaline solution, therefore, hydroxo species are formed by reactions such as:

$$[AuCl_4]^- + H_2O \rightleftharpoons [AuCl_3(OH)]^- + H^+ + Cl^- \quad \beta_1 = 10^{-6.22}\ M^2$$

$$[AuCl_4]^- + 2H_2O \rightleftharpoons [AuCl_2(OH)_2]^- + 2H^+ + 2Cl^- \quad \beta_2 = 10^{-13.26}\ M^4$$

The species $[AuCl(OH)_3]^-$ and $[Au(OH)_4]^-$ are probably also formed but the corresponding stability constants have not been determined [88].

(c) Iodide complexes of gold(III)

Salts of the complex ion $[AuI_4]^-$ cannot be prepared in pure form from aqueous solution since partial reduction to $[AuI_2]^-$

takes place [85]. The equilibrium constant is given below [90].

$[AuI_2]^- + I_2 \rightleftharpoons [AuI_4]^- \quad K = 2 \times 10^2\ M^{-1}$

Pure $Et_4N[AuI_4]$ can be prepared by reaction of $Et_4N[AuCl_4]$ with anhydrous liquid HI. It decomposes slowly in air in the solid state and rapidly on dissolution in organic solvents [85].

The complex ion $[Au(CN)_2I_2]^-$ is very much more stable than $[AuI_4]^-$ itself, as evidenced by the equilibrium constant [4].

$[Au(CN)_2]^- + I_2 \rightleftharpoons [Au(CN)_2I_2]^- \quad K = 1.3 \times 10^4\ M^{-1}$

REFERENCES

1 H. T. S. Britton and E. N. Dodd, *J. Chem. Soc.*, (1935) 100.
2 L. H. Jones and J. M. Smith, *J. Chem. Phys.*, 41 (1964) 2507.
3 L. H. Jones, *Inorg. Chem.*, 3 (1964) 1581.
4 M. H. Ford-Smith, J. J. Habeed and J. H. Rawsthorne, *J. Chem. Soc., Dalton Trans.*, (1972) 2116.
5 J. M. Smith, L. H. Jones, I. K. Kressin and R. A. Peneman, *Inorg. Chem.*, 4 (1965) 369.
6 A. E. Remick, *J. Am. Chem. Soc.*, 69 (1947) 94.
7 A. Sacco and M. Freni, *Gazz. Chim. Ital.*, 86 (1956) 195.
8 L. H. Skibsted and J. Bjerrum, *Acta Chem. Scand., Ser. A*, 28 (1974) 740.
9 M. Weishaupt and J. Straehle, *Z. Naturforsch. B*, 31B (1976) 554.
10 B. P. Block and J. C. Bailar, *J. Am. Chem. Soc.*, 73 (1951) 4722.
11 W. H. Baddley, F. Basolo, H. B. Gray, C. Nolting and A. J. Poë, *Inorg. Chem.*, 2 (1963) 921.
12 C. F. Weick and F. Basolo, *Inorg. Chem.*, 5 (1966) 576.
13 M. S. Kharasch and H. Isbell, *J. Am. Chem. Soc.*, 53 (1931) 3059.
14 Yu. N. Kukushkin and L. I. Zorina, *Zh. Neorg. Khim.*, 12 (1967) 568.
15 C. S. Gibson and W. M. Colles, *J. Chem. Soc.*, (1931) 2407.
16 P. K. Monaghan and R. J. Puddephatt, *Inorg. Chim. Acta*, 15 (1975) 231.
17 D. B. Dell'Amico and F. Calderazzo, *Gazz. Chim. Ital.*, 103 (1973) 1099.
18 F. Calderazzo and D. B. Dell'Amico, *J. Organomet. Chem.*, 76 (1974) C59.
19 R. Hüttel and A. Konietzny, *Chem. Ber.*, 106 (1973) 2098.
20 R. J. Dubois, J. Hagimassy, A. C. Noble and F. D. Popp, *J. Heterocycl. Chem.*, 3 (1966) 377.
21 C. M. Harris, *J. Chem. Soc.*, (1959) 682.
22 C. M. Harris and T. N. Lockyer, *J. Chem. Soc.*, (1959) 3083.
23 T. Mundorf and K. Dehnicke, *Z. Anorg. Allg. Chem.*, 408 (1974) 146.
24 R. J. Charlton, C. M. Harris, H. Patil and N. C. Stephenson, *Inorg. Nucl. Chem. Lett.*, 2 (1966) 409.
25 W. T. Robinson and E. Sinn, *J. Chem. Soc., Dalton Trans.*, (1975) 726.
26 N. F. Borkett and M. I. Bruce, *Inorg. Chim. Acta*, 12 (1975) L33.
27 T. Inazu, *Bull. Chem. Soc. Jpn.*, 39 (1966) 1065.
28 S. Yamada and K. Yamanouchi, *Bull. Chem. Soc. Jpn.*, 43 (1970) 1744.
29 D. Malone and B. Sen, *J. Inorg. Nucl. Chem.*, 35 (1973) 2114.

30 P. Bamfield and P. A. Mack, *J. Chem. Soc. C*, (1968) 1961.
31 E. B. Fleischer and A. Laszlo, *Inorg. Nucl. Chem. Lett.*, 5 (1969) 373; R. Timkovich and A. Tulinksy, *Inorg. Chem.* (1977) 962.
32 P. S. K. Chia and S. E. Livingstone, *Aust. J. Chem.*, 21 (1968) 339.
33 G. A. Barclay, C. M. Harris and J. V. Kingston, *Chem. Ind.* (1965) 227.
34 F. G. Mann and D. Purdie, *J. Chem. Soc.*, (1940) 1235.
35 B. T. Heaton and R. J. Kelsey, *Inorg. Nucl. Chem. Lett.*, 11 (1975) 363.
36 R. J. Puddephatt and P. J. Thompson, *J. Chem. Soc., Dalton Trans.*, (1975) 1810.
37 M. F. Perutz and O. Weisz, *J. Chem. Soc.*, (1946) 438.
38 G. Bandoli, D. A. Clemente, G. Marangoni and L. Cattalini, *J. Chem. Soc., Dalton Trans.*, (1973) 886.
39 V. F. Duckworth and N. C. Stephenson, *Inorg. Chem.*, 8 (1969) 1661.
40 C. M. Harris and R. S. Nyholm, *J. Chem. Soc.*, (1957) 63; L. F. Warren and M. A. Bennett, *Inorg. Chem.*, 15 (1976) 3126.
41 M. C. Muller, *Gold Bull.*, 7 (1974) 39; R. J. H. Clark, M. C. Muller and R. S. Nyholm, *J. Chem. Soc., Dalton Trans.*, to be published.
42 C. C. Addison, G. S. Brownlee and N. Logan, *J. Chem. Soc., Dalton Trans.*, (1972) 1440.
43 C. D. Garner and S. C. Wallwork, *J. Chem. Soc. A.*, (1970) 3092.
44 A. Dunand and R. Gerdil, *Acta Crystallogr., Sect. B*, B31 (1975) 370.
45 A. R. Latham, V. C. Mascall and H. B. Gray, *Inorg. Chem.*, 4 (1965) 788.
46 R. A. Potts, *J. Inorg. Nucl. Chem.*, 34 (1972) 1749.
47 A. C. Fabretti, G. C. Pellacani, G. Peyronel and B. Scapinelli, *J. Inorg. Nucl Chem.*, 36 (1974) 1067.
48 A. C. Fabretti, G. C. Franchini, G. C. Pellacani and G. Peyronel, *Inorg. Nucl. Chem. Lett.*, 12 (1976) 163.
49 G. C. Pellacani, *Inorg. Chim. Acta*, 12 (1975) L3.
50 K. C. Dash and H. Schmidbaur, *Chem. Ber.*, 106 (1973) 1221.
51 E. E. Aynsley, N. N. Greenwood and J. B. Leach, *Chem. Ind. (London)*, (1966) 379.
52 P. J. Hendra and N. Sadasivan, *J. Chem. Soc.*, (1965) 2063.
53 F. Coletta, R. Ettore and A. Gambaro, *Inorg. Nucl. Chem. Lett.*, 8 (1972) 667.
54 H. J. A. Blaauw, R. J. F. Nivard and G. J. M. van der Kerk, *J. Organomet. Chem.*, 2 (1964) 236.
55 P. T. Beurskens, H. J. A. Blaauw, J. A. Cras and J. J. Steggerda, *Inorg. Chem.*, 7 (1968) 805.
56 P. T. Beurskens, J. A. Cras and J. G. M. van der Linden, *Inorg. Chem.*, 9 (1970) 475.
57 J. G. M. van der Linden, *Rec. Trav. Chim. Pays-Bas*, 90 (1971) 1027.
58 S. Akerstrom, *Ark. Kemi*, 14 (1959) 403.
59 T. Takiguchi, M. Abe, K. Kurosaki, E. Asada and M. Nakagoma, *Kogyo Kagaku Zasshi*, 70 (1967) 1182.
60 J. G. M. van der Linden and W. P. M. Nijssen, *Z. Anorg. Allg. Chem.*, 392 (1972) 93.
61 B. Lorenz, R. Kirmse and E. Hoyer, *Z. Anorg. Allg. Chem.*, 378 (1970) 144.
62 N. Sonada and T. Tanaka, *J. Inorg. Nucl. Chem.*, 35 (1973) 1145.
63 V. V. K. Rao and A. Mueller, *Z. Chem.*, 10 (1970) 197.
64 A. Mueller, V. V. K. Rao and G. Klinksiek, *Chem. Ber.*, 104 (1971) 1892.
65 A. Davison, D. V. Howe and E. T. Shawl, *Inorg. Chem.*, 6 (1967) 458.
66 J. H. Enemark and J. A. Ibers, *Inorg. Chem.*, 7 (1968) 2636.
67 A. Davison, N. Edelstein, A. H. Maki and R. H. Holm, *Inorg. Chem.*, 2 (1963) 1227.
68 T. J. Bergendahl and J. H. Waters, *Inorg. Chem.*, 14 (1975) 2556.
69 J. G. M. van der Linden and H. G. J. de Roer, *Inorg. Chim. Acta*, 5 (1971) 254.
70 J. G. M. van der Linden, *J. Inorg. Nucl. Chem.*, 34 (1972) 1645.
71 J. H. Noordik and P. T. Beurskens, *J. Cryst. Mol. Struct.*, 1 (1971) 339.

72 J. H. Noordik, T. W. Hummelink and J. G. M. van der Linden, *J. Coord. Chem.*, 2 (1973) 185.
73 J. A. Cras, J. H. Noordik, P. T. Beurskens and A. M. Verhoeven, *J. Cryst. Mol. Struct.*, 1 (1971) 155.
74 J. H. Noordik, *Cryst. Struct. Commun.*, 2 (1973) 81.
75 A. G. Sharpe, *J. Chem. Soc.*, (1949) 2901.
76 A. G. Sharpe, *J. Chem. Soc.*, (1950) 2907.
77 R. Hoppe and W. Klemm, *Z. Anorg. Allg. Chem.*, 268 (1952) 364.
78 A. J. Edwards and G. R. Jones, *J. Chem. Soc., A.* (1969) 1936.
79 N. Bartlett and P. L. Robinson, *J. Chem. Soc.*, (1961) 3417.
80 J. C. Bowles and D. Hall, *J. Chem. Soc., Chem. Commun.*, (1971) 1523.
81 P. S. Gomm and A. E. Underhill, *Inorg. Nucl. Chem. Lett.*, 10 (1974) 309.
82 P. G. Jones, J. J. Guy and G. M. Sheldrick, *Acta Crystallogr., Sect. B*, 31 (1975) 2687.
83 M. R. Caira, L. R. Nassimbeni and A. L. Rodgers, *Acta Crystallogr., Sect. B*, 31 (1975) 1112.
84 (a) J. C. M. T. Eijndhoven and G. C. Verschoor, *Mater. Res. Bull*, 9 (1974) 1667.
84 (b) H. Ohtaki and M. Maeda, *Asahi Garasu Kogyo Gijutsu Shoreikai Kenkyu Hokuku* 28 (1976) 221; *Chem Abstr.*, 86 (1977) 181120.
85 J. L. Ryan, *Inorg. Chem.*, 8 (1969) 2058.
86 A. J. Hall and D. P. N. Satchell, *J. Chem. Soc., Chem. Commun.*, (1976) 163.
87 V. I. Dubinskii, V. M. Shul'man and B. I. Peshchevitskii, *Zh. Neorg. Khim.*, 13 (1968) 54.
88 L. Carlsson and G. Lundgren, *Acta Chem. Scand.*, 21 (1967) 819.
89 B. I. Peshchevitskii, V. I. Belevantsev and N. V. Kurbatova, *Zh. Neorg. Khim.*, 16 (1971) 1898.
90 Å. Håkansson and L. Johansson, *Chem. Scr.*, 7 (1975) 201.

GOLD(V) COMPLEXES

The chemistry of gold(V) is limited to the fluoride AuF_5 and to the complex ion $[AuF_6]^-$. The chemistry of AuF_5 was discussed in Chapter 2, so that we shall be concerned only with the hexafluoroaurates(V) here.

The first gold(V) complex to be prepared was $[Xe_2F_{11}][AuF_6]$, formed by reaction of a mixture of gold(III) fluoride and xenon difluoride with fluorine at 400 °C and a pressure of 100 p.s.i. [1].

$$2\,XeF_2 + AuF_3 + 5F_2 \rightarrow [Xe_2F_{11}][AuF_6]$$

Oxidation by fluorine is also used to prepare $Cs[AuF_6]$ from $Cs[AuF_4]$ [1], and $[O_2][AuF_6]$ from gold powder and oxygen [1, 2].

The powerful oxidative fluorinating agent KrF_2 can also be used to form the $[AuF_6]^-$ ion at room temperature [3].

$$7\,KrF_2 + 2Au \xrightarrow{HF} 2[KrF][AuF_6] + 5Kr$$

Other salts of the $[AuF_6]^-$ ion can be prepared by displacement of one cation by another [1, 3].

$$CsF + [Xe_2F_{11}][AuF_6] \xrightarrow{110\,^{\circ}C} 2XeF_6 + Cs[AuF_6]$$

$$[KrF][AuF_6] + O_2 \rightarrow [O_2][AuF_6] + Kr + \tfrac{1}{2}F_2$$

Alternatively, reaction of a suitable fluoride with AuF_5 can give hexafluoroaurates(V) [2, 3].

$$2\,XeF_2 + AuF_5 \xrightarrow{HF} [Xe_2F_3][AuF_6]$$

$$NOF + AuF_5 \xrightarrow{HF} [NO]^+[AuF_6]$$

Heating the complex $[Xe_2F_{11}][AuF_6]$ to 110 °C leads to evolution of XeF_6 and the complex $[XeF_5][AuF_6]$ remains [4]. Also known are the salts $K[AuF_6]$, $[IF_6][AuF_6]$ [5] and $[BrF_6][AuF_6]$ [5, 6].

The most interesting property of the hexafluoroaurates(V) is the decomposition of the $[O_2]^+$ or $[KrF]^+$ salts to give AuF_5 [2, 3]. Displacement of AuF_5 also takes place on reaction with SbF_5 [5].

$$[AuF_6]^- + SbF_5 \rightarrow [SbF_6]^- + AuF_5$$

The crystal and molecular structure of $[Xe_2F_{11}][AuF_6]$ has been determined by X-ray diffraction, and the $[AuF_6]^-$ was shown to have octahedral stereochemistry with r(AuF) 185—190 pm. A very slight distortion from regular symmetry is perhaps due to weak co-ordination of two of the fluorine atoms of $[AuF_6]^-$ to xenon atoms of the counter-ion. The Raman spectra [1, 3] and ^{197}Au Mössbauer spectra [4] of various $[AuF_6]^-$ salts suggest that the ion has a regular octahedral structure, though the absence of ^{197}AuF coupling (^{197}Au, $I = \frac{3}{2}$) in the ^{19}F n.m.r. spectrum of $[XeF_5][AuF_6]$ in hydrogen fluoride solvent suggests that there may be some distortion of the $[AuF_6]^-$ ion in this medium [3]. A regular octahedral structure is of course expected since Au(V) has the low-spin t_{2g}^6 electron configuration in both AuF_5 and $[AuF_6]^-$.

Gold(V) complexes have also been suggested as reaction intermediates. For example the reaction of $[AuMe_3(PMe_3)]$ with methyl iodide to give ethane and *cis*-$[AuIMe_2(PMe_3)]$ was suggested to proceed *via* the gold(V) complex $[AuIMe_4(PMe_3)]$, but this is now considered unlikely [7].

REFERENCES

1 K. Leary and N. Bartlett, *J. Chem. Soc., Chem. Commun.*, (1972) 903.
2 M. J. Vasile, T. J. Richardson, F. A. Stevie and W. E. Falconer, *J. Chem. Soc., Dalton Trans.*, (1976) 351.
3 J. H. Holloway and G. J. Schrobilgen, *J. Chem. Soc., Chem. Commun.*, (1975) 623.
4 G. Kaindl, K. Leary and N. Bartlett, *J. Chem. Phys.*, 59 (1973) 5050.
5 N. Bartlett and K. Leary, *Rev. Chim. Miner.*, 13 (1976) 82.
6 V. B. Sokolov, V. N. Prusakov, A. V. Ryzhkov and Yu. B. Drobyshevskii, *Dokl. Akad. Nauk SSSR*, 229 (1976) 884.
7 A. Shiotani and H. Schmidbaur, *J. Organomet. Chem.*, 37 (1972) C24.

CHAPTER 7

ORGANOGOLD CHEMISTRY

Great advances in our knowledge of organogold chemistry have been made in recent years, including the synthesis of such complex ions as $[AuMe_2]^-$ and $[AuMe_4]^-$, and of many olefin complexes of gold. We shall begin by describing complexes with gold—carbon σ-bonds, and examine the π-bonded derivatives later. The σ-bonded derivatives are the easier to prepare and may even exist as intermediates in living systems. Thus, it has been shown that methyl-vitamin B_{12} reacts with a mixture of $[AuCl_2]^-$ and $[AuCl_4]^-$ to give initially a methylgold derivative [1].

σ-BONDED ORGANOGOLD(I) COMPLEXES

(a) Dialkylaurates(I) and related species

These complexes, which are the simplest alkylgold(I) complexes yet prepared [2], can be formed by reactions such as:

$$[AuMe(PPh_3)] + MeLi \rightarrow Li^+ [AuMe_2]^- + PPh_3$$

The complex can be isolated if the ligand PMDT($=Me_2NCH_2$-$CH_2N(Me)CH_2CH_2NMe_2$) is added to form a complex with the Li^+ ion [3]. Thus, [Li(PMDT)] $[AuMe_2]$ has m.p. 120—123°C, and the anion has been shown by vibrational spectroscopy to contain a linearly coordinated gold(I) atom.

Complexes with mixed alkyl groups can be prepared *in situ* by similar reactions but they have not yet been isolated [2].

$$[AuMe(PPh_3)] + {}^tBuLi \longrightarrow Li^+ [Au(Me)\,{}^tBu]^- + PPh_3$$

$$[Au\,{}^tBu(PPh_3)] + MeLi \nearrow$$

In a similar way the complexes $Bu_4N^+[Au(C_6F_5)_2]^-$ and $Bu_4N^+[Au(C_6F_5)X]^-$, where X = halogen, can be prepared (tht = tetrahydrothiophen) [4].

$$[Au(C_6F_5)(tht)] + [Bu_4N]^+X^- \rightarrow [Bu_4N]^+[Au(C_6F_5)X]^- + tht$$

Some particularly stable derivatives of this kind are prepared by reactions of gold(I) complexes with trimethylphosphine ylid, $Me_3P{=}CH_2$ and similar species [5—7].

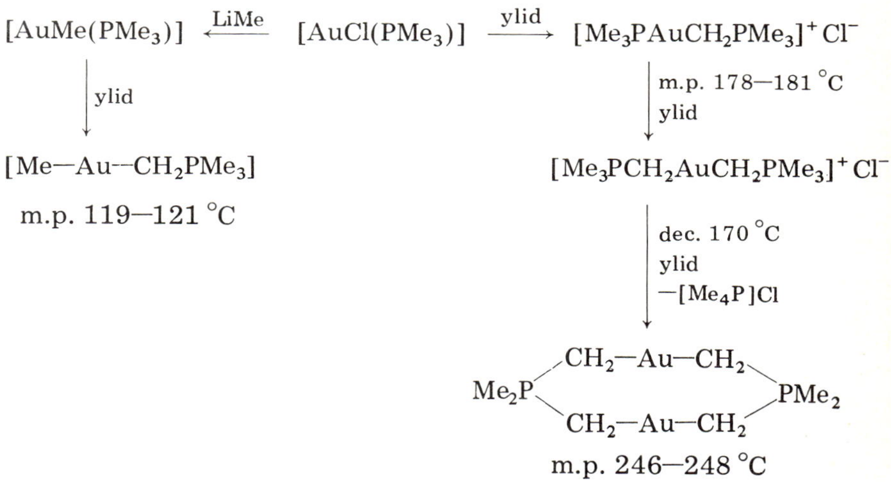

The complexes shown below can be prepared in a similar way from suitable phosphine ylid precursors [7b].

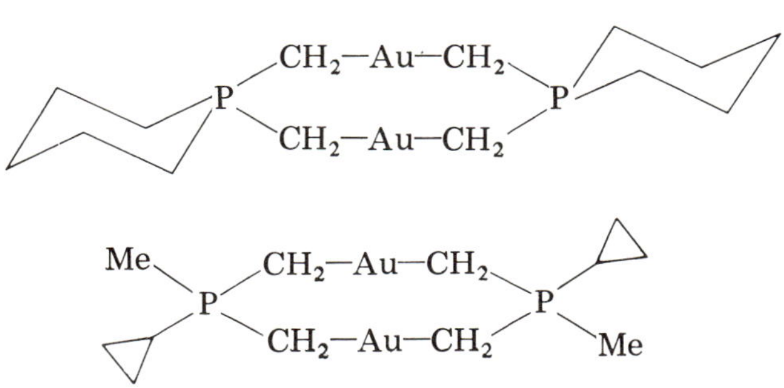

The complex $[Au\{CH(COPh)PPh_3\}_2]^+[AuCl_4]^-$ derived from benzoyltriphenylphosphine ylid is also of this general type [7a].

All of these complexes with gold(I) bound to two carbon atoms are very easily oxidised to organogold(III) species, as described later.

(b) Complexes of the type [RAuL]

More complexes of the general form [RAuL], where L is a "stabilising" ligand which is most commonly a tertiary phosphine, are known than for all other organogold(I) complexes. They were first reported in 1959 to be formed by reaction of a gold(I) halide with an organolithium reagent [9].

$$[AuCl(PEt_3)] + MeLi \rightarrow [AuMe(PEt_3)] + LiCl$$

This is still the most widely used preparative route to these complexes, though use of Grignard reagents is sometimes preferred [14, 19].

$$[AuCl(PPh_3)] + {}^tBuMgCl \rightarrow [Au({}^tBu)(PPh_3)] + MgCl_2$$

Some of the compounds which have been prepared by these and other methods are given in Table 7.1. The compounds are monomeric and diamagnetic with linearly co-ordinated gold(I) atoms. The dipole moments of the complexes are lower than those of similar halide complexes {e.g., $[AuMe(PEt_3)]$ 5.5 D, $[AuMe(PPh_3)]$ 5.6 D, $[AuCl(PEt_3)]$ 8.4 D} indicating that the Au—C bond has low polarity [9]. (1 D = 3.3356 × 10^{-20} C m.)

An unusual route to benzylgold(I) compounds is illustrated by the reaction of Et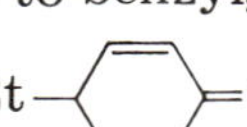$=CH_2$ with $[(Ph_3PAu)_3O]^+$ to give

[Et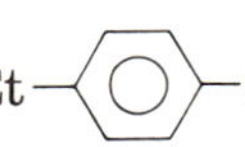$CH_2Au(PPh_3)]$ [25a].

The gold—carbon σ-bond is probably a strong bond, though no bond energy determinations have been made for gold(I) complexes. This strong Au—C bond, coupled with the low affinity of gold for oxygen-donor ligands, leads to some unusual

TABLE 7.1

Complex[a]	m.p./°C	Ref.
$[AuMe(PMe_3)]$	70—71	8
$[AuMe(PEt_3)]$	62	9
$[AuMe(PMe_2Ph)]$	30	10
$[AuMe(PMePh_2)]$	69—70	11
$[AuMe(PPh_3)]$	175	9
$[AuMe\{P(OMe)_3\}]$	0—2	12
$[AuEt(PPh_3)]$	150—152 dec.	13
$[Au^iPr(PPh_3)]$	150—151 dec.	14
$[Au^tBu(PPh_3)]$	139—142 dec.	14
$[AuPh(PPh_3)]$	164 dec.	9
$[Au(C_6Cl_4H)(PPh_3)]$		15a
$[Au(C_6F_5)(PPh_3)]$	170 dec.	15b
$[Au(C_6Cl_5)(PPh_3)]$	211 dec.	15b
$[Au(C_6Br_5)(PPh_3)]$	130 dec.	15b
$[Au(C_6F_5)(AsPh_3)]$	182 dec.	16
$[Au(CF_3)(PPh_3)]$	150 dec.	17
$[Au(CH_2SiMe_3)(PPh_3)]$	111—113	18
$[Au(CH_2Ph)(PPh_3)]$		19
$[Au(CH{=}CH_2)(PPh_3)]$		19
$[Au(4\text{-}FC_6H_4)(PPh_3)]$	154	20
$[Au(3\text{-}FC_6H_4)(PPh_3)]$	149—151	21
$[Au(4\text{-}FC_6H_4)(4FC_6H_4NC)]$	153—155 dec.	20
$[Au(2\text{-}Me_2NC_6H_4)(PPh_3)]$	172—174 dec.	22
$[Au(CF_2CF_2Me)(PMePh_2)]$	153—153.5	23
$[Au(C_5H_5)(PPh_3)]$	>100 dec.	24
$[Au(C{\equiv}CPh)(PPh_3)]$	163—164 dec.	25
$[Au(C{\equiv}CPh)(^iPrNH_2)]$	>115 dec.	25
$[Au(C{\equiv}CPh)(BuNC)]$	40.5—41.5	25

[a] All complexes are white.

structures. Thus, gold(I) acetylacetonates contain C-bonded, rather than the usual O-bonded, acetylacetonate ligand [26].

$$[AuCl(PPh_3)] + Tl(acac) \rightarrow [Au\{CH(COMe)_2\}(PPh_3)] + TlCl$$

Similarly, gold(I) derivatives of monoketones and aldehydes are also C-bonded. These compounds can be prepared in a number of ways. For example reaction of $[(Ph_3PAu)_3O]^+$-$[BF_4]^-$ with acetophenone or vinyl n-butyl ether gives $[Au(CH_2COPh)(PPh_3)]$ or $[Au(CH_2CHO)(PPh_3)]$, respectively, and oxidation of vinyl(triphenylphosphine)gold(I) with potassium permanganate in acetone solution gives [Au(CH_2-COMe) (PPh_3)] [27—29]. Another example is the ylid derivative [Au{CH(COPh)PPh_3}PPh_3] [7a].

(c) Cyclopentadienylgold(I) complexes

Reaction of (dicyclopentadiene)chlorogold(I) with cyclopentadienylsodium, C_5H_5Na, gives an explosive yellow compound AuC_5H_5. This is undoubtedly polymeric and reacts with triphenylphosphine to give [Au(C_5H_5) (PPh_3)], which can also be prepared by the reaction [30]:

$$AuCl(PPh_3) + C_5H_5Na \rightarrow NaCl + [Au(C_5H_5)(PPh_3)]$$

Infrared and n.m.r. studies indicate that the cyclopentadienyl group is σ-bonded to gold [30—32]. However, at room temperature the hydrogen atoms of the C_5H_5 group appear equivalent in the n.m.r spectrum since the molecule is fluxional, with the gold atom moving rapidly from one carbon atom to another of the η^1-C_5H_5 group.

Ferrocenylgold(I) complexes can be prepared by the reactions [33, 34]:

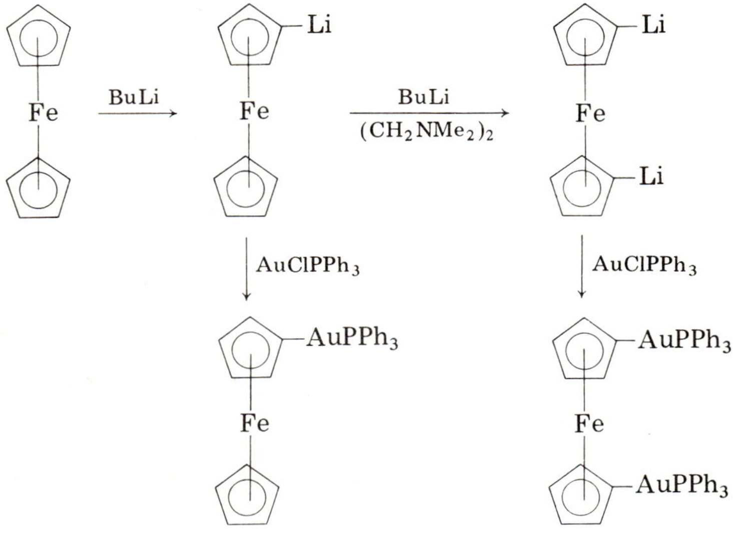

The manganese derivatives $[Mn(CO_3)\{C_5H_4Au(PPh_3)\}]$ and $[Mn(CO)_2(PPh_3)\{C_5H_4Au(PPh_3)\}]$ can be prepared in an analogous way [35, 36].

(d) Alkynylgold(I) complexes

Alkynylgold(I) complexes have been known since 1900 when the explosive gold(I) acetylide, Au_2C_2, was reported [37]. The more stable phenylethynylgold(I) is precipitated as a yellow powder when phenylacetylene and sodium acetate are added to an aqueous solution containing $[AuBr_2]^-$. The complex AuCCPh is probably polymeric, but one isomer of the similar complex $AuCC^tBu$ exists as a tetramer which probably has the structure shown in Fig. 7.1.

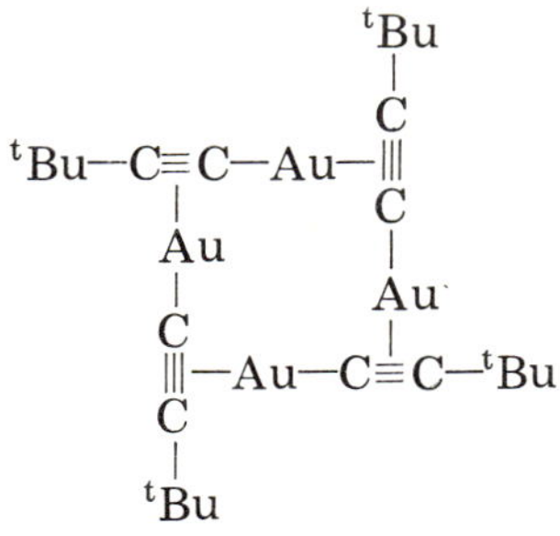

Fig. 7.1. The proposed structure of 3,3-dimethylbut-1-ynylgold(I).

Studies of the infrared spectra of these complexes indicate that the gold—alkyne π-bond, which causes this oligomerisation, is weak and the polymers are easily broken down by addition of other ligands.

$$(AuCCPh)_n + n\,Et_3As \rightarrow n\,[Et_3As{-}Au{-}C{\equiv}CPh]$$

Other ligands such as ammonia, pyridine, tertiary phosphines and isocyanides give similar complexes [25]. The complex $[Au(^iPrNH_2)(C{\equiv}CPh)]$ has been shown to contain a linearly co-ordinated gold(I) centre, though association through hydrogen bonding by the amine ligand is present [38].

Alkynylgold(I) complexes can also be prepared from gold(I) iodide and potassium acetylide, KC_2H, in liquid ammonia [39]. An ammonia complex of Au_2C_2 is obtained first, but with excess of KC_2H this reacts further to give the complex $K[Au(C{\equiv}CH)_2]$ and, in the presence of acetylene, $K_2[HC{\equiv}C{-}Au{-}C{\equiv}C{-}Au{-}C{\equiv}CH]$.

The complexed alkynylgold(I) species generally have high thermal stability and low chemical reactivity. They are, like other derivatives of gold(I) with electronegative σ-bonded species (e.g., trifluoromethyl and pentafluorophenyl derivatives), considerably more stable than the methyl- or ethylgold(I) complexes. An alternative preparative route involves cleavage of a methylgold bond by a terminal alkyne [23, 40].

$$[AuMe(PPh_3)] + PhC{\equiv}CH \rightarrow [Au(C{\equiv}CPh)(PPh_3)] + CH_4$$

(e) Complexes with bridging ligands

The polymeric complexes $[Au(C{\equiv}CR)_n]$ can be considered as special cases of complexes in which a σ-bonding ligand for gold(I) contains a second donor site through which it can bond to another gold atom and hence cause oligomerisation. Another example is given by the cyclopentadienylgold(I) polymer, $[(AuC_5H_5)_n]$.

A simpler example is found in 2-pyridylgold(I) which is formed thus [41]:

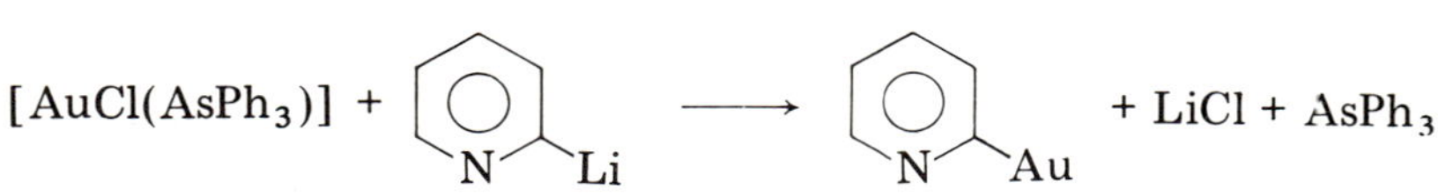

The product is suggested to be a trimer with the structure shown in Fig. 7.2.

The alkoxy(alkylimino)methylgold(I) complexes can be prepared by the reaction [42]:

$$[AuCl(SMe_2)] + RNC + R'OH + KOH \rightarrow AuC(OR'){=}NR + KCl + H_2O + Me_2S$$

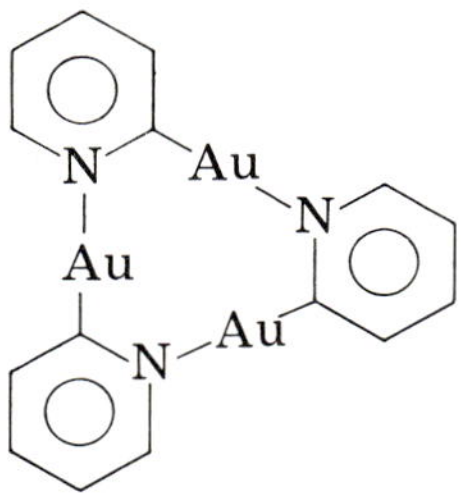

Fig. 7.2. The proposed structure of 2-pyridylgold(I).

The compounds are also trimeric and so probably have the structure shown in Fig. 7.3. Each gold atom can be oxidised in turn without breaking down the trimeric structure [42].

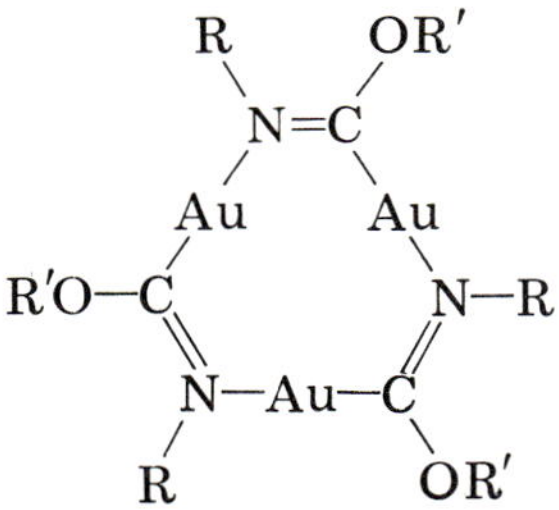

Fig. 7.3. The proposed structure of alkoxy(alkylimino)methylgold(I) complexes.

Other likely examples of oligomers of this kind are the compounds 1-methylbenzimidazol-2-ylgold(I) [43],

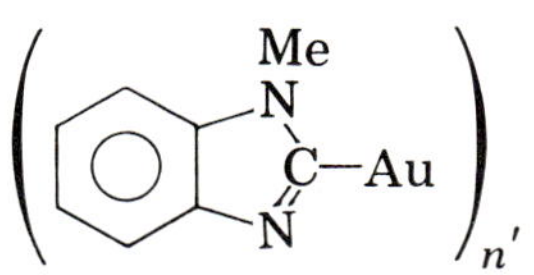

and 2-(dimethylaminomethyl)phenyl-gold(I) [22, 44] which can be formed by the reactions:

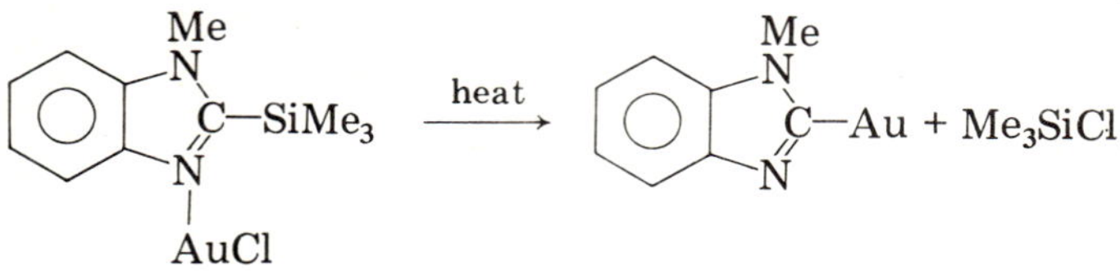

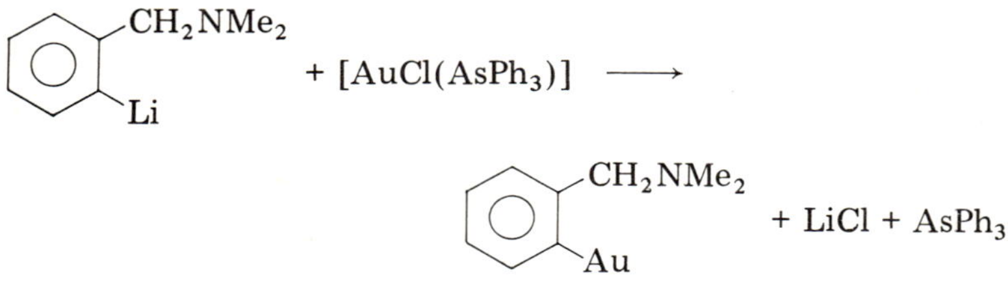

(f) Complexes of gold(I) with bridging carbon atoms

Reaction of ferrocenyl(triphenylphosphine)gold(I) with a deficiency of $H[BF_4]$ takes place according to the equation [45, 46]:

$$2[(C_5H_5)Fe(C_5H_4AuPPh_3)] + H[BF_4] \rightarrow [(C_5H_5)FeC_5H_4Au_2(PPh_3)_2]^+ [BF_4]^- + (C_5H_5)_2Fe$$

The complex is important as an example of a gold(I) complex with a bridging organic group and has the structure [47] shown in Fig. 7.4.

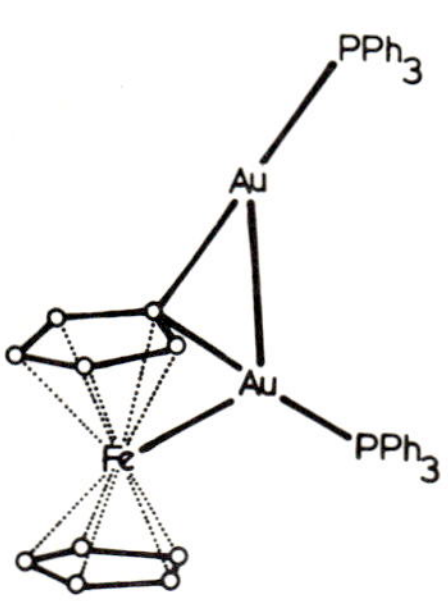

Fig. 7.4. The structure of the ion $[(\eta\text{-}C_5H_5)Fe\{\eta\text{-}C_5H_4(AuPPh_3)_2\}]^+$.

Each C—Au—P group is linear and there is also evidence for a gold—gold bond (r(AuAu) = 276.8 pm) and, for one of the gold atoms, a bond to iron also [47].

A similar complex is formed by a cyclopentadienylmanganese derivative [48].

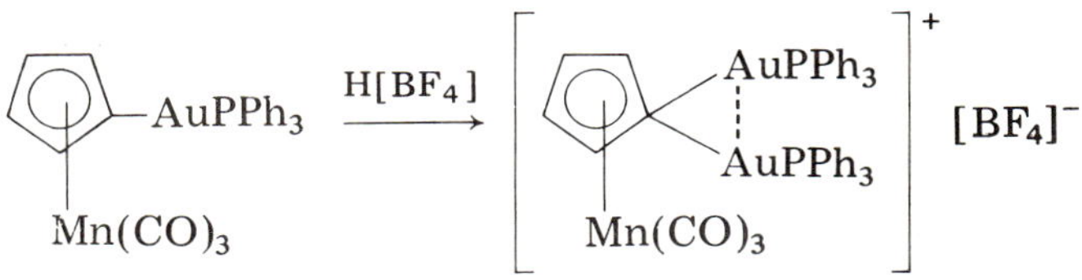

More important is the observation that complexes of this type are formed by phenyl- and vinylgold(I) complexes as illustrated below [49, 50]:

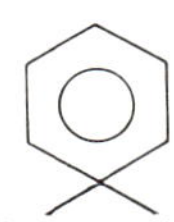

$PhAuPPh_3 + H[BF_4] \rightarrow [Ph_3PAu \cdots AuPPh_3]^+ [BF_4]^-$

$Ph_2C{=}CHAuPPh_3 + H[BF_4] \rightarrow [Ph_2C{=}CH(AuPPh_3)_2]^+ [BF_4]^-$

The extent of gold—gold interaction in these compounds is not known, since no X-ray structural determinations have been made. For the vinylgold derivative, it is suggested that much of the positive charge is on the β-carbon atom, through the resonance form $[Ph_2C^+{-}CH(AuPPh_3)_2]$.

These complexes are electron-deficient and the bridging aryl or vinyl group is lost on treatment with a tertiary phosphine or with chloride [45, 46, 49, 50].

$[Ph(AuPPh_3)_2]^+ [BF_4]^- \xrightarrow{Ph_3P} [PhAuPPh_3] + [Au(PPh_3)_2]^+ [BF_4]^-$

$[Ph(AuPPh_3)_2]^+ [BF_4]^- \xrightarrow{Cl^-} [PhAuPPh_3] + [AuCl(PPh_3)]$

Other complexes which are believed to contain bridging aryl groups are $Ph_6Au_2Zn_2$, prepared from $[Au_2Cl_6]$ and diphenylzinc, and $Ar_4Au_2Li_2$, $Ar_4Au_2Cu_2$, $Ar_4Cu_4Au_2(CF_3SO_3)_2$ and $Ar_4Ag_4Au_2(CF_3SO_3)_2$, where Ar is 2-(dimethylaminomethyl)-phenyl. Some proposed structures are [44, 51, 52] shown in Fig. 7.5.

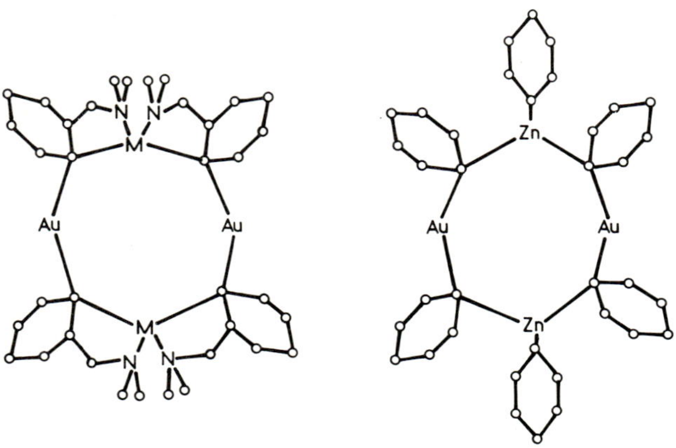

(a) (b)

Fig. 7.5. The proposed structures of (a) $Ar_4Au_2M_2$ where Ar = 2-$Me_2NCH_2C_6H_4$ and M=Cu or Li, and (b) $Ph_6Au_2Zn_2$.

The following complex derived from a phosphine ylid is unusual in containing two gold atoms bound to the same carbon atom, but in which the bridge is not electron-deficient [52b].

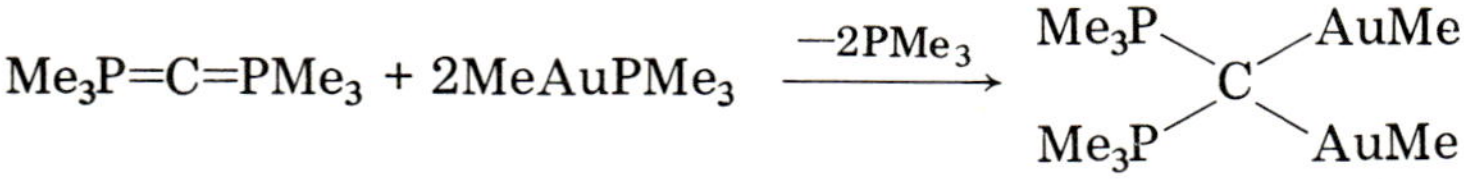

(g) Gold(I) ketenide

Gold(I) ketenide, $Au_2C{=}C{=}O$, may be prepared by reaction of chloro(1,4-oxathian)gold(I) with ketene in acetonitrile in the presence of 2,6-dimethylpyridine, L, and triethylamine [53]. Formation occurs by the reactions:

$$[AuCl(C_4H_8OS)] + L \rightleftharpoons [AuClL] + C_4H_8OS$$

$$2[AuClL] + CH_2{=}C{=}O + 2Et_3N \rightarrow Au_2(C{=}C{=}O)L + L + 2[Et_3NH]^+Cl^-$$

The base L is readily lost to give the ketenide $Au_2C{=}C{=}O$ as a yellow, explosive powder.

The structure of this compound is not known but it almost certainly contains ketenide units bridging between gold(I) atoms to give a polymeric structure. Copper and silver ketenides are also known. The compound decomposes to gold and carbon suboxide, C_3O_2, and reacts with phenylacetylene to give ketene and phenylethynylgold(I) [53].

REACTIONS OF ORGANOGOLD(I) COMPLEXES

(a) Thermal and photochemical decomposition of alkylgold(I) complexes

Methyl(triphenylphosphine)gold(I) decomposes quantitatively at about 100 °C either as the solid or in solution according to the equation [9, 14]:

$$2[AuMe(PPh_3)] \rightarrow C_2H_6 + 2Au + 2PPh_3$$

Similarly, ethyl(triphenylphosphine)gold(I) gives largely n-butane by coupling of the ethyl groups, but the similar isopropyl- and tertiary-butylgold(I) complexes give more complex products. Thus, from isopropyl(triphenylphosphine)gold(I) the chief volatile products are propane and propene arising from disproportionation of the propyl groups. A kinetic study of the decomposition of $[AuMe(PPh_3)]$ has shown that the decomposition follows first-order kinetics, is unaffected by the presence of oxygen but is retarded by added triphenylphosphine. Thus, the mechanism shown in the following equations is indicated [14]:

$$[AuMe(PPh_3)] \rightleftharpoons [AuMe] + PPh_3$$

$$[AuMe] + [AuMe(PPh_3)] \longrightarrow C_2H_6 + 2Au + PPh_3$$

This coupling of organic groups by decomposition of organogold(I) complexes is also observed for 2-pyridylgold(I) which gives gold and 2,2′-bipyridine at 150 °C. Decomposition of substituted derivatives can be useful in the synthesis of the corresponding substituted 2,2′-bipyridine [41].

In some cases the coupling is catalysed by copper(II) chloride [54].

$$2AuC{\equiv}CPh + 2CuCl_2 \rightarrow PhC{\equiv}C{-}C{\equiv}CPh + 2CuCl + 2AuCl$$

Most organogold compounds are photosensitive but only for methyl(triphenylphosphine)gold(I) has the mechanism of photolysis been investigated. Photolysis in $CDCl_3$ ultimately gives methyl radicals according to the scheme (S = singlet, T = triplet excited state):

$$[AuMe(PPh_3)] \xrightarrow{h\nu} [AuMe(PPh_3)]^S \longrightarrow [AuMe(PPh_3)]^T$$

$$[AuMe(PPh_3)]^T \xrightarrow{CDCl_3} [AuCl(PPh_3)] + [(Me + CDCl_2)]^T$$

The methyl radicals can then react in a number of ways to give CH_3D, C_2H_6, CH_3CDCl_2 and CH_3CCl_3 [55].

(b) Ligand substitution in organogold(I) complexes

In the complexes [AuRL], it is often possible to displace one ligand L by another without cleavage of the gold—carbon bond;

$$[Au(C{\equiv}CPh)AsEt_3] + 4\text{-}MeC_6H_4NC \rightarrow [Au(C{\equiv}CPh)(4\text{-}MeC_6H_4NC)] + AsEt_3$$

From reactions such as this, the following series of ligands, in order of affinity for gold(I), $PEt_3 > P(OPh)_3 > 4\text{-}MeC_6H_4NC >$

$AsEt_3 > SbEt_3 > NH_3$, RNH_2, was obtained [25]. Similarly, by displacement reactions on the complexes [AuMeL] the series of ligands $Ph_3P > Et_3P > Me_3P > (MeO)_3P > Ph_3As$ was obtained. This displacement reaction offers a convenient route to complexes with bidentate ligands [8, 12].

$$2\,[AuMe(PMe_3)] + Me_2PCH_2CH_2PMe_2 \rightarrow \underset{\textstyle AuMe}{Me_2P}CH_2CH_2\underset{\textstyle AuMe}{PMe_2}$$

(c) Reactions of alkylgold(I) complexes with acids

Many protic reagents will cleave the gold—carbon bond in organogold(I) complexes. Methyl- [9, 23], phenyl- [56, 57], vinyl- [50], and ferrocenylgold [33, 35] compounds all react as illustrated by the following reactions:

$$[AuMe(PPh_3)] + HCl \rightarrow [AuCl(PPh_3)] + CH_4$$

$$2\,[AuMe(PPh_3)] + (CF_3)_2C(OH)_2 \rightarrow [(CF_3)_2C(OAuPPh_3)_2] + 2CH_4$$

$$[AuPh(PPh_3)] + MeCO_2H \rightarrow [Au(O_2CMe)(PPh_3)] + C_6H_6$$

$$[Au(CH{=}CH_2)(PPh_3)] + PhCO_2H \rightarrow [Au(O_2CPh)(PPh_3)] + C_2H_4$$

Cleavage of the vinylgold bond takes place with retention of stereochemistry about the C=C bond [58].

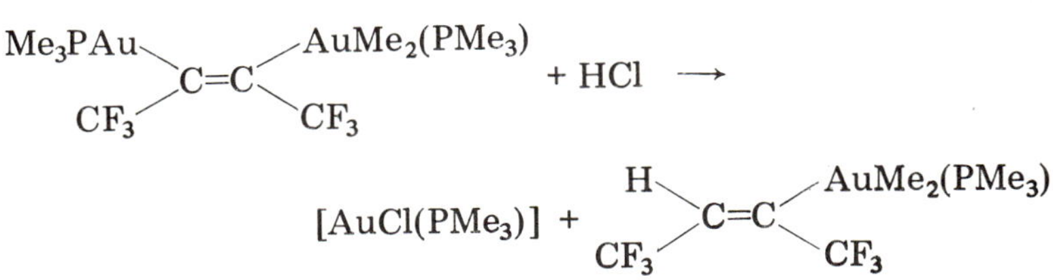

In some cases, it has been suggested that the acid cleavage reactions occur by protonation of the gold centre to give a hydridogold(III) species which then undergoes reductive elimination of alkane, as illustrated in the equations:

$$[AuMe^tBu]^- + PPh_3 + H^+ \longrightarrow [AuHMe^tBu(PPh_3)]$$

$$[AuHMe^tBu(PPh_3)] \longrightarrow [Au^tBu(PPh_3)] + CH_4 \quad \text{or} \quad [AuMe(PPh_3)] + C_4H_{10}$$

$$[AuMe(PMe_3)] + PhSH \longrightarrow [AuHMe(SPh)(PMe_3)]$$

$$[AuHMe(SPh)(PMe_3)] \longrightarrow [Au(SPh)(PMe_3)] + CH_4$$

If these hydridogold(III) intermediates do exist, they are too short-lived to be detected directly [2, 11].

Even very weak acids such as terminal alkynes [23, 40] and diphenylarsine [58b] can cleave the gold—carbon bond in methylgold(I) complexes:

$$[AuMe(PMe_2Ph)] + CF_3C{\equiv}CH \rightarrow [Au(C{\equiv}CCF_3)(PMe_2Ph)] + CH_4$$

$$[AuMe(PMe_2Ph)] + Ph_2AsH \rightarrow [\{Au(AsPh_2)\}_n] + CH_4 + PMe_2Ph$$

(d) Reactions with acyl halides

Acyl halides can react with organogold(I) complexes in a number of ways. The simplest reactions give the gold(I) halide and a ketone [56]:

$$[AuPh(PPh_3)] + MeCOCl \rightarrow [AuCl(PPh_3)] + MeCOPh$$

With functionally substituted alkylgold(I) complexes, 1,4-cleavage rather than the usual 1,2-cleavage may occur [29]:

$$[Au(CH_2COMe)(PPh_3)] + MeCOCl \rightarrow [AuCl(PPh_3)] + MeCO_2C(Me){=}CH_2$$

In some cases more complex reactions take place. For example, reaction of benzoyl chloride with [AuPh(PPh$_3$)] gives [AuCl(PPh$_3$)] along with benzene, benzaldehyde and biphenyl rather than the expected benzophenone. It is thought that the initial reaction is an electron-transfer from gold to give $[AuPh(PPh_3)]^{+\cdot}$ $[PhCOCl]^{-\cdot}$, rather than direct electrophilic substitution at gold. The intermediate then breaks down to give the final products [59].

(e) Reactions with metal halides

Mercury(II) chloride reacts rapidly with methyl- [13], ethyl- [13], phenyl- [56], and vinylgold(I) [50, 60] complexes to cleave the gold—carbon bond according to the equation:

$$[AuRL] + HgCl_2 \rightarrow [AuClL] + RHgCl$$

In most cases the reactions follow the rate law, rate = $k[AuRL][HgX_2]$ and an S_E2 mechanism of substitution at the carbon atom was suggested [13]. An alternative oxidative addition—reductive elimination sequence has also been proposed [60].

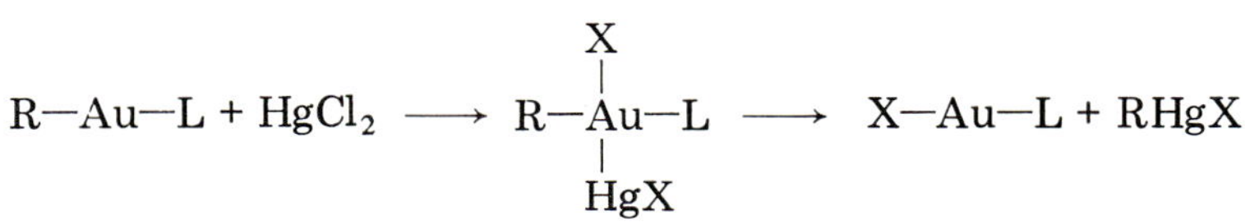

With the complex $[Au\{C(CN)(CO_2Et)Bu\}(PPh_3)]$, cleavage by mercury(II) salts in the polar dimethylsulphoxide-1,4-dioxane solvent mixture proceeds by the S_E1 mechanism, involving rate-determining formation of the carbanion.

Cleavage of the vinylgold bond takes place with retention of stereochemistry about the C=C bond [61]:

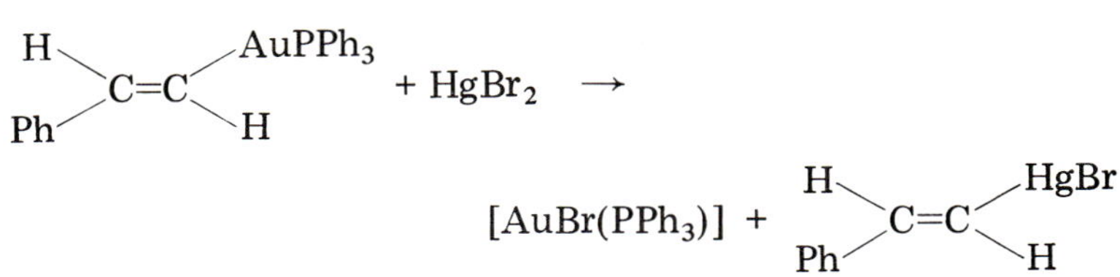

Platinum(II) and palladium(II) halides also react with organogold(I) complexes as seen in the following equations [57, 62]:

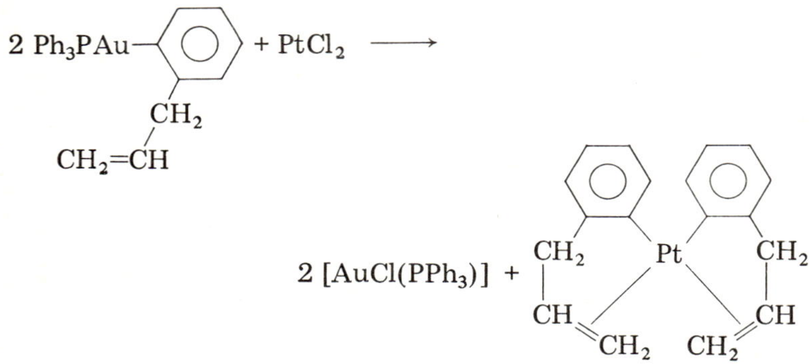

$$[AuMe(PMe_2Ph)] + cis\text{-}[PdCl_2(PMe_2Ph)_2] \longrightarrow [AuCl(PMe_2Ph)] + trans\text{-}[PdClMe(PMe_2Ph)_2]$$

With methylplatinum(II) complexes an equilibrium may be reached as shown in the equation:

$$[AuMe(PMe_2Ph)] + trans\text{-}[PtXMe(PMe_2Ph)_2] \overset{K}{\rightleftharpoons} [AuX(PMe_2Ph)] + cis\text{-}[PtMe_2(PMe_2Ph)_2]$$

The equilibrium constant K is very much dependent on the halogen X, with K for $X = I > Br > Cl$ [62].

Alkylgold(I) complexes also react readily with gold(III) halides [10, 63, 64].

$$[AuMe(PPh_3)] + cis\text{-}[AuMe_2I(PPh_3)] \rightarrow [AuI(PPh_3)] + [AuMe_3(PPh_3)]$$

The reactions take place with retention of stereochemistry at the gold(III) centre as shown by the reaction [17, 64]:

$$CF_3\text{—}\underset{\underset{\displaystyle I}{|}}{\overset{\overset{\displaystyle Me}{|}}{Au}}\text{—}PMe_3 + [AuMe(PMe_3)] \rightarrow CF_3\text{—}\underset{\underset{\displaystyle Me}{|}}{\overset{\overset{\displaystyle Me}{|}}{Au}}\text{—}PMe_3 + [AuI(PMe_3)]$$

With halides of thallium(III) or gold(III) oxidation of the organogold(I) complex rather than cleavage of the gold—carbon bond may occur [62, 65].

$$[Au(C_6Cl_5)(AsPh_3)] + TlCl_3 \rightarrow cis\text{-}[AuCl_2(C_6Cl_5)(AsPh_3)] + TlCl$$

(f) Reactions with halogens

Usually, halogens react with alkylgold(I) complexes to give the corresponding alkyl halide and either a gold(I) or gold(III) halide [23, 29, 56].

$$[AuPh(PPh_3)] + Br_2 \longrightarrow PhBr + [AuBr(PPh_3)] \xrightarrow{Br_2} [AuBr_3(PPh_3)]$$

Occasionally some coupling of the organic groups occurs. Thus reaction of ferrocenyl(triphenylphosphine)gold(I) with bromine gave 15% biferrocenyl along with bromoferrocene (55%) [66].

If the gold—carbon bond in the organogold(I) complex is particularly strong, oxidation to a gold(III) complex may occur. This reaction was first discovered with pentafluorophenylgold(I) complexes [20].

$$[Au(C_6F_5)(PPh_3)] + Br_2 \rightarrow [AuBr_2(C_6F_5)(PPh_3)]$$

The reaction has been found to be quite general for perhalogenoarylgold(I) complexes, although the course of reaction is dependent on the halogen and the particular gold(I) complex as illustrated below [15b, 65, 67].

$$[Au(C_6F_5)(AsPh_3)] + X_2 \xrightarrow{X = Cl} [AuCl_2(C_6F_5)(AsPh_3)]$$

$$[Au(C_6F_5)(AsPh_3)] + X_2 \xrightarrow{X = Br\ or\ I} [AuX(AsPh_3)] + C_6F_5X$$

$$[Au(C_6Cl_5)(AsPh_3)] + X_2 \xrightarrow{X = Cl,\ Br\ or\ I} [AuX_2(C_6Cl_5)(AsPh_3)]$$

Interestingly, the oxidation by chlorine gives *trans*-$[AuCl_2(C_6X_5)(PPh_3)]$ which can then slowly rearrange to the *cis*-isomer, while oxidation by thallium(III) chloride gives the *cis*-isomer directly [15b, 65].

Similarly, the ylid derivative $[Au(CH_2PMe_3)_2]Br$ is oxidised by iodine to $[AuI_2(CH_2PMe_3)_2]Br$ [68], but particularly interesting are the reactions of the dimeric ylid derivatives which, with a deficiency of halogen, give stable gold(II) complexes [5—7].

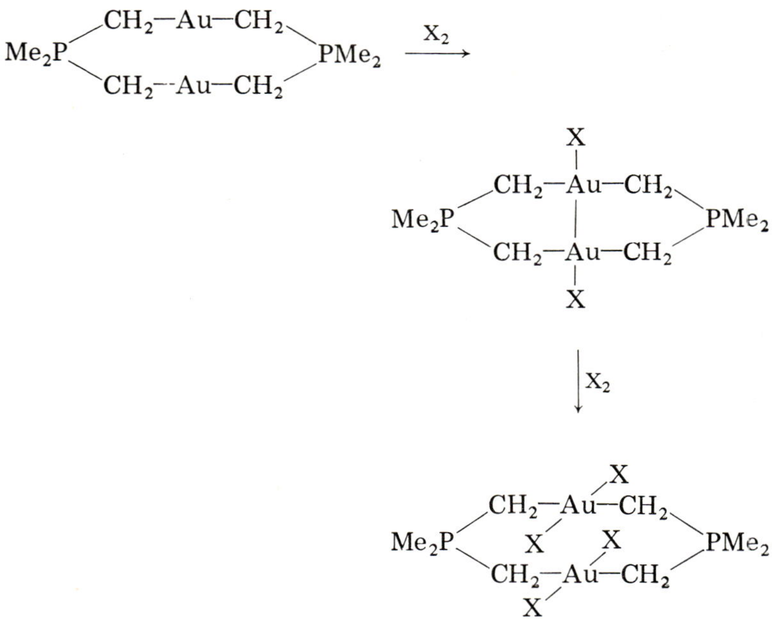

It is possible that cleavage of the gold—carbon bond in organogold(I) complexes by halogens may take place, not by direct electrophilic substitution, but by an oxidative-addition reductive-elimination sequence. Evidence for this mechanism has been found in at least one case by isolation of an unstable gold(III) intermediate [57].

$$[Au(C_6H_4CH{=}CH_2)(PPh_3)] + Br_2 \xrightarrow{-40^\circ C} [AuBr_2(C_6H_4CH{=}CH_2)(PPh_3)]$$

$$[AuBr_2(C_6H_4CH{=}CH_2)(PPh_3)] \xrightarrow{20^\circ C} [AuBr(PPh_3)] + BrC_6H_4CH{=}CH_2$$

(g) Reactions with alkyl halides

The initial reaction of a methylgold(I) complex [AuMeL], L = tertiary phosphine, with methyl iodide occurs by oxidative addition [10, 63].

$$[AuMeL] + MeI \rightarrow cis\text{-}[AuMe_2IL]$$

However, since this product reacts rapidly with [AuMeL], the isolated products are $[AuMe_3L]$ and [AuIL].

$$cis\text{-}[AuMe_2IL] + [AuMeL] \rightleftharpoons [AuMe_3L] + [AuIL]$$

If these products are left in solution, in methyl iodide a further reaction takes place slowly. The nature of this reaction is dependent on the ligand L. Thus, when L is triphenylphosphine, the complex $[AuMe_3L]$ undergoes reductive elimination of ethane to reform [AuMeL] so that, after further reaction of this with methyl iodide, the overall reaction can be written [60, 63]:

$$[AuMe(PPh_3)] + MeI \rightarrow [AuI(PPh_3)] + C_2H_6$$

However, when L is trimethylphosphine or dimethylphenylphosphine, the complex $[AuMe_3L]$ is stable to reductive elimination and the trace of [AuMeL] formed by reaction with [AuIL] reacts with methyl iodide to give finally *cis*-$[AuMe_2IL]$ [10, 63].

Reaction of ethyl(triphenylphosphine)gold(I) with methyl iodide gives butane in the early stages of reaction, probably by the sequence of reactions [63].

$$[AuEt(PPh_3)] + MeI \longrightarrow [AuIMeEt(PPh_3)]$$

$$[AuIMeEt(PPh_3)] \xrightarrow{[AuEt(PPh_3)]} [AuMeEt_2(PPh_3)] + [AuI(PPh_3)]$$

$$[AuMe(PPh_3)] + C_4H_{10} \longleftarrow [AuMeEt_2(PPh_3)] + [AuI(PPh_3)]$$

The complexes [AuMeL] also react with trifluoromethyl iodide and the nature of the products is again dependent on the ligand L [17].

$$[AuMeL] + CF_3I \xrightarrow{L=PMe_3} [AuMe_2(CF_3)L] + IAuL$$

$$[AuMeL] + CF_3I \xrightarrow{L=PPh_3} [Au(CF_3)L] + MeI$$

The reactions give a useful route to trifluoromethylgold complexes.

Dialkylaurates(I) are even more reactive to oxidative addition reactions, and the reactions with alkyl halides give a useful synthetic route to trialkylgold(III) complexes [2].

$$[AuCl(PMe_3)] + 2MeLi \longrightarrow Li[AuMe_2] + PMe_3 + LiCl$$

$$Li[AuMe_2] \xrightarrow{MeI} [AuMe_3(PMe_3)] + LiI$$

Mixed alkylgold(III) complexes can be obtained in a similar way.

$$Li[AuMe_2] + PPh_3 + EtI \rightarrow trans\text{-}[AuMe_2Et(PPh_3)] + LiI$$

Prepared in this way, the *trans*-isomers are formed predominantly, but *cis*-isomers can be prepared by the alternative method [2]:

$$Li[AuMe(CH_2CMe_3)] + PPh_3 + MeI \rightarrow cis\text{-}[AuMe_2(CH_2CMe_3)(PPh_3)] + LiI$$

A methylgold(II) complex is prepared by reaction of methyl iodide with a dimeric ylid complex [6].

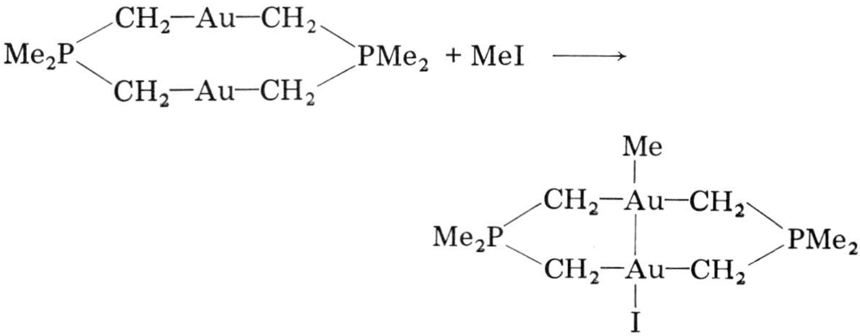

(h) Reactions with unsaturated reagents

Fluorinated alkenes react with methylgold(I) complexes under ultraviolet irradiation to give fluoroalkylgold(I) complexes by insertion into the methylgold bond [23].

$[AuMe(PPh_3)] + CF_2{=}CF_2 \rightarrow [Au(CF_2CF_2Me)(PPh_3)]$

A free-radical mechanism is likely for this reaction, but it is also possible that rearrangement of an intermediate tetrafluoroethylene complex of gold $[AuMe(PPh_3)(C_2F_4)]$ may be responsible. This mechanism is made more reasonable by the isolation of an unstable tetracyanoethylene complex of gold [40].

$[AuMe(PMe_2Ph)] + (CN)_2C{=}C(CN)_2 \longrightarrow$

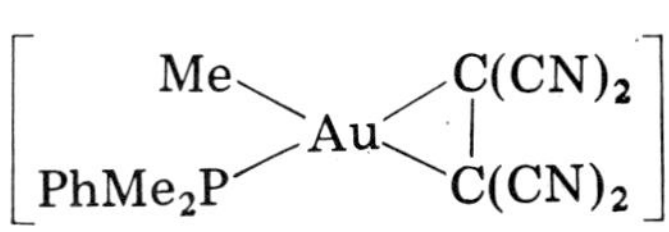

Reactions with alkynes are more complex [23, 40, 58]. Thus, reaction with hexafluorobut-2-yne gives a mixed oxidation state Au(I)—Au(III) complex.

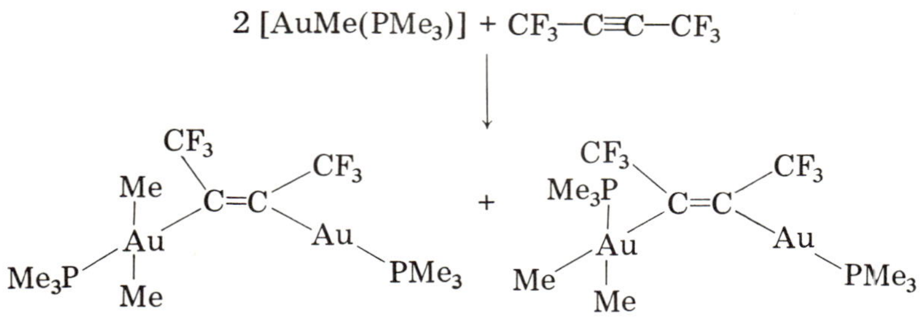

The reaction involves transfer of a methyl group from one gold atom to another and formation of a *cis*-bridging C_4F_6 unit. From [AuMe(PMe_3)], the complexes with both *trans* and *cis* stereochemistry at the gold(III) centre are formed, but with PMe_2Ph or $PMePh_2$ ligands only the *trans*-isomer is formed. These gold(I)—gold(III) complexes decompose slowly in the presence of hexafluorobut-2-yne by reductive elimination at the gold(III) centre. One of two decomposition routes is followed:

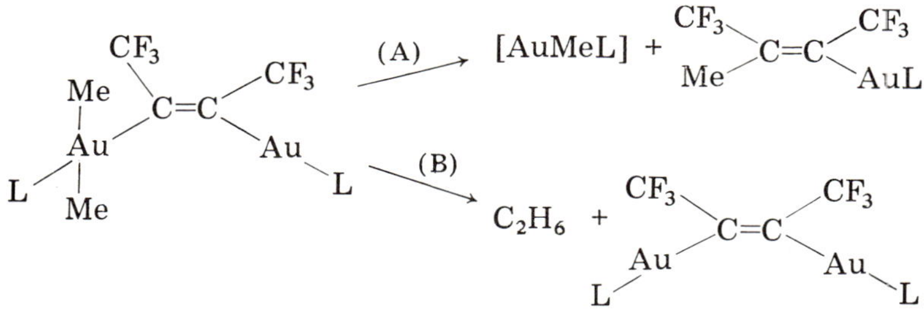

Route (A) gives the product of insertion of the acetylene into the methylgold bond, and the *cis*-stereochemistry about the C=C bond is retained during the reductive elimination reaction [58]. Route (B) also gives a product with *cis*-stereochemistry about the C=C bond [23, 69].

In similar reactions, but-2-yne with [AuMe(PPh_3)] gives [Ph_3PAuC(Me)=C(Me)AuPPh_3], and dimethyl acetylenedicarboxylate with [AuMe(PMe_2Ph)] gives [Au{C(CO_2Me)=C(CO_2Me)Me}(PMe_2Ph)] and *cis*-[AuMe_2{C(CO_2Me)=C(CO_2Me)Me}(PMe_2Ph)] but no intermediates have been isolated in these cases [23, 40].

A complex containing two gold atoms is formed by reaction of hexafluoroacetone with $[AuMe(PPh_3)]$.

$$2\,[AuMe(PPh_3)] + (CF_3)_2CO \longrightarrow C_2H_6 + Ph_3PAuO-\underset{CF_3}{\overset{CF_3}{\underset{|}{\overset{|}{C}}}}-AuPPh_3$$

Finally, sulphur dioxide reacts with methylgold(I) [40] and arylgold(I) [57, 70] complexes by insertion into the gold—carbon bond with formation of *S*-sulphinatogold(I) complexes.

$$[AuMe(PMe_3)] + SO_2 \rightarrow Me-\underset{O}{\overset{O}{\underset{\|}{\overset{\|}{S}}}}-Au-PMe_3$$

With the methylgold(I) complexes, bright red intermediates, believed to be SO_2 complexes, are formed at low temperatures [40].

The reverse of this kind of insertion reaction can be used to prepare organogold(I) complexes, for example in the decarboxylation reaction [23]:

$$Au(O_2CC_6F_5)(PPh_3) \xrightarrow{\text{heat}} [Au(C_6F_5)(PPh_3)] + CO_2$$

σ-BONDED ORGANOGOLD(III) COMPLEXES

It will again be convenient to subdivide these compounds according to the number of gold—carbon bonds present, beginning with those containing the maximum number of four Au—C bonds.

(a) Tetra-alkylaurate(III) and related complexes

The simplest complex of this type is $Li^+[AuMe_4]^-$, prepared in ether solution by the reaction [3]:

$$[AuMe_3(PPh_3)] + LiMe \rightarrow Li[AuMe_4] + PPh_3$$

After solvation of the lithium ion by the ligand PMDT = $Me_2NCH_2CH_2N(Me)CH_2CH_2NMe_2$, the complex [Li(PMDT)] $[AuMe_4]$ can be isolated. It melts with decomposition at 86—88 °C, and, on the basis of its vibrational spectra, is thought to contain a square-planar $[AuMe_4]^-$ ion [3].

The analogous pentafluorophenylgold(III) complex can be prepared directly from $K[AuCl_4]$ [4]:

$$K[AuCl_4] + 4LiC_6F_5 + [Bu_4N]^+Br^- \rightarrow [Bu_4N][Au(C_6F_5)_4] + 4LiCl + KBr$$

Compounds containing four Au—C bonds can also be prepared from dimethylgold chloride and trimethylphosphine ylid [5]:

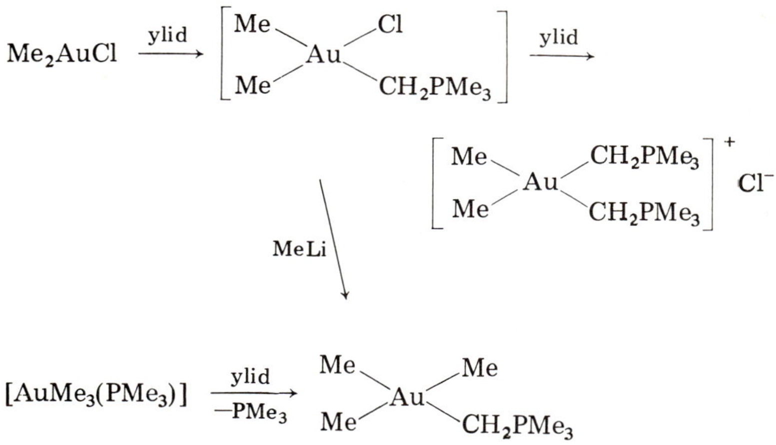

The sulphur ylid derivatives $[Me_3AuCH_2SMe_2]$ and $[Me_3AuCH_2SOMe_2]$ have also been prepared by displacement of triphenylphosphine from $[Me_3AuPPh_3]$ by the corresponding sulphur ylid, and the molecular structure of the sulphoxonium ylid derivative has been determined by X-ray methods [70b].

Another unusual compound of this type is the tetrazolatogold(III) complex prepared by the reaction [71]:

$[Ph_4As][Au(N_3)_4] + 4MeNC \longrightarrow$

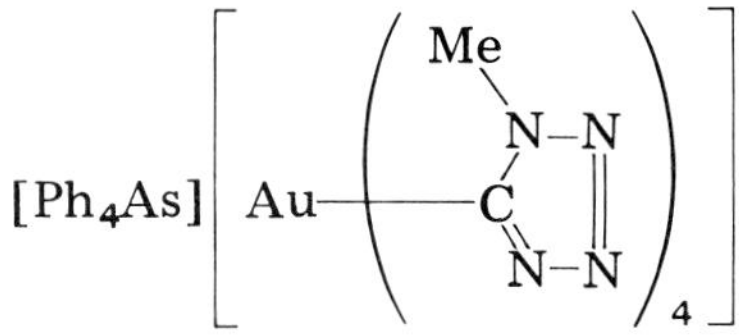

(b) Trialkylgold(III) and related complexes

A compound formulated as trimethylgold is formed on treating gold(III) bromide or dimethylgold bromide with methyllithium at —65 °C in ether [72].

$$Au_2Br_6 + 6LiMe \rightarrow 2AuMe_3 + 6LiBr$$

The product is difficult to characterise since it decomposes to gold, methane and ethane at about —40 °C. It is now thought more likely to be $Li[AuMe_3Br]$ rather than the etherate $[AuMe_3(OEt_2)]$ [3]. Similarly, the compound originally suggested to be $[Au(C_6F_5)_3(OEt_2)]$ [20] is more likely to be $[Au(C_6F_5)_3Br]^-$ [4].

More stable complexes can be obtained if "trimethylgold" is treated with nitrogen, phosphorus or arsenic donor ligands at low temperatures. The products are of the general type $[AuMe_3L]$ and some examples are given in Table 7.2.

Trimethylgold complexes can also be prepared by ligand displacement reactions [8, 9].

$$[Me_3AuNH_2CH_2CH_2NH_2AuMe_3] + 2PMe_3 \rightarrow 2[AuMe_3(PMe_3)] + NH_2CH_2CH_2NH_2$$

$$[AuMe_3(PMe_3)] + PEt_3 \rightarrow [AuMe_3(PEt_3)] + PMe_3$$

However, the most convenient synthesis involves the oxidation of dimethylaurate(I), prepared *in situ*, with methyl halides [2]:

$$AuBr(PPh_3) + 2LiMe + MeI \rightarrow AuMe_3(PPh_3) + LiBr + LiI$$

TABLE 7.2

Compounds with 3 Au—C σ-bonds

Compound	m.p./°C	Ref.
$[Me_3AuNH_2CH_2CH_2NH_2AuMe_3]$	94—98 dec.	72
$[AuMe_3(NH_2CH_2Ph)]$	51.5—53	72
$[AuMe_3(PMe_3)]$	23	9
$[AuMe_3(PMe_2Ph)]$	oil	10, 62
$[AuMe_3(PMePh_2)]$	63—65	11
$[AuMe_3(PPh_3)]$	120 dec.	9
$[AuMe_3(AsMe_3)]$	*ca.* 0	73
$[AuMe_3(PEt_3)]$	—5	8
$[Au(C_6F_5)_3(PPh_3)]$	243—246 dec.	20
cis-$[Au(C_6F_5)_2(C_6Br_5)(PPh_3)]$	170 dec.	15b
trans-$[AuMe_2Et(PPh_3)]$		2, 74
cis-$[AuMe_2Et(PPh_3)]$		74
cis-$[AuMe_2(C_5H_5)(PPh_3)]$	99 dec.	75
cis-$[AuMe_2(CD_3)(PPh_3)]$		75
trans-$[AuMe_2(CF_3)(PMe_3)]$	97—100	17
cis-$[AuMe_2(C{\equiv}CCF_3)(PMe_2Ph)]$		40
$Bu_4N[Au(C_6F_5)_3Br]$		4
cis-$[AuMe_2(CH_2PMe_3)Cl]$	107—110	5, 68
cis-$[AuMe_2(CH_2PMe_3)(PMe_3)]Cl$	111 dec.	5, 68

This method has also been used to prepare compounds with different alkyl groups on the same gold atom:

$$[AuBr(PPh_3)] + 2LiMe + Me_2CHBr \rightarrow trans\text{-}[AuMe_2(CHMe_2)(PPh_3)] + 2LiBr$$

For the preparation of the similar t-butyl complex, it is necessary to use the mixed methyl(t-butyl)aurate(I):

$$[AuMe(PPh_3)] + Li^tBu + MeI \rightarrow trans\text{-}[AuMe_2{}^tBu(PPh_3)] + LiI$$

The *cis*-isomers can be prepared by stereospecific substitution reactions (R = CD_3, Et, Pr, or C_5H_5) [74, 75]:

$$cis\text{-}[AuMe_2I(PPh_3)] + LiR \rightarrow cis\text{-}[AuMe_2R(PPh_3)] + LiI$$

Substitution in a monoarylgold(III) complex is possible in one case [15b], but attempts to alkylate $[AuBr_3(PPh_3)]$ lead to partial reduction to gold(I) complexes.

$$cis\text{-}[AuBr_2(C_6Br_5)(PPh_3)] + 2C_6F_5MgBr \rightarrow cis\text{-}[Au(C_6F_5)_2(C_6Br_5)(PPh_3)] + 2MgBr_2.$$

(c) Dialkylgold(III) and related complexes

The unit *cis*-AuR_2^+ is very much favoured and a large number of such complexes are known. They may be anionic, neutral or cationic as typified by the complexes *cis*-$[AuMe_2Cl_2]^-$, *cis*-$[AuMe_2Cl(PMe_3)]$ and *cis*-$[AuMe_2(PMe_3)_2]^+$. The alternative *trans*-stereochemistry is found only in the unusual ylid complexes (X = halogen) [6]:

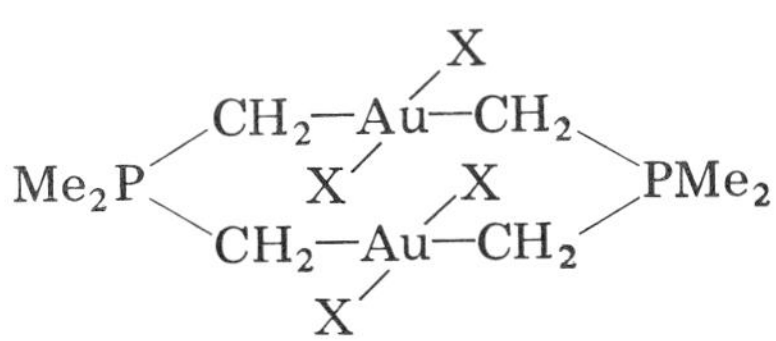

Reasons for this very strong preference for the *cis*-stereochemistry have been proposed. Thus, the orbitals available for bonding by gold(III) are, in order of decreasing stability, $5d_{x^2-y^2}$, 6s, $6p_x$ and $6p_y$. It is supposed that, in complexes such as *cis*-$[AuMe_2Cl_2]^-$, the strongly σ-bonding methyl groups will demand the lower energy orbitals on gold, i.e., the 5d and 6s orbitals, leaving the chloride ligands to overlap with the 6p orbitals. This is only possible if the *cis*-isomer is formed, since the ds hybrid orbitals have lobes directed at 90° to one another [76]. The theory correctly predicts that the Au—Cl bonds will be weaker than in $[AuCl_4]^-$ in which regular dsp^2 hybridisation is assumed, and is also consistent with the higher energy of ligand-to-metal charge transfer transitions in $[Au_2Me_4(\mu\text{-}Br)_2]$ than in $[Au_2Br_6]$. More generally, it has been suggested that two soft ligands will have a destabilising effect on one another when in mutually *trans* positions and when attached to a "class b" or soft metal ion. The effect is termed anti-symbiosis and it again predicts that two soft methyl ligands will prefer to be mutually *cis* when bound to the soft gold(III) centre [77].

(d) Dialkylgold halides

Diethylgold bromide was the first organogold compound to be prepared [78]. It was subsequently shown to be dimeric with halogen bridges [79].

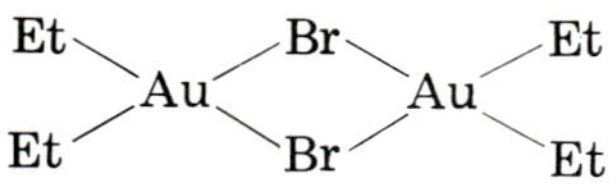

The complexes are usually prepared from a suitable gold(III) halide and a Grignard reagent in ether solution. Initially, hydrated tetrabromoauric(III) acid was used but, since the associated water destroys much of the Grignard reagent, anhydrous gold(III) halides or suitable complexes such as $[AuCl_3(py)]$ or $[AuBr_2(py)_2]Br$ are preferred [80].

$$[Au_2Br_6] + 4MeMgBr \rightarrow [Au_2Me_4(\mu\text{-}Br)_2] + 4MgBr_2$$

If a Grignard reagent derived from an alkyl iodide is used, the corresponding iodogold complex is always formed by halogen exchange. Organolithium reagents have not been used extensively since they react further to give trialkylgold(III) or tetra-alkylaurate(III) species. However, further reaction of these compounds with gold(III) halides [72] or with acids [72, 81] can give dialkylgold(III) compounds.

$$4''AuMe_3'' + [Au_2Br_6] \rightarrow 3[Au_2Me_4(\mu\text{-}Br)_2]$$

$$2''AuMe_3'' + 2HCl \rightarrow [Au_2Me_4(\mu\text{-}Cl)_2] + 2CH_4$$

Some of the known dialkylgold halides are given in Table 7.3, and a more extensive list has been published [82]. A few compounds derived from diGrignard reagents are known; for example the Grignard reagent derived from 1,5-dibromopentane gives a pentamethylenegold complex in low yield [89]:

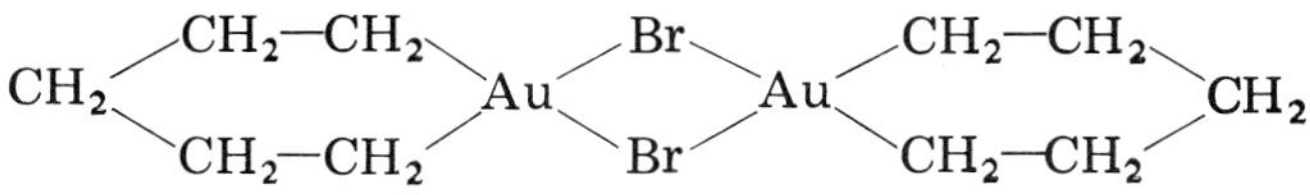

The pseudohalides are generally prepared by metathesis from the corresponding dialkylgold halides. The preferred method for the dimethylgold derivatives is to treat dimethylgold(III) iodide with silver nitrate solution. After filtration of the silver iodide formed, an aqueous solution of *cis*-$[AuMe_2(OH_2)_2]^+$ is obtained. This is then treated with a sodium pseudohalide, whereupon the required dimethylgold pseudohalide precipitates [83, 84].

The dialkylgold pseudohalides are of three structural types. The azide and cyanate compounds are dimeric with a bridging

TABLE 7.3

Dialkylgold halides and pseudohalides, R_2AuX

Compound	m.p./°C	Ref.
$(Me_2AuCl)_2$	71—72	72
$(Me_2AuBr)_2$	68—69	80
$(Me_2AuI)_2$	78.5	80
$(Me_2AuCN)_4$	140 dec.	83
$(Me_2AuSCN)_2$	73—74	84
$(Me_2AuSeCN)_2$	92 dec.	83
$(Me_2AuNCO)_2$	75 dec.	83
$(Me_2AuN_3)_2$	explodes	85
$(Et_2AuCl)_2$	48	86
$(Et_2AuBr)_2$	58	78
$(Et_2AuI)_2$	67—68	87
$(Et_2AuCN)_4$	103—105	86, 88
$(Pr_2AuCl)_2$	liq.	86
$\{(PhCH_2)_2AuCl\}$	70—110 dec.	86
$\{(Me_3CCH_2)_2AuBr\}_2$	121—123	72

nitrogen atom, thus giving four-membered rings as with the dialkylgold halides [84, 85].

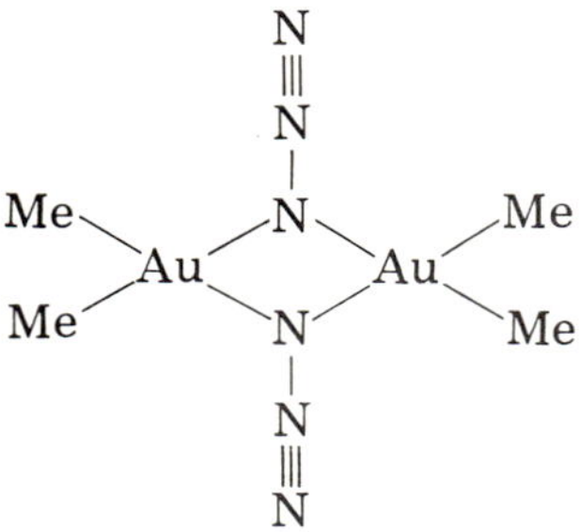

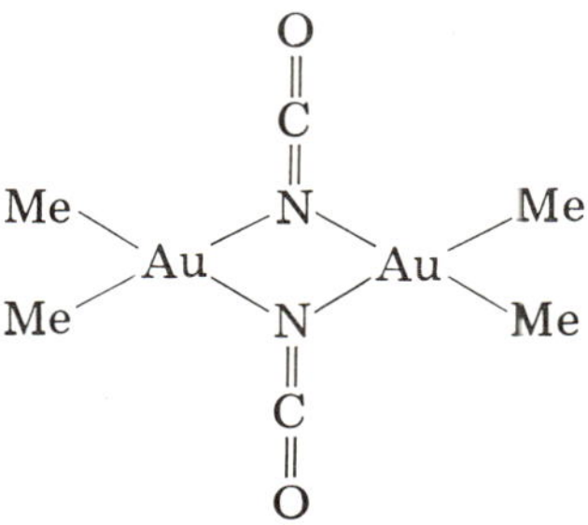

The dialkylgold thiocyanates and selenocyanates are also dimeric but bridge through the nitrogen and sulphur or selenium atoms, respectively. The n.m.r spectra thus contain two methylgold signals due to the methyl groups *trans* to nitrogen and sulphur or selenium [83, 84].

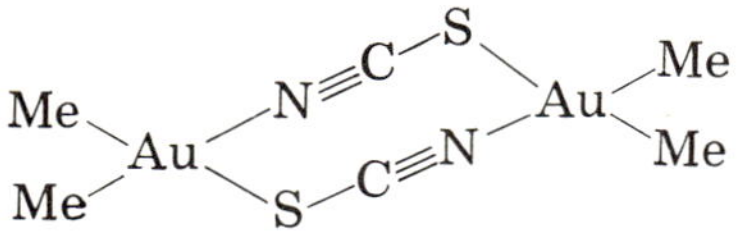

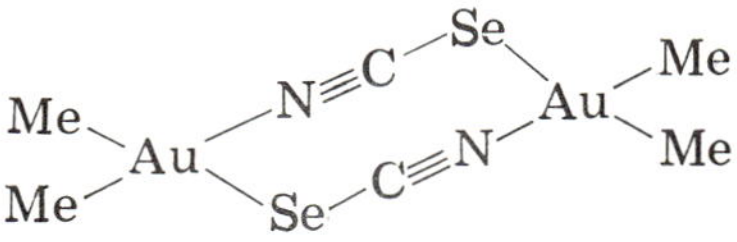

In contrast the dialkylgold cyanides are tetrameric as shown by molecular weight determinations [83, 88] and, in one case, by X-ray structure determination [90]. The tetrameric structures are necessary because cyanide can form only linear bridges, and this is confirmed by the structure determination for dipropylgold cyanide. The most symmetrical structure for dimethylgold cyanide is shown in Fig. 7.6(a). This would be expected to give two methylgold signals in the n.m.r. spectrum due to the methyl groups *trans* to carbon and nitrogen, but four such signals are actually observed. It is suggested that there is disorder in the cyanide bridges giving structures such as Fig. 7.6(b).

The dialkylgold cyanides decompose readily on being heated, and intermediates of formula $Au_2R_2(CN)_2$ can be isolated. These are probably best formulated as polymeric

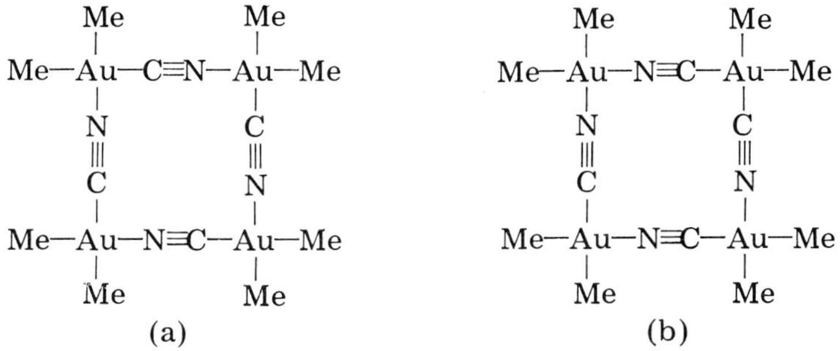

Fig. 7.6. The structure of dimethylgold cyanide.

$R_2AuCN \cdot AuCN$, that is as mixed gold(I)—gold(III) complexes [82, 88].

$$2R_2AuCN \rightarrow R_2Au_2(CN)_2 + R{-}R \rightarrow 2AuCN + 2R{-}R$$

In addition to the neutral halogen-bridged complexes a few anionic species are known. Thus, reaction of dimethylgold(III) halides with excess of halide gives the corresponding *cis*-$[AuMe_2X_2]^-$, which can be isolated as the tetraphenylarsonium or caesium salts [91].

(e) Complexes with Group VI donor ligands

The aquo complex *cis*-$[AuMe_2(OH_2)_2]^+$ has been studied in some detail [92]. In acidic aqueous solution the lifetime of the bound water molecules is only about 2×10^{-5} s at 25 °C due to exchange with solvent water [93]. In aqueous solution hydrolysis occurs according to the equations [94]:

$$2[AuMe_2(OH_2)_2]^+ \rightleftharpoons [\{AuMe_2(OH)\}_2] + 2H_3O^+$$

$$[\{AuMe_2(OH)\}_2] + 2OH^- \rightleftharpoons 2cis\text{-}[AuMe_2(OH)_2]^-$$

Although dimethylgold hydroxide is dimeric in aqueous solution, in organic solvents and in the solid state it is a tetramer with the unique structure shown in Fig. 7.7 [95]. In this structure, the two methyl groups on each gold atom are not equivalent and give rise to two signals in the n.m.r. spectrum.

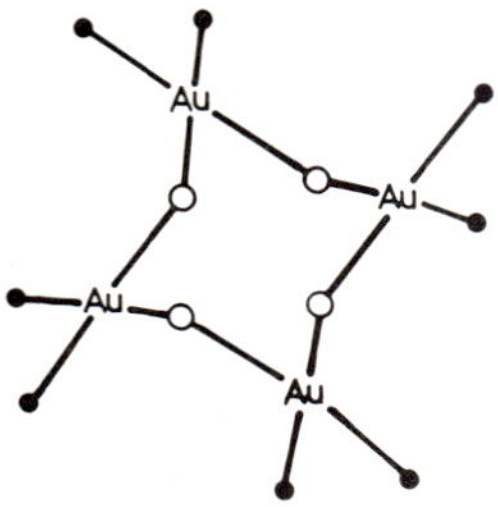

Fig. 7.7. The structure of dimethylgold hydroxide (closed circles represent methyl groups, open circles represent hydroxyl groups).

The formation of this tetrameric structure for dimethylgold hydroxide is unexpected, particularly since trimethylsiloxydimethylgold [96] and the hydroxide [97] shown in Fig. 7.8 have the more usual dimeric structures.

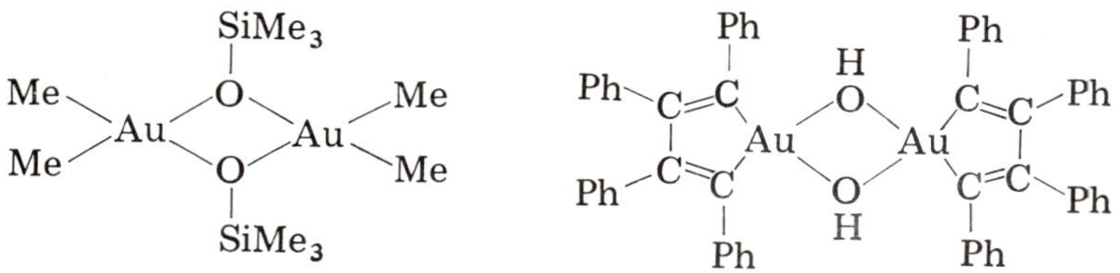

Fig. 7.8. The structures of $(Me_2AuOSiMe_3)_2$ and $(Ph_4C_4AuOH)_2$.

Dimethylgold acetate is also dimeric. It is prepared by the reaction [98, 99]:

$$(Me_2AuBr)_2 + 2AgOAc \rightarrow (Me_2AuOAc)_2 + 2AgBr$$

The proposed structures of this and of bis(diethylgold)oxalate [100] are shown in Fig. 7.9.

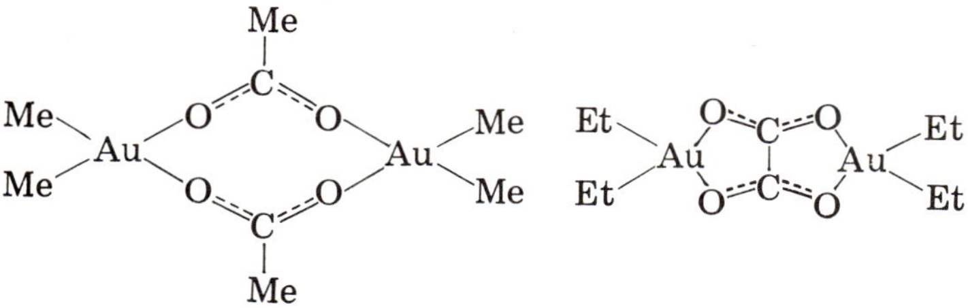

Fig. 7.9. The proposed structures of dimethylgold acetate and oxalate.

Bis(diethylgold)phthalate probably has a structure similar to the oxalate, but with other dicarboxylic acids more highly associated structures may be formed. Thus, malonic acid gives a complex containing four diethylgold units, $[(Et_2Au)_4\{CH_2(CO_2)_2\}_2]$, while terephthalic acid gives an insoluble polymer [100].

Salts of inorganic acids, such as the sulphate, are also known [86, 101].

$$2(Et_2AuBr)_2 + 2Ag_2SO_4 \rightarrow [\{Et_2Au)_2SO_4\}_2] + 4AgBr$$

This is also dimeric, as is the corresponding monophenylphosphate [102]. The diphenylphosphate, $[\{Et_2AuO_2P(OPh)_2\}_2]$, is also dimeric but the phosphate $(Et_2Au)_3PO_4$ is an insoluble polymer. The particular structure adopted is determined by the need to give square-planar gold(III) centres by bridging of the anion, and this is achieved in different ways depending on the particular anion.

Also known are dimethylgold acetylacetonate [80, 103] and 8-hydroxyquinolinate [104] and several complexes of $Me_2Au(III)$ with salicylaldimine Schiff bases [105].

The simplest compounds derived from sulphur or selenium donor ligands are formed from "trimethylgold" [58b, 72],

$2''Me_3Au'' + 2\ PhSeH \longrightarrow$ 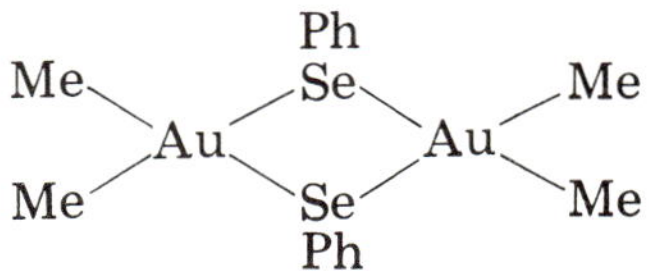 $+ 2CH_4$

However, the usual route to such compounds is by metathesis from the dialkylgold halides [98, 99].

$$[(Me_2AuBr)_2] + 2[EtOCS_2]K \rightarrow 2KBr + 2[Me_2AuS_2COEt]$$

The dithiocarbamates, which are monomeric, can also be prepared by the reaction [106]:

$$[AuBr_2(S_2CNEt_2)] + 2MeMgBr \rightarrow [AuMe_2(S_2CNEt_2)] + 2MgBr_2$$

The structures of some of the known complexes are given in Fig. 7.10.

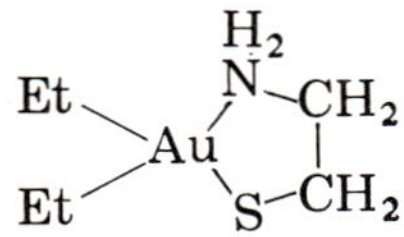

(ref. 107)

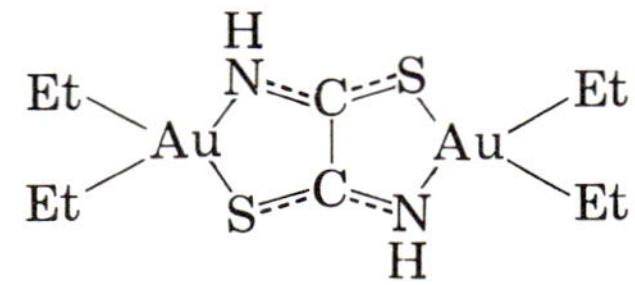

(ref. 107)

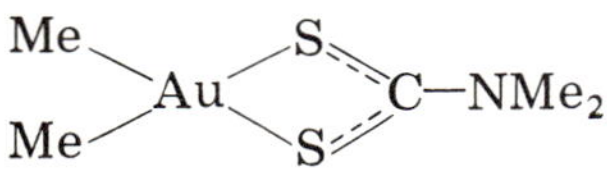

(refs. 98, 99, 106)

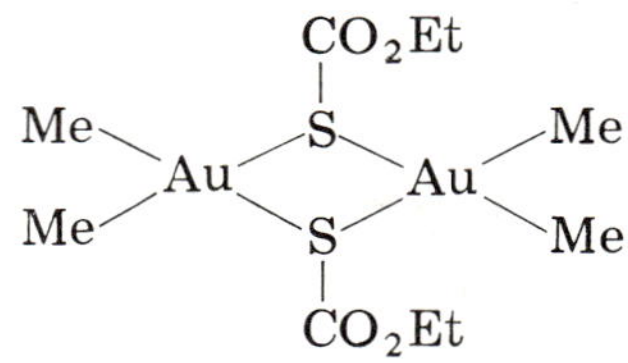

(ref. 98)

Fig. 7.10. Structures of some dialkylgold thiolates.

The halide bridges in the dialkylgold halides can be split by dialkylsulphides and dialkylselenides to give complexes, *cis*-$[AuR_2X(SR'_2)]$ [80, 108].

$(Me_2AuCl)_2 + 2Me_2Se \rightarrow 2$*cis*-$[AuMe_2Cl(SeMe_2)]$

With chelate ligands, such as $MeSCH_2CH_2SMe$, bridging compounds are formed with the dialkylgold halides but with the weakly co-ordinating nitrate ion an ionic complex can be isolated as shown [108].

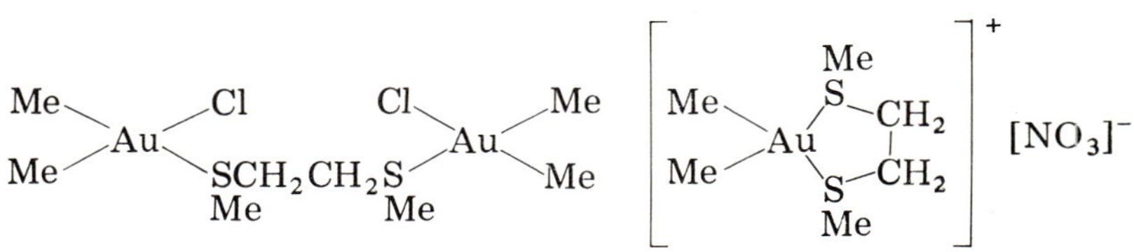

(f) Complexes with Group V donor ligands

Dimethylgold(III) iodide dissolved in liquid ammonia is present as *cis*-$[AuMe_2(NH_3)_2]^+I^-$, but in pyridine only the neutral *cis*-$[AuMe_2I(py)]$ is formed [109]. For a solution of the latter complex the n.m.r spectrum contains two signals due to the methyl groups *trans* to iodide and pyridine, but the signals coalesce if the solution is heated or has pyridine added to it, due to a ligand-exchange reaction which is rapid on the n.m.r. time scale [109].

With the chelate ligands 2,2′-bipyridine and 1,10-phenanthroline, the dimethylgold halides react to give the ionic *cis*-$[AuMe_2(bipy)]^+$ *cis*-$[AuMe_2Cl_2]^-$ or *cis*-$[AuMe_2(phen)]^+$ $[AuMe_2Cl_2]^-$, respectively [110]. In contrast, 2,7-dimethyl-1,8-naphthyridine gives only a 1:1 complex, which undergoes a rapid fluxional process as shown in the equation (X = Cl,Br, I,NCO,SCN,SeCN or CN):

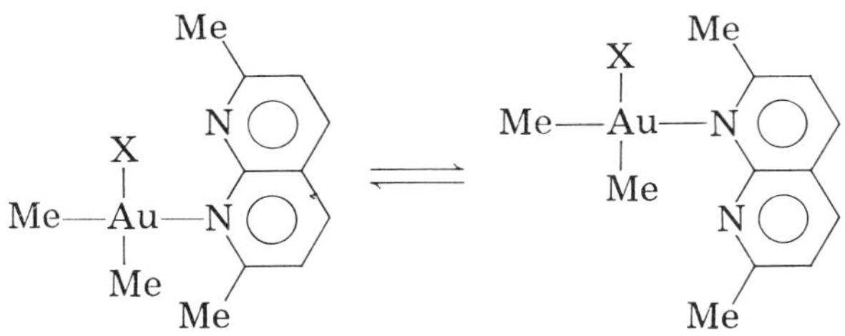

The effect is to make the methyl groups of the ligand effectively equivalent on the n.m.r. time scale, while the two methylgold groups remain non-equivalent [110].

If the compounds $(Me_2AuX)_2$ contain a strong bridging ligand X, then addition of a further ligand may be unfavourable [110, 111]. For example the following systems exist at equilibrium in solution (1,8-naph = 2,7-dimethyl-1,8-naphthyridine):

$$2\,cis\text{-}[AuMe_2(SCN)(py)] \rightleftharpoons [\{AuMe_2(SCN)\}_2] + 2py$$

$$4\,cis\text{-}[AuMe_2(CN)(1,8\text{-naph})] \rightleftharpoons [\{AuMe_2(CN)\}_4] + 4(1,8\text{-naph})$$

Reactions of dimethylgold halides with tridentate nitrogen- and phosphorus-donor ligands have also been investigated, but there is no evidence for formation of five-co-ordinate species [112].

Dimethylgold(III) halides and pseudohalides form complexes with tertiary phosphines and arsines [75, 111, 113]:

$$[(Me_2AuCl)_2] + 2AsPh_3 \rightarrow 2\,cis\text{-}[AuMe_2Cl(AsPh_3)]$$

With tertiary phosphines, displacement of the halide from gold may also occur [113, 114].

$$cis\text{-}[AuMe_2Cl(PMe_3)] + PMe_3 \rightarrow cis\text{-}[AuMe_2(PMe_3)_2]^+Cl^-$$

A slightly different sequence of reactions occurs with chelate phosphine ligands [111, 113]:

$$[(Me_2AuCl)_2] + Ph_2PCH_2CH_2PPh_2 \longrightarrow$$

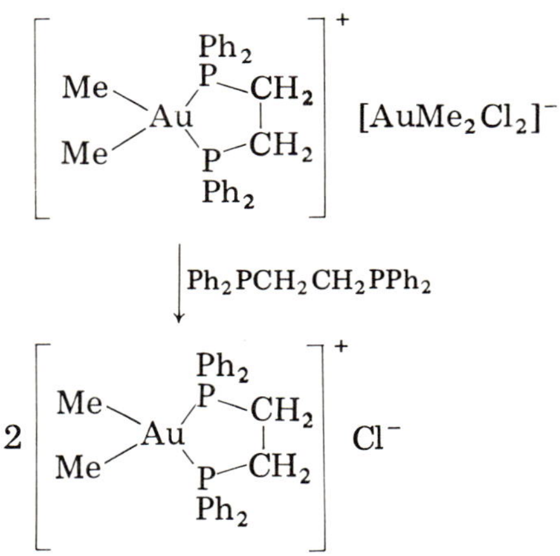

Complexes of the form *cis*-$[AuR_2XL]$ can also be prepared by reaction of $[AuR_3L]$ with acids [17], metal halides [13] or halogens [2]. The redistribution reaction is a useful synthetic route to these complexes [62].

$$2[AuMe_3(PMe_2Ph)] + [AuBr_3(PMe_2Ph)] \rightarrow 3\,cis\text{-}[AuMe_2Br(PMe_2Ph$$

Oxidation of gold(I) complexes has been used to prepare pentafluorophenylgold(III) complexes [115, 116].

$$[AuCl(PPh_3)] + (C_6F_5)_2TlBr \rightarrow cis\text{-}[Au(C_6F_5)_2Cl(PPh_3)] + TlBr$$

The stereochemistry of the product was confirmed by an X-ray structure determination, and the structure of $[Au(C_6F_5)(PPh_3)]$, formed by reduction with hydrazine, has also been determined [117].

Reaction of *cis*-$[Au(C_6F_5)_2Cl(PPh_3)]$ with silver perchlorate gives the corresponding gold perchlorate [16, 118].

$$cis\text{-}[Au(C_6F_5)_2Cl(PPh_3)] + AgClO_4 \rightarrow cis\text{-}[Au(C_6F_5)_2(ClO_4)(PPh_3)]$$

A wide range of derivatives can then be formed by displacement of the labile perchlorate by other ligands.

$$cis\text{-}[Au(C_6F_5)_2(ClO_4)(PPh_3)] + Ph_3PO \rightarrow cis\text{-}[Au(C_6F_5)_2(PPh_3)(OPPh_3)]^+ [ClO_4]^-$$

$$cis\text{-}[Au(C_6F_5)_2(ClO_4)(AsPh_3)] + KCN \rightarrow cis\text{-}[Au(C_6F_5)_2(CN)(PPh_3)] + K[ClO_4]$$

In addition to the tertiary phosphine complexes, it is also possible to prepare a derivative of diphenylphosphine by the reaction [58b]:

$$2\,[AuMe_3(PMe_2Ph)] + 2Ph_2PH \rightarrow$$

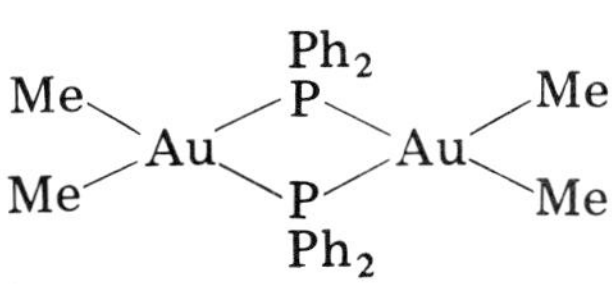

$$+ 2CH_4 + 2PMe_2Ph$$

(g) Complexes with one gold—carbon bond

Benzene reacts with anhydrous gold(III) chloride in dry carbon tetrachloride solution to give phenylgold(III) dichloride which is dimeric [119—121].

$$2Au_2Cl_6 + 2C_6H_6 \rightarrow (PhAuCl_2)_2 + 2H[AuCl_4]$$

The product is formed as a brown suspension, and reaction must be stopped by addition of ether after a short time or further reaction with benzene occurs. The reaction involves electrophilic substitution by gold(III) at the aromatic ring, and is generally known as an auration reaction by analogy with the better known mercuration of arenes. In accord with this view, auration of alkylbenzenes gives very largely the corresponding 4-alkylphenylgold(III) dichloride by *para*-substitution.

The scope of the auration reactions appears to be limited by the need for a potentially vacant co-ordination site on gold to be present as in $[Au_2Cl_6]$. Thus, $H[AuCl_4]$ or $[Au_2Cl_6]$ in donor solvents will not react with benzene [119]. Also, if the arene has substituents which can co-ordinate to gold, then simple co-ordination compounds are formed [119, 122]:

$$2PhCN + [Au_2Cl_6] \rightarrow 2[AuCl_3(PhCN)]$$

However, decomposition of (azobenzene)trichlorogold(III) gives 2-chloroazobenzene and a mechanism involving initial *ortho*-auration of azobenzene is thought likely [123, 124]:

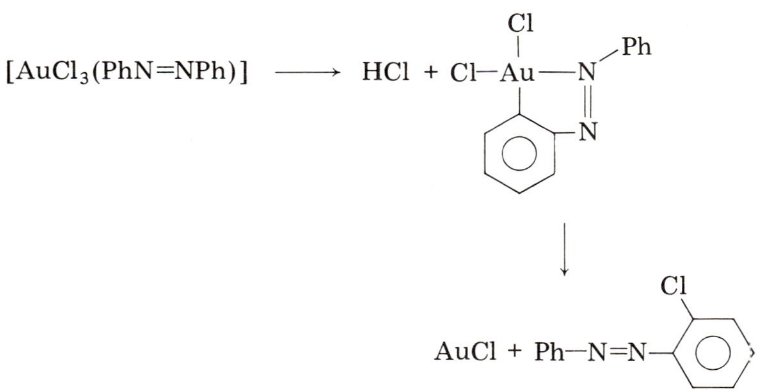

A second route to arylgold(III) complexes is by reaction of an arylhydrazine with tetrachloroaurate(III) [125]:

$$2[Et_4N][AuCl_4] + PhNHNH_2 \cdot HCl \rightarrow [Et_4N][AuCl_2] + [Et_4N][PhAuCl_3] + N_2 + 4HCl$$

Functionally substituted arylgold complexes can also be prepared in this way. Thus, 4-nitrophenylhydrazine gives the corresponding 4-nitrophenyltrichloroaurate(III). The ion $[PhAuCl_3]^-$ can also be obtained by reaction of chloride with dimeric $(PhAuCl_2)_2$, while neutral ligands with $(PhAuCl_2)_2$ give such complexes as *cis*-$[AuCl_2PhL]$ where L = SPr_2, PPh_3 or pyridine [120, 126].

Auration of an alkene occurs in the bromination of 2-vinylphenylphosphine or 2-allylphenylphosphine complexes of gold(I) [127].

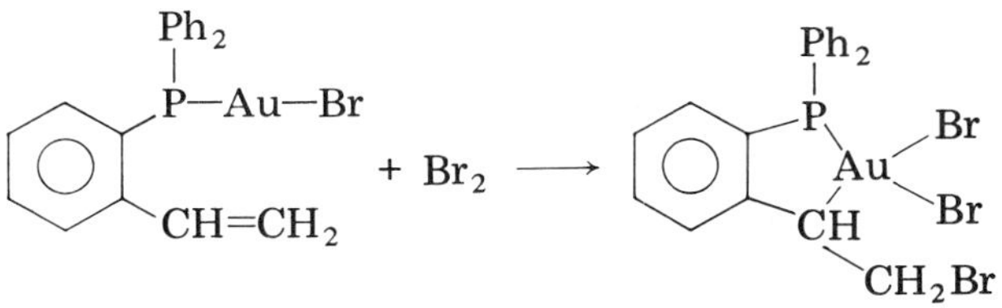

Reaction of this complex with methanol causes a ring expansion to occur, as proved by an X-ray structure determination of the product [128].

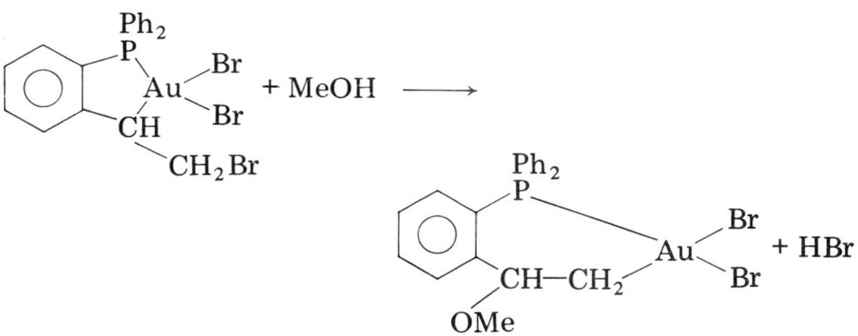

Reaction of organogold(I) complexes with halogens usually results in cleavage of the gold—carbon bond, but with perhaloarylgold(I) complexes oxidation without cleavage often occurs.

$$[Au(C_6Cl_5)(AsPh_3)] + Cl_2 \rightarrow trans\text{-}[AuCl_2(C_6Cl_5)(AsPh_3)]$$

Here, *trans*-addition occurs but oxidation with thallium(III) chloride gives the *cis*-isomer.

$$[Au(C_6Cl_5)(AsPh_3)] + TlCl_3 \rightarrow cis\text{-}[AuCl_2(C_6Cl_5)(AsPh_3)] + TlCl$$

Even with the perhaloarylgold(I) complexes cleavage of the gold—carbon bond rather than oxidation sometimes occurs.

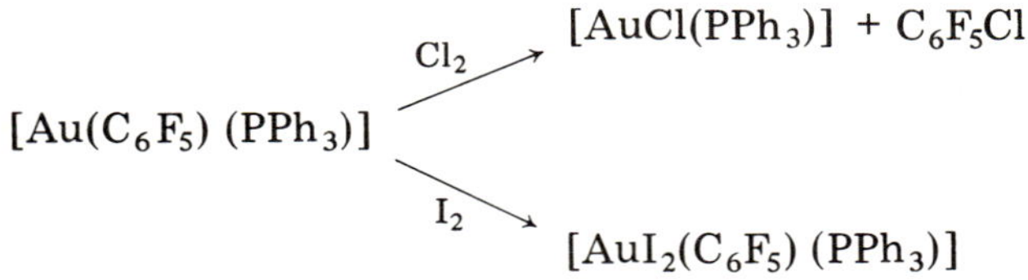

Whether cleavage or oxidation occurs is dependent on the aryl group, the neutral ligand and the particular halogen [15b, 65, 67].

No simple alkylgold dihalides are known. Reaction of dimethylgold(III) bromide with bromine gives a complex $[Au_2Br_4Me_2]$ but this is known to have the structure

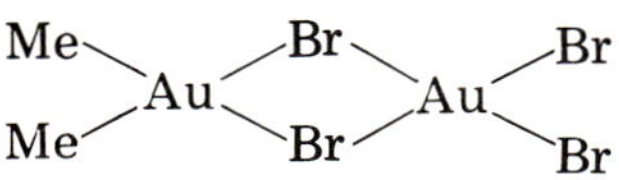

and is thus a dimer of dimethylgold bromide with gold(III) bromide [80, 129, 129b]. In confirmation, the similar $[Au_2Br_4Et_2]$ reacts with bromide to give $[(Et_2AuBr)_2]$ and $[AuBr_4]^-$.

REACTIONS OF ORGANOGOLD(III) COMPLEXES

Many reactions of organogold(III) complexes have been described above, but it will be useful to consider here a few reactions, so that comparisons between complexes with different numbers of gold—carbon bonds can be made.

(a) Isomerisation of organogold(III) complexes

Simple *cis-trans* isomerisation has been observed in the mono-arylgold(III) [15b] and trialkylgold(III) [74] complexes, but not with the dialkylgold(III) complexes for which the *cis*-stereochemistry is almost always found.

trans-$[AuCl_2(C_6Br_5)(PPh_3)]$ → *cis*-$[AuCl_2(C_6Br_5)(PPh_3)]$

trans-$[AuMe_2Et(PPh_3)]$ ⇌ *cis*-$[AuMe_2Et(PPh_3)]$

Elevated temperatures are usually needed to effect the isomerisation. Since the rate of the latter reaction is reduced by the presence of triphenylphosphine, a rearrangement of a three co-ordinate T-shaped intermediate is thought likely [74].

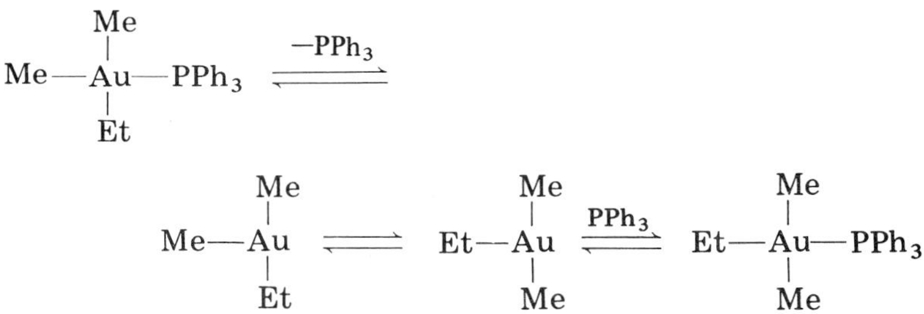

A second form of isomerisation is involved in the rearrangement of a t-butylgold(III) to an iso-butylgold(III) complex [74]:

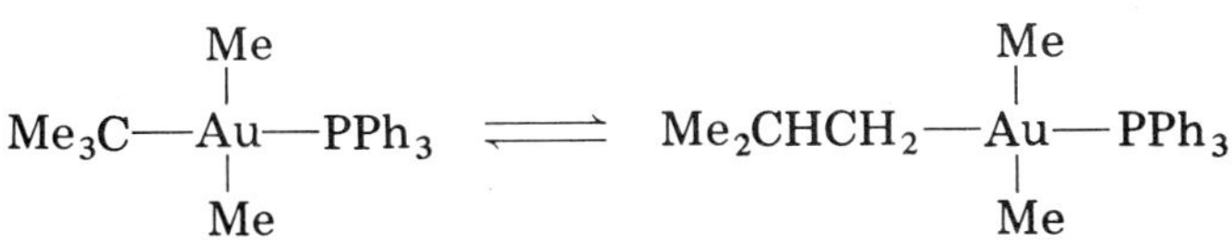

This reaction is also retarded by added triphenylphosphine, and a mechanism involving dissociation of a ligand is suggested:

trans-$[AuMe_2(^tBu)(PPh_3)]$ ⇌ $[AuMe_2{}^tBu] + PPh_3$

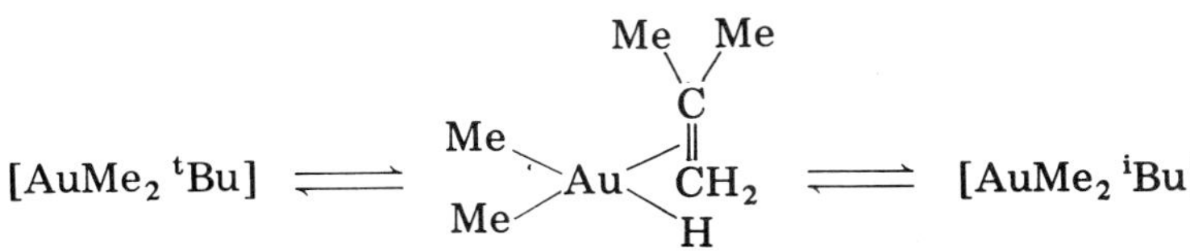

$[AuMe_2{}^iBu] + PPh_3$ ⇌ *trans*-$[AuMe_2{}^iBu(PPh_3)]$

(b) *Reactions with acids, halogens, metal halides and unsaturated reagents*

The dialkylgold halides and related complexes are inert to these reagents. Thus, reactions with halogens take place only slowly at room temperature [80], while with acids and metal halides no reaction occurs.

In contrast, the trialkylgold(III) and tetraalkylaurate(III) complexes are reactive, as illustrated by the following reactions which occur readily at room temperature [2, 5, 13, 62]:

$$[AuMe_3(CH_2PMe_3)] + HCl \rightarrow cis\text{-}[AuMe_2Cl(CH_2PMe_3)] + CH_4$$

$$[AuMe_3(PPh_3)] + I_2 \rightarrow cis\text{-}[AuMe_2I(PPh_3)] + MeI$$

$$[AuMe_3(PPh_3)] + HgCl_2 \rightarrow cis\text{-}[AuMe_2Cl(PPh_3)] + MeHgCl$$

The complexes $[AuMe_3L]$ and $[AuMe_4]^-$ contain mutually *trans* methyl groups, and these methylgold bonds are relatively weak [76, 77]. It is always one of these bonds that is cleaved on reaction with electrophilic reagents.

The reaction of *trans*-$[AuMe_2(CF_3)(PMe_3)]$ with HI takes place with retention of stereochemistry at the gold centre [17]:

$$trans\text{-}[AuMe_2(CF_3)(PMe_3)] + HI \longrightarrow CH_4 + \left[\begin{array}{c} I \\ | \\ CF_3{-}Au{-}PMe_3 \\ | \\ Me \end{array}\right]$$

Organogold(III) complexes are inert to reactions with alkenes and alkynes, though organogold(I) complexes do react. The difference is attributed to the lower ability of gold(III) to form alkene or alkyne complexes. However, sulphur dioxide does react with trimethylgold(III) complexes [40].

$$[AuMe_3(PMe_3)] + SO_2 \rightarrow cis\text{-}[AuMe_2(-\overset{\overset{O}{\|}}{\underset{\underset{O}{\|}}{S}}Me)(PMe_3)]$$

Even in liquid sulphur dioxide, the resulting dimethylgold(III) complex does not undergo further reaction rapidly.

Both nitrogen monoxide and nitrogen dioxide react with $[AuMe_3(PMe_2Ph)]$ to cleave a methylgold bond and give *cis*-$[AuMe_2(NO_2)(PMe_2Ph)]$ in each case [130].

(c) Reduction of organogold(III) complexes

Thermal decomposition of organogold(III) complexes generally occurs by reductive elimination to give a gold(I) complex, which may decompose further to metallic gold. The reaction is common to all organogold(III) complexes, though the products vary from complex to complex. Arylgold(III) dihalides decompose to give the aryl halide and gold(I) halide [119].

$$[(PhAuCl_2)_2] \rightarrow 2PhCl + 2AuCl$$

The reaction is accelerated in the presence of water [121].

Decomposition of dialkylgold(III) complexes usually leads to coupling of the alkyl groups. The reactions occur for cationic [114], neutral [88] and anionic complexes [102, 110].

$$cis\text{-}[AuMe_2(PPh_3)_2]^+ \longrightarrow [Au(PPh_3)_2]^+ + C_2H_6$$

$$[(Et_2AuCN)_4] \longrightarrow 2\,Et_2Au_2(CN)_2 + 2C_4H_{10}$$

$$\downarrow$$

$$4AuCN + 4C_4H_{10}$$

$$cis\text{-}[AuMe_2Cl_2]^- \longrightarrow [AuCl_2]^- + C_2H_6$$

For the cationic complexes the bulk of the ligand appears to accelerate the reductive elimination, since *cis*-$[AuMe_2(PPh_3)_2]^+$ decomposes at room temperature while the similar *cis*-$[AuMe_2(PMe_3)_2]^+$ does not [114].

Different reactions are observed with a few pentafluorophenylgold(III) complexes, where a gold—carbon bond may remain intact. Thus decomposition of *cis*-$[Au(C_6F_5)_2(NO_3)(PPh_3)]$

gives [Au(C_6F_5)(PPh_3)] on reductive elimination [16]. The same product is formed by reduction of *cis*-[Au(C_6F_5)$_2$Cl(PPh_3)] with hydrazine [115].

The decomposition of trialkylgold(III) complexes also leads to coupling of alkyl groups [9, 72].

$$[AuMe_3(PPh_3)] \rightarrow [AuMe(PPh_3)] + C_2H_6$$

The reaction is an intramolecular process in solvents such as dimethylsulphoxide. Thus, decomposition of a mixture of [AuMe$_3$(PPh_3)] and [Au(CD_3)$_3$(PPh_3)] gave C_2H_6 and C_2D_6 but very little of the mixed CH_3CD_3, though CH_3CD_3 is formed in non-polar solvents like benzene. Further, it is shown that *cis*-alkyl groups are eliminated by comparing the products of decomposition of *cis*- and *trans*-dimethyl(ethyl)gold(III) complexes.

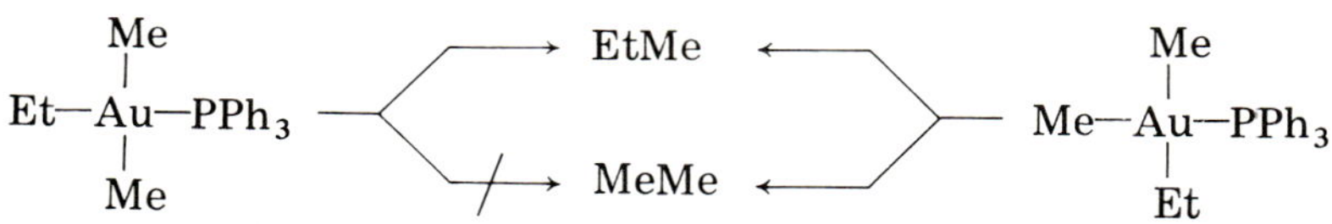

The failure of the *trans*-isomer to give significant amounts of ethane indicates a strong preference for *cis*-elimination. Finally, it is shown that the reactions are strongly retarded by added triphenylphosphine. It is suggested therefore that predissociation of the tertiary phosphine occurs, and that reductive elimination then occurs from the resulting complex [R_3Au] [74].

A concerted mechanism is indicated by the observation that reductive elimination from a vinylgold(III) complex occurs with retention of configuration about the C=C bond [58].

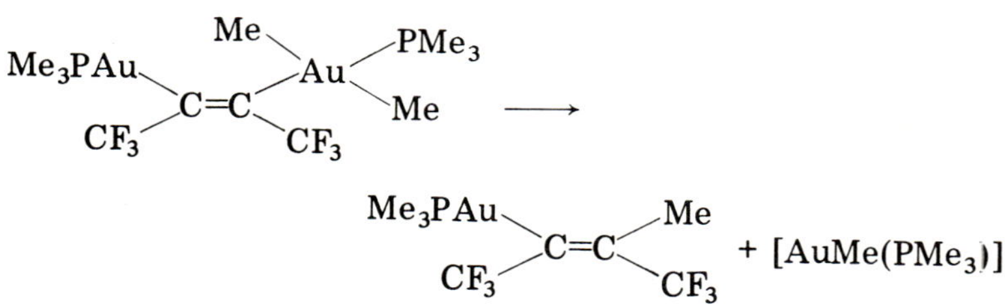

Of the complexes with four Au—C bonds, $[AuMe_4]^-$ is stable to reductive elimination at room temperature [3], but some complexes derived from trimethylphosphine ylid do decompose by coupling of alkyl groups [5].

$$[AuMe_3(CH_2PMe_3)] \rightarrow [AuMe(CH_2PMe_3)] + C_2H_6$$

In one case the enthalpy change on decomposition has been determined [6].

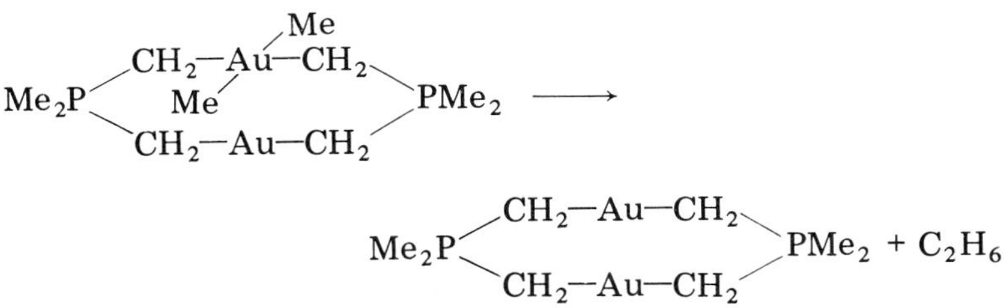

For this reaction $\Delta H = -38$ kJ mol^{-1}. If it is assumed that the only contributions to this are from cleavage of two Au—Me bonds and formation of a C—C bond [E(C—C) = 351 kJ mol^{-1}], then the gold—methyl bond energy can be calculated to be ca. 157 kJ mol^{-1}. This is similar to the methyl—platinum bond energy and higher than the methyl—mercury bond energy. Thus the gold—carbon σ-bond appears to be a strong bond.

Finally, it is worth noting that the decomposition of ethylgold(I) and ethylgold(III) complexes by coupling of alkyl groups is not typical of other ethyl derivatives of transition metals. These more usually decompose by the β-elimination mechanism:

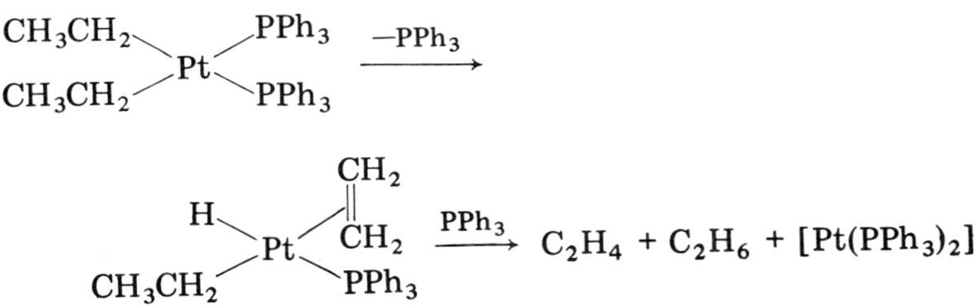

The reluctance of ethylgold complexes to decompose in this way, may be due to the weaker metal—hydride and metal—alkene bonds formed by gold compared with platinum for example.

CARBENE COMPLEXES OF GOLD

Most carbene complexes of gold have been prepared by reactions of isocyanide complexes of gold with alcohols or amines [131—135].

$$[4\text{-MeC}_6\text{H}_4\text{NCAuCl}] + \text{MeOH} \rightarrow [4\text{-MeC}_6\text{H}_4\text{NH(MeO)C—Au—Cl}]$$

$$[4\text{-MeC}_6\text{H}_4\text{NCAuCl}] + 4\text{-MeC}_6\text{H}_4\text{NH}_2 \rightarrow [(4\text{-MeC}_6\text{H}_4\text{NH})_2\text{-C—Au—Cl}]$$

In these complexes, there is restricted rotation about the C—N bonds, and as a result geometrical isomers can be formed as shown [133]:

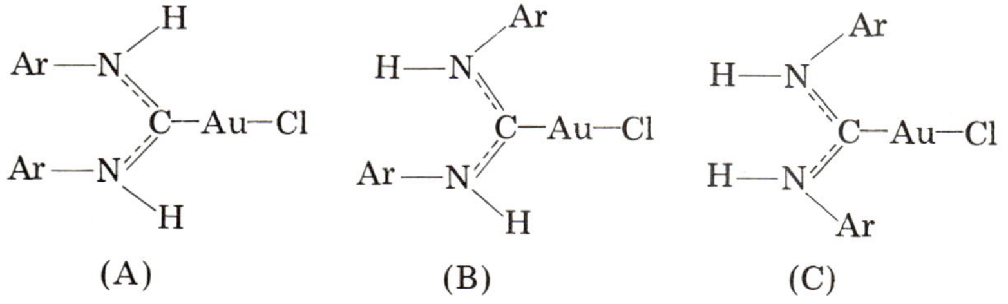

The *cis—trans* isomer (B) predominates. The products obtained from reactions with alkylamines contain additional amine [132, 133]:

$$[^t\text{BuNCAuCl}] + {}^t\text{BuNH}_2 \rightarrow ({}^t\text{BuNH})_2\text{CAuCl} \cdot {}^t\text{BuNH}_2$$

Complexes of this kind have been formulated as three co-ordinate monomers [132] or as four co-ordinate chloride-bridged dimers [133].

Bis(carbene)gold(I) complexes can be prepared from the corresponding bis(isocyanide)gold(I) complex [132, 134].

$$[(4\text{-MeC}_6\text{H}_4\text{NC})_2\text{Au}]^+ + 2\text{EtOH} \rightarrow [\{4\text{-MeC}_6\text{H}_4\text{NH(EtO)C}\}_2\text{Au}]^+$$

$$[({}^t\text{BuNC})_2\text{Au}]^+ + 2\text{Et}_2\text{NH} \rightarrow [\{{}^t\text{BuNH(Et}_2\text{N)C}\}_2\text{Au}]^+$$

Complexes of this kind can also be prepared directly from $H[AuCl_4]$, an alkyl isocyanide and an amine [131]. The sequence of reactions probably involves formation of an isocyanide complex of gold(III), reduction by excess of isocyanide to the gold(I) complex $[Au(CNR)_2]^+$ and then reaction with the amine to give the carbene complexes. It is thought that these *bis*-carbene complexes again have planar structures due to C=N multiple bonding. Hence, many isomers are possible as a result of restricted rotation about the C—N and Au—C bonds. The compounds with methyl substituents are thought to have the structures [131]:

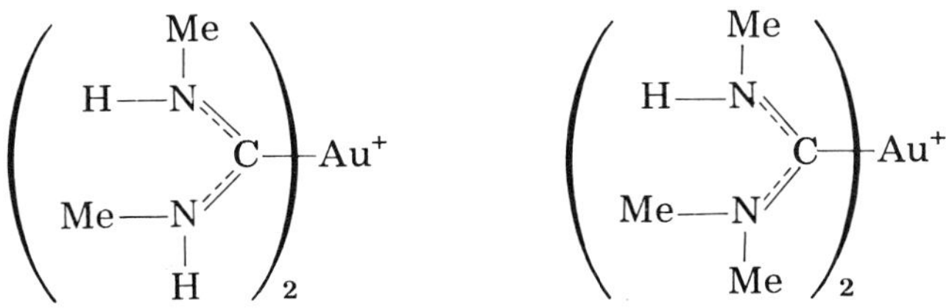

Reaction of 2- or 3-hydroxyalkyl isocyanides with gold(III) halides leads to formation of cyclic carbene complexes by intramolecular attack of the hydroxy group on the co-ordinated isocyanide [139].

Some of the known carbene complexes of gold are listed in Table 7.4.

These carbene complexes can be deprotonated by reaction with base to give iminomethylgold complexes. The reactions are reversible, so that reaction of the iminomethylgold complexes with acids gives an alternative synthetic route to the carbene complexes of gold [42, 131, 133, 138].

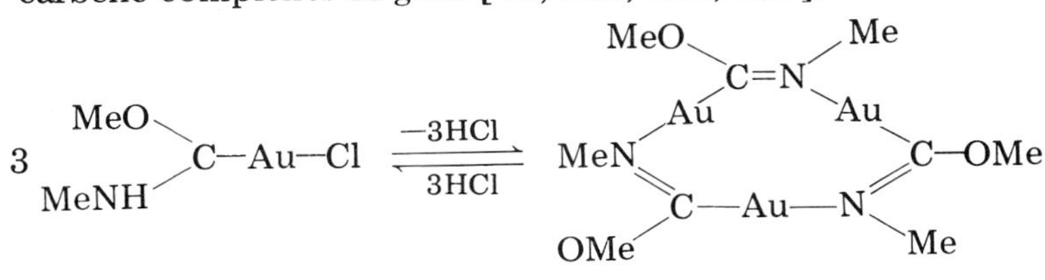

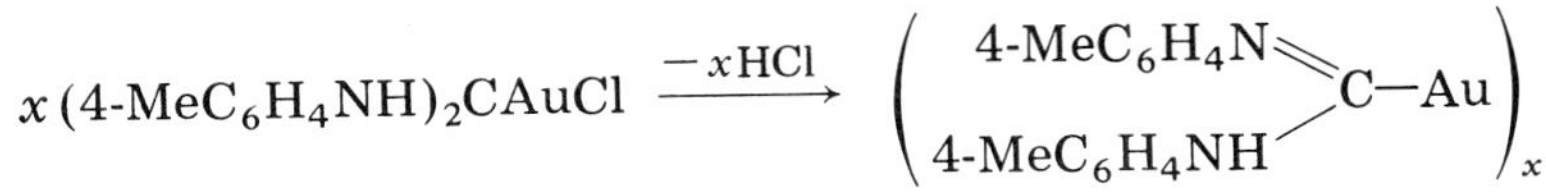

TABLE 7.4

Carbene complexes of gold

Complex	m.p./°C	Ref.
$[AuCl\{C(OMe)NHMe\}]$	129—130	42, 131
$[AuCl\{C(OMe)NHCH_2Ph\}]$		135
$[AuCl\{C(OMe)NHC_6H_4Me\text{-}4\}]$	167	135
$[AuCl\{C(NHC_6H_4Me\text{-}4)_2\}]$	162	133
$[AuCl\{C(NH^iPr)(NH^tBu)\}(NH_2{}^tBu)]$		132
$[Au\{C(NHMe)_2\}_2][PF_6]$		131
$[Au\{C(NH^iPr)_2\}_2][PF_6]$		132
$[Au\{C(NHC_6H_4Me\text{-}4)_2\}_2][ClO_4]$	180	134
$[Au\{C(OEt)NHC_6H_4Me\text{-}4\}_2][ClO_4]^a$	185—186	134
$[Au\{C(OEt)NHC_6H_4Me\text{-}4\}_2][ClO_4]^a$	153	134
$[Au\{\overline{CN(Ph)CH_2CH_2N}Ph\}_2][BF_4]$	218—225 dec.	136
$[AuI_2\{C(NHMe\}_2][PF_6]$		131
$[AuI_2\{C(NHC_6H_4Me\text{-}4)_2\}_2][ClO_4]$	124	137
$[AuI_2\{C(OEt)NHC_6H_4Me\text{-}4\}_2][ClO_4]$	162 dec.	137

[a] Geometrical isomers.

Particularly interesting is the observation that the carbene ligands can be displaced from gold by ligands such as cyanide, triphenylphosphine and methyl isocyanide [131, 132, 134]. The displacement is accompanied by proton transfer.

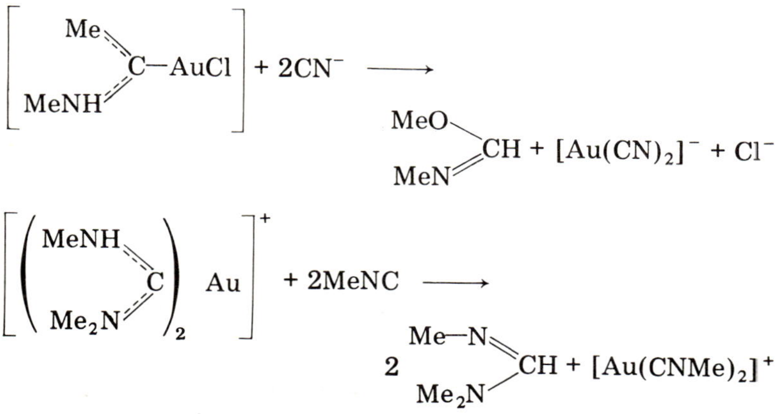

It is interesting that H[$AuCl_4$] acts as a catalyst for the addition of amines to isocyanides to give formamidine derivatives.

From the results described above, it is clearly likely that the reaction proceeds by the following sequence: first an isocyanide complex of gold(I) is formed, which reacts with the amine to give a carbene complex of gold(I). The carbene ligand is then displaced by isocyanide to give the formamidine derivative and regenerate the isocyanide complex, thus completing the catalytic cycle.

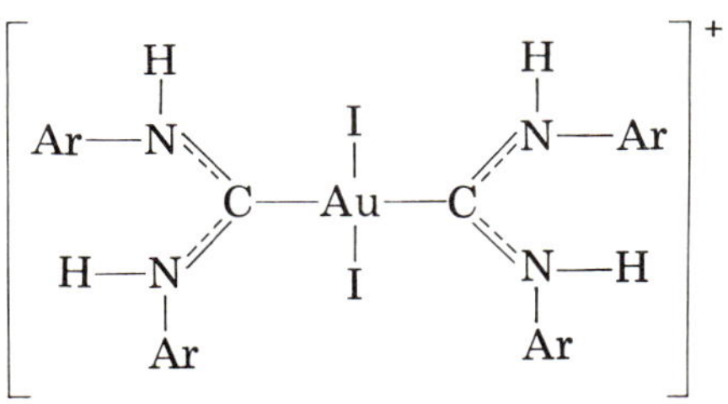

$[ClO_4]^-$

(Ar = 4-MeC_6H_4)

Fig. 7.11. The structure of a gold(III) carbene complex.

The carbene complexes of gold(I) can be oxidised by halogens to give carbene complexes of gold(III) [131, 137].

$$[Au\{C(NHMe)_2\}_2]^+ + I_2 \rightarrow [AuI_2\{C(NHMe)_2\}_2]^+$$

The structure of one such complex has been determined by X-ray methods and is shown in Fig. 7.11.

The stereochemistry at gold(III) is *trans*-square planar, and at the carbene carbon atoms trigonal planar. The C—N bonds are relatively short indicative of multiple bonding due to overlap between the $2p_z$ orbital of the carbene carbon atom and the filled 2p orbitals of the nitrogen atoms. The Au—C bond lengths are 208 pm [140].

A quite different route to carbene complexes of gold is based on cleavage of the C=C bond of electron-rich alkenes [141].

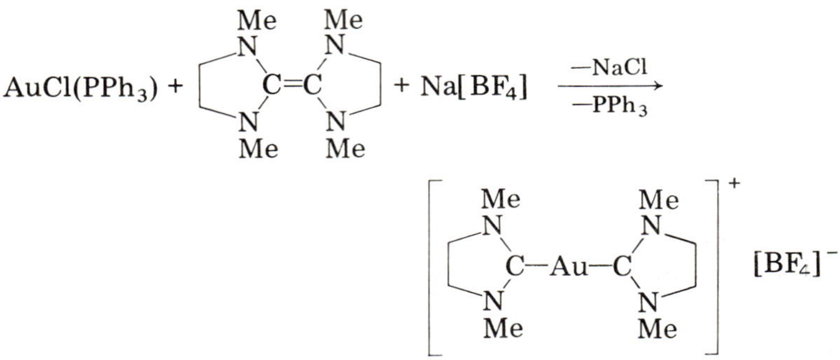

ALKENE AND ALKYNE COMPLEXES OF GOLD

Alkene complexes of gold in general have rather low thermal stability as illustrated by the decomposition temperatures of such complexes given in Table 7.5. The simplest route to alkene complexes of gold(I) is to treat a gold(I) halide with the alkene [142, 143].

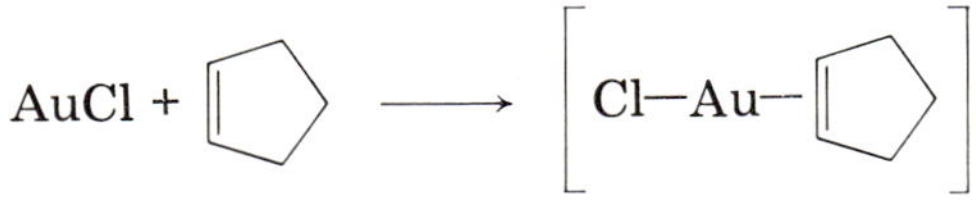

The same method can be used to prepare alkyne complexes [146, 147]

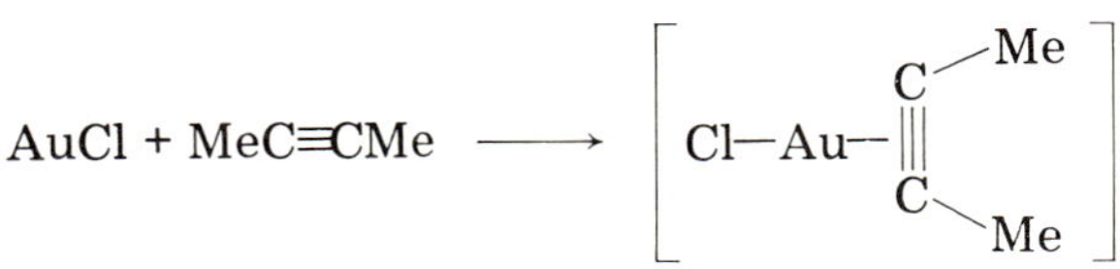

$$AuBr + 2\ \text{cyclo-octyne} \longrightarrow [AuBr(\text{cyclo-octyne})_2]$$

To minimise decomposition of the products, the reactions are carried out at low temperature.

TABLE 7.5

Alkene and alkyne complexes of gold

Complex[a]	Dec./°C	Ref.
[AuCl(cyclopentene)]	55—60	142
[AuCl(cyclohexene)]	60	142
[AuCl(cycloheptene)]	93—98	142
[AuCl(cyclo-octene)]	93—96	142
[AuCl(octene-1)]	0	143
[AuCl(decene-1)]	0	143
[AuCl(dodecene-1)]	23—24	143
[AuCl(hexadecene-1)]	43—45	143
[AuCl(cyclo-octatetraene)]	−20	144
[AuCl(hexamethyl-Dewarbenzene)]	139—145	145
[AuCl(dimethylacetylene)]	0—5	146
[AuCl(diphenylacetylene)]	75—78	146
[AuBr(cyclo-octyne)$_2$]	135—143	147
$(AuCl)_2$(1,5-cyclo-octadiene)	110—114	142, 148, 149
[$AuCl_2$(1,5-cyclo-octadiene)]Cl		149
[AuMe(PMe_2Ph)(tetracyanoethylene)]	45—47	40

[a] All complexes are white or yellow.

Reactions of alkenes and alkynes with gold(III) halides are often complex. It is thought that the first step is formation of an alkene or alkyne complex of gold(III). With 1,5-cyclo-octadiene, COD, such a complex can be isolated [149].

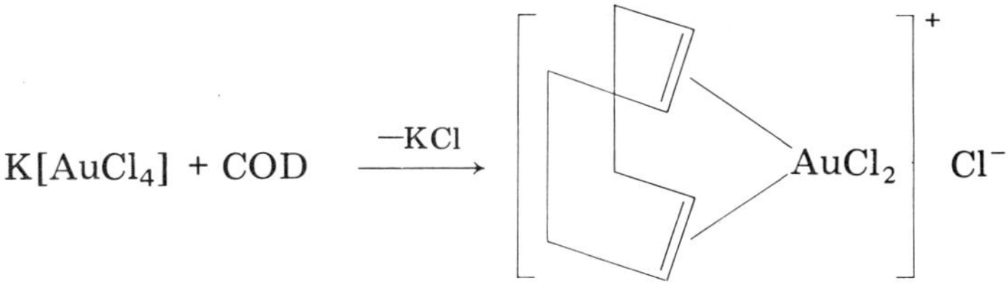

However, even this complex decomposes to give first a mixed gold(I)—gold(III) complex {Au(COD)Cl_2} and then the gold(I) complex [$(AuCl)_2$(COD)] [142, 148, 149]. With other alkenes, reduction of gold with parallel formation of chlorine adducts of the alkenes occurs rapidly even at low temperatures. The reactions occur in several stages. First, the complex [$AuCl_3$-(alkene)] is assumed to be formed, but this decomposes to give gold(I)—gold(III) complexes such as [Au_2Cl_4(alkene)] and alkene dichloride. Further, alkene can react to give [Au_2Cl_4-(alkene)$_n$] where n = 2 or 3, but these complexes react further to give finally [AuCl(alkene)] and more alkene dichloride. For example, propene gives over 90% 1,2-dichloropropane [150]. *trans*-Chlorination of the alkene occurs predominantly [144, 150]. Alternatively, it is possible that direct chlorination by the gold(III) halide occurs [144, 145].

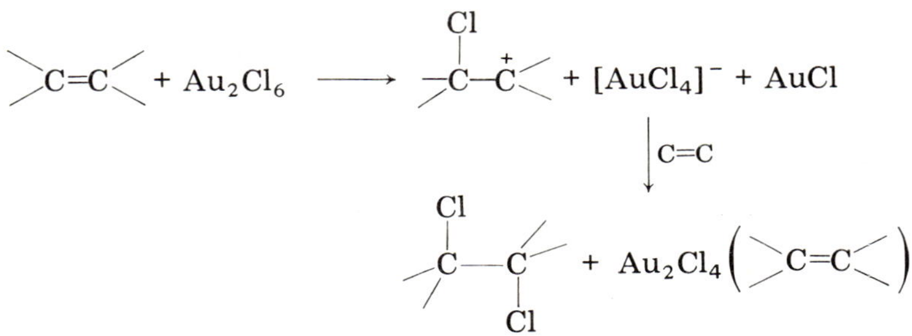

This is likely in the reaction of hexamethyl Dewarbenzene with Au_2Cl_6, when a stable chlorinated carbocation is formed [145].

The reaction of gold(III) chloride with but-2-yne is particularly interesting. The overall reaction is given by the equation:

$$4MeC{\equiv}CMe + Au_2Cl_6 \longrightarrow$$

$$[Au(MeC{\equiv}CMe)_2]^+[AuCl_4]^- +$$

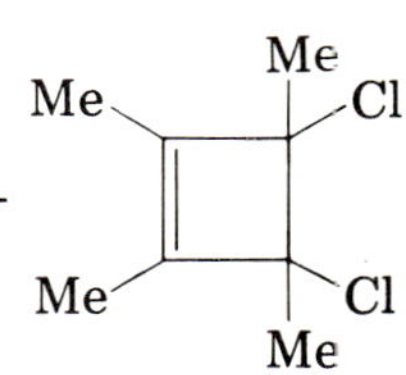

The first step involves formation of the π-complexes [Au_2Cl_6(MeC≡CMe)] and [$Au_2Cl_6(MeC{\equiv}CMe)_2$]. These can then rearrange to give σ-bonded chlorovinylgold(III) complexes such as

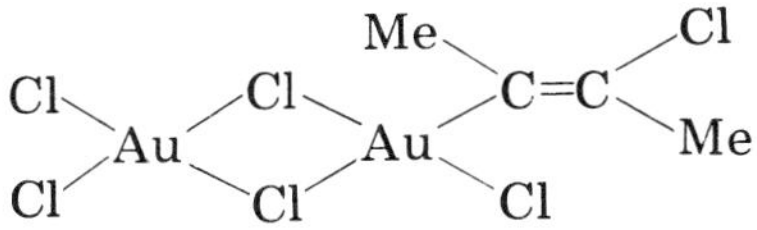

Further reaction with but-2-yne, possibly by insertion of the alkyne into the Au—C σ-bond followed by reductive cyclisation, gives the dichlorocyclobutene derivative [146] and the alkyne complex of gold(I).

Although many alkene and alkyne complexes of gold have been prepared in recent years, the structures of many of the complexes are still not known. Various spectroscopic and chemical studies indicate that the gold—alkene or gold—alkyne bond is considerably weaker than similar bonds to platinum in platinum(II) complexes. In particular, the complexes dissociate to give free alkene or alkyne in solution [146, 147].

$$AuCl + MeC{\equiv}CMe \rightleftharpoons [AuCl(MeC{\equiv}CMe)] \quad K = 6.3 \ l\,mol^{-1}$$

The thermal stability of complexes $[AuCl(RCH{=}CH_2)]$ increases as the chain length of the alkyl group R increases [143].

A rather different type of alkene complex is formed by addition of tetracyanoethylene to methylgold(I) complexes. The compounds can be considered as three-co-ordinate complexes of gold(I) or as gold(III) complexes with a metallocyclopropane ring as shown [40].

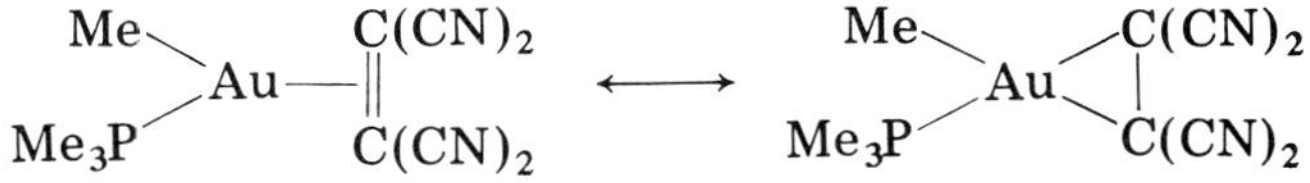

REFERENCES

1 R. J. P. Williams, G. Agnes, S. Bendle, H. A. O. Hill and F. R. Williams, *J. Chem. Soc., Chem. Commun.* (1971) 809.

2 A. Tamaki and J. K. Kochi, *J. Organomet. Chem.*, 51 (1973) C39; *J. Chem. Soc., Dalton Trans.* (1973) 2620.

3 G. W. Rice and R. S. Tobias, *Inorg. Chem.*, 14 (1975) 2402; 15 (1976) 489.
4 R. Uson, A. Laguna and J. Vicente, *J. Chem. Soc., Chem. Commun.*, (1976) 353; R. Uson, A. Laguna and J. Vicente, *J. Organomet. Chem.*, 131 (1977) 471.
5 H. Schmidbaur, *Acc. Chem. Res.*, 8 (1975) 62.
6 H. Schmidbaur and R. Franke, *Inorg. Chim. Acta*, 13 (1975) 85.
7 (a) H. Schmidbaur, J. R. Mandl, A. Frank and G. Huttner, *Chem. Ber.*, 109 (1976) 466; H. Schmidbaur and H. P. Scherm, *Chem. Ber.*, 110 (1977) 1576.
(b) P. A. Arnup and M. C. Baird, *Inorg. Nucl. Chem. Lett.*, 5 (1969) 65.
8 H. Schmidbaur and A. Shiotani, *Chem. Ber.*, 104 (1971) 2821.
9 G. Galvin, G. E. Coates and C. Parkin, *Chem. Ind.*, (1959) 1628; G. E. Coates and C. Parkin, *J. Chem. Soc.* (1963) 421.
10 A. Johnson and R. J. Puddephatt, *J. Organomet. Chem.*, 85 (1975) 115.
11 A. Johnson and R. J. Puddephatt, *J. Chem. Soc., Dalton Trans.*, (1975) 115.
12 H. Schmidbaur and R. Franke, *Chem. Ber.*, 105 (1972) 2985.
13 B. J. Gregory and C. K. Ingold, *J. Chem. Soc. B*, (1969) 276.
14 A. Tamaki and J. K. Kochi, *J. Organomet. Chem.*, 61 (1973) 441.
15 (a) S. Numata, H. Kurosawa and R. Okawara, *J. Organomet. Chem.*, 102 (1975) 259.
(b) R. Uson and A. Laguna, *Synth. React. Inorg. Met.-Org. Chem.*, 5 (1975) 17.
16 R. Uson, A. Laguna and J. Buil, *J. Organomet. Chem.*, 85 (1975) 403.
17 A. Johnson and R. J. Puddephatt, *J. Chem. Soc., Dalton Trans.*, (1976) 1360.
18 B. Wozniak, J. D. Ruddick and G. Wilkinson, *J. Chem. Soc. A*, (1971) 3116.
19 A. N. Nesmeyanov, E. G. Perevalova, V. V. Krivikh, A. N. Kosina, K. I. Grandberg and E. I. Smyslova, *Izv. Akad. Nauk. SSSR, Ser. Khim.*, (1972) 653; *Chem. Abstr.*, 77 (1972) 101818.
20 L. G. Vaughan and W. A. Sheppard, *J. Am. Chem. Soc.*, 91 (1969) 6161; *J. Organomet. Chem.*, 22 (1970) 739.
21 D. I. Nichols, *J. Chem. Soc. A* (1970) 1216.
22 G. van Koten and J. G. Noltes, *J. Organomet. Chem.*, 80 (1974) C56.
23 C. M. Mitchell and F. G. A. Stone, *J. Chem. Soc., Dalton Trans.*, (1972) 102.
24 R. Hüttel, U. Raffay and H. Reinheimer, *Angew. Chem. Int. Ed.*, 6 (1967) 862.
25 G. E. Coates and C. Parkin, *J. Chem. Soc.*, (1962) 3220.
(a) V. I. Rozenberg, R. I. Gorbacheva, E. I. Smyslova, K. I. Grandberg, V. A. Nikanorov, Y. G. Bundel and O. A. Reutov, *Dokl. Akad. Nauk. SSSR*, 225 (1975) 1082.
26 D. Gibson, B. F. G. Johnson, J. Lewis and C. Oldham, *Chem. Ind.*, (1966) 342; D. Gibson, *Coord. Chem. Rev.*, 4 (1969) 225.
27 A. N. Nesmeyanov, K. I. Grandberg, V. P. Dyadchenko, D. A. Lemenovskii and E. G. Perevalova, *Izv. Akad. Nauk. SSSR, Ser. Khim.*, (1974) 1206; A. N. Nesmeyanov, E. G. Perevalova, D. A. Lemenovskii, V. P. Dyadchenko and K. I. Grandberg, *Izv. Akad. Nauk. SSSR, Ser. Khim.*, (1974) 1661.
28 A. N. Nesmeyanov, E. G. Pervalova, V. P. Dyadchenko and K. I. Grandberg, *Izv. Akad. Nauk. SSSR, Ser. Khim.*, (1974) 2872.
29 A. N. Nesmeyanov, K. I. Grandberg, E. I. Smyslova and E. G. Perevalova, *Izv. Akad. Nauk. SSSR, Ser. Khim.*, (1974) 2872.
30 R. Hüttel, U. Raffay and H. Reinheimer, *Angew. Chem. Int. Ed.*, 6 (1967) 862.
31 C. H. Campbell and M. L. H. Green, *J. Chem. Soc. A*, (1971) 3282.
32 G. Ortaggi, *J. Organomet. Chem.*, 80 (1974) 275.
33 A. N. Nesmeyanov, E. G. Perevalova, D. A. Lemenovskii, A. N. Kosina and K. I. Grandberg, *Izv. Akad. Nauk. SSSR, Ser. Khim.*, (1969) 2030.
34 E. G. Perevalova, D. A. Lemenovskii, O. B. Afanasova, V. P. Dyadchenko, K. I. Grandberg and A. N. Nesmeyanov, *Izv. Akad. Nauk. SSSR, Ser. Khim.*, (1972) 2594.

35 A. N. Nesmeyanov, K. I. Grandberg, T. V. Baukova, A. N. Kosina and E. G. Perevalova, *Izv. Akad. Nauk. SSSR, Ser. Khim.*, (1969) 2032.
36 A. N. Nesmeyanov, E. G. Perevalova, T. V. Baukova and K. I. Grandberg, *Izv. Akad. Nauk. SSSR, Ser. Khim.*, (1973) 2641.
37 A. Matthews and L. L. Waters, *J. Am. Chem. Soc.*, 22 (1900) 108.
38 P. W. R. Corfield and H. M. M. Shearer, *Acta Crystallogr.*, 23 (1967) 156.
39 R. Nast and U. Kirner, *Z. Anorg. Allg. Chem.*, 330 (1964) 311.
40 A. Johnson, R. J. Puddephatt and J. L. Quirk, *J. Chem. Soc., Chem. Commun.*, (1972) 938; A. Johnson and R. J. Puddephatt, *J. Chem. Soc., Dalton Trans.*, (1977) 1384.
41 L. G. Vaughan, *J. Am. Chem. Soc.*, 92 (1970) 730; *U.S. Pat.*, 3,755,329, 1973; *Chem. Abstr.*, 79 (1973) 115732.
42 G. Minghetti and F. Bonati, *Inorg. Chem.*, 13 (1974) 1600; A. L. Balch and D. J. Doonan, *J. Organomet. Chem.*, 131 (1977) 137.
43 P. Jutzi and H. Heusler, *J. Organomet. Chem.*, 114 (1976) 265.
44 J. G. Noltes, *J. Organomet. Chem.*, 100 (1975) 177.
45 E. G. Perevalova, D. A. Lemenovskii, K. I. Grandberg and A. N. Nesmeyanov, *Dokl. Akad. Nauk. SSSR*, 202 (1972) 93.
46 A. N. Nesmeyanov, E. G. Perevalova, O. B. Ofanasova and K. I. Grandberg, *Izv. Akad. Nauk. SSSR, Ser. Khim.*, (1974) 484; (1975) 477.
47 V. G. Andrianov, Yu. T. Struchkov and E. R. Rossinskaya, *J. Chem. Soc., Chem. Commun.*, (1973) 338.
48 A. N. Nesmeyanov, E. G. Perevalova, T. V. Baukova and K. I. Grandberg, *Izv. Akad. Nauk. SSSR, Ser. Khim.*, (1974) 866.
49 K. I. Grandberg, T. V. Baukova, E. G. Perevalova and A. N. Nesmeyanov, *Dokl. Akad. Nauk. SSSR*, 206 (1972) 1355.
50 V. I. Sokolov, K. I. Grandberg, V. V. Bashilov, D. A. Lemenovskii, P. V. Petrovskii and O. A. Reutov, *Dokl. Akad. Nauk. SSSR*, 214 (1974) 393; K. I. Grandberg, E. I. Smyslova and A. I. Kosina, *Izv. Akad. Nauk. SSSR, Ser Khim.*, (1973) 2787.
51 P. W. J. de Graaf, J. Boersma and G. J. M. van der Kerk, *J. Organomet. Chem.*, 78 (1974) C19; 127 (1977) 391.
52 (a) G. van Koten and J. G. Noltes, *J. Organomet. Chem.*, 82 (1974) C53; G. van Koten, J. T. B. H. Zastrzebski and J. G. Noltes, *J. Chem. Soc., Chem. Commun.*, (1977) 203.
(b) H. Schmidbaur and O. Gasser, *Angew. Chem. Int. Ed.*, 15 (1976) 502.
53 E. T. Blues, D. Bryce-Smith, I. W. Lawston and G. D. Wall, *J. Chem. Soc., Chem. Commun.*, (1974) 513; E. T. Blues, D. Bryce-Smith and I. W. Lawston, *Gold Bull.*, 9 (1976) 88; D. Bryce-Smith and E. T. Blues, *Ger. Pat.*, 2,336,396, 1974, *Chem. Abstr.*, 80 (1974) 108672.
54 A. M. Sladkov and I. R. Gol'ding, *Dokl. Akad. Nauk. SSSR*, 200 (1971) 132.
55 P. W. N. M. van Leeuwen, R. Kaptein, R. Huis and C. F. Roobeek, *J. Organomet. Chem.*, 104 (1976) C44.
56 E. G. Perevalova, T. V. Baukova, E. I. Goryunov and K. I. Grandberg, *Izv. Akad. Nauk. SSSR, Ser. Khim.*, (1970) 2148.
57 M. Aresta and G. Vasapollo, *J. Organomet. Chem.*, 50 (1973) C51.
58 (a) J. A. J. Jarvis, A. Johnson and R. J. Puddephatt, *J. Chem. Soc., Chem. Commun.*, (1973) 373.
(b) R. J. Puddephatt and P. J. Thompson, *J. Organomet. Chem.*, 117 (1976) 395.
59 E. G. Perevalova, D. A. Lemenovskii, T. V. Baukova, E. I. Smyslova, K. I. Grandberg and A. N. Nesmeyanov, *Dokl. Akad. Nauk. SSSR*, 206 (1972) 883.
60 A. Shiotani and H. Schmidbaur, *J. Organomet. Chem.*, 37 (1972) C24.
61 A. N. Nesmeyanov, E. G. Perevalova, M. V. Ovchinnikov and K. I. Grandberg, *Izv. Akad. Nauk. SSSR*, (1975) 2282.
62 R. J. Puddephatt and P. J. Thompson, *J. Chem. Soc., Dalton Trans.*, (1975) 1810.

63 A. Tamaki and J. K. Kochi, *J. Organomet. Chem.*, 40 (1972) C81; 64 (1974) 411.
64 G. W. Rice and R. S. Tobias, *J. Organomet. Chem.*, 86 (1975) C37.
65 R. Uson, A. Laguna and J. Vicente, *J. Organomet. Chem.*, 86 (1975) 415.
66 E. G. Perevalova, D. A. Lemenovskii, K. I. Grandberg and A. N. Nesmeyanov, *Dokl. Akad. Nauk. SSSR*, 199 (1971) 832.
67 R. Uson, A. Laguna and J. Vicente, *J. Organomet. Chem.*, 104 (1976) 401.
68 H. Schmidbaur and R. Franke, *Inorg. Chim. Acta.*, 13 (1975) 79.
69 C. J. Gilmore and P. Woodward, *J. Chem. Soc., Chem. Commun.*, (1971) 233.
70 (a) J. Bailey and M. J. Mays, *J. Organomet. Chem.*, 63 (1973) C24; M. J. Mays and J. Bailey, *J. Chem. Soc., Dalton Trans.*, (1977) 578.
(b) J. P. Fackler Jr., and C. Paparizos, *J. Am. Chem. Soc.*, 99 (1977) 2363.
71 W. Beck, K. Burger and W. P. Fehlhammer, *Chem. Ber.*, 104 (1971) 1816.
72 H. Gilman and L. A. Woods, *J. Am. Chem. Soc.*, 70 (1948) 550.
73 C. F. Shaw and R. S. Tobias, *Inorg. Chem.*, 12 (1973) 965.
74 A. Tamaki, S. A. Magennis and J. K. Kochi, *J. Am. Chem. Soc.*, 96 (1974) 6140; S. Komiya, T. A. Albright, R. Hoffmann and J. K. Kochi, *J. Am. Chem. Soc.*, 98 (1976) 7255.
75 S. W. Krauhs, G. C. Stocco and R. S. Tobias, *Inorg. Chem.*, 10 (1971) 1365.
76 R. S. Tobias, *Inorg. Chem.*, 9 (1970) 1296.
77 R. G. Pearson, *Inorg. Chem.*, 12 (1973) 712.
78 V. J. Pope and C. S. Gibson, *J. Chem. Soc.*, 91 (1907) 2061.
79 A. Buraway, C. S. Gibson, G. C. Hampson and H. M. Powell, *J. Chem. Soc.*, (1937) 1690.
80 F. H. Brain and C. S. Gibson, *J. Chem. Soc.*, (1939) 762.
81 E. G. Perevalova, K. I. Grandberg, D. A. Lemenovskii and T. V. Baukova, *Izv. Akad. Nauk. SSSR, Ser. Khim.*, (1971) 2077.
82 B. Armer and H. Schmidbaur, *Angew. Chem. Int. Ed.*, 9 (1970) 101.
83 F. Stocco, G. C. Stocco, W. M. Scovell and R. S. Tobias, *Inorg. Chem.*, 10 (1971) 2639.
84 W. M. Scovell, G. C. Stocco and R. S. Tobias, *Inorg. Chem.*, 9 (1970) 2682.
85 W. Beck, W. P. Fehlhammer, P. Pöllman and R. S. Tobias, *Inorg. Chim. Acta*, 2 (1968) 467.
86 M. S. Kharasch and H. S. Isbell, *J. Am. Chem. Soc.*, 53 (1931) 2701.
87 C. S. Gibson and J. L. Simonsen, *J. Chem. Soc.*, (1930) 2531.
88 A. Buraway, C. S. Gibson and S. Holt, *J. Chem. Soc.* (1935) 1024.
89 A. Buraway and C. S. Gibson, *J. Chem. Soc.*, (1936) 324.
90 R. F. Phillips and H. M. Powell, *Proc. Roy. Soc., Ser. A.*, 173 (1939) 147.
91 W. M. Scovell and R. S. Tobias, *Inorg. Chem.*, 9 (1970) 945.
92 M. G. Miles, G. E. Glass and R. S. Tobias, *J. Am. Chem. Soc.*, 88 (1966) 5738.
93 G. E. Glass, W. B. Schwabacher and R. S. Tobias, *Inorg. Chem.*, 7 (1968) 2471.
94 S. J. Harris and R. S. Tobias, *Inorg. Chem.*, 8 (1969) 2259.
95 G. E. Glass, J. H. Konnert, M. G. Miles, D. Britton and R. S. Tobias, *J. Am. Chem. Soc.*, 90 (1968) 1131.
96 H. Schmidbaur and M. Bergfeld, *Inorg. Chem.*, 5 (1966) 2069.
97 M. Peteau-Boisdenghian, J. Meunier-Piret and M. van Meerssche, *Cryst. Struct. Commun.*, 4 (1975) 375.
98 M. Bergfeld and H. Schmidbaur, *Chem. Ber.*, 102 (1969) 2408.
99 D. I. Nichols and A. S. Charleston, *J. Chem. Soc., A* (1969) 2581.
100 C. S. Gibson and W. P. Weller, *J. Chem. Soc.*, (1941) 102.
101 R. V. G. Ewens and C. S. Gibson, *J. Chem. Soc.*, (1941) 109.
102 M. E. Foss and C. S. Gibson, *J. Chem. Soc.*, (1949) 3075.
103 G. E. Glass and R. S. Tobias, *J. Organomet. Chem.*, 15 (1968) 481.
104 E. Rivarola, G. C. Stocco, B. L. Pepe and R. Barbieri, *J. Organomet. Chem.*, 14 (1968) 467.
105 K. S. Murray, B. E. Reichert and B. O. West, *J. Organomet. Chem.*, 61 (1973) 451.

106 H. J. A. Blaauw, R. J. F. Nivard and G. J. M. van der Kerk, *J. Organomet. Chem.*, 2 (1964) 236.
107 R. V. G. Ewens and C. S. Gibson, *J. Chem. Soc.*, (1949) 431.
108 H. Schmidbaur and K. C. Dash, *Chem. Ber.*, 105 (1972) 3662.
109 H. Hagnauer, G. C. Stocco and R. S. Tobias, *J. Organomet. Chem.*, 46 (1972) 179.
110 H. Schmidbaur and K. C. Dash, *J. Am. Chem. Soc.*, 95 (1973) 4855.
111 G. C. Stocco and R. S. Tobias, *J. Am. Chem. Soc.*, 93 (1971) 5057.
112 G. C. Stocco, L. Pellerito and N. Bertazzi, *Inorg. Chim. Acta*, 12 (1975) 67.
113 A. Shiotani and H. Schmidbaur, *Chem. Ber.*, 104 (1971) 2838.
114 C. F. Shaw, J. W. Lundeen and R. S. Tobias, *J. Organomet. Chem.*, 51 (1973) 365.
115 R. S. Nyholm and P. Royo, *J. Chem. Soc., Chem. Commun.*, (1969) 421.
116 R. Uson, P. Royo and A. Laguna, *J. Organomet. Chem.*, 69 (1974) 361.
117 R. W. Baker and P. Pauling, *J. Chem. Soc., Chem. Commun.*, (1969) 745; *J. Chem. Soc., Dalton Trans.*, (1972) 2264.
118 R. Uson, A. Laguna and J. L. Sanjoaquin, *J. Organomet. Chem.*, 80 (1974) 147.
119 M. S. Kharasch and H. S. Isbell, *J. Am. Chem. Soc.*, 53 (1931) 3053; M. S. Kharasch and T. M. Beck, *J. Am. Chem. Soc.*, 56 (1934) 2057.
120 K. S. Liddle and C. Parkin, *J. Chem. Soc., Chem. Commun.*, (1972) 26.
121 P. W. J. de Graaf, J. Boersma and G. J. M. van der Kerk, *J. Organomet. Chem.*, 105 (1976) 399.
122 F. Calderazzo and D. B. Dell'Amico, *J. Organomet. Chem.*, 76 (1974) C59.
123 R. Hüttel and A. Konietzny, *Chem. Ber.*, 106 (1973) 2098.
124 P. K. Monaghan and R. J. Puddephatt, *Inorg. Chim. Acta*, 15 (1975) 231.
125 P. Braunstein, *J. Chem. Soc., Chem. Commun.*, (1973) 851; P. Braunstein and R. J. H. Clark, *Inorg. Chem.*, 13 (1974) 2224.
126 M. McPartlin and A. J. Markwell, *J. Organomet. Chem.*, 57 (1973) C25.
127 M. A. Bennett, K. Hoskins, W. R. Kneen, R. S. Nyholm, P. B. Hitchcock, R. Mason, G. B. Robertson and A. D. C. Towl, *J. Am. Chem. Soc.*, 93 (1971) 4591.
128 M. A. Bennett, K. Hoskins, W. R. Kneen, R. S. Nyholm, R. Mason, P. B. Hitchcock, G. B. Robertson and A. D. C. Towl, *J. Am. Chem. Soc.*, 93 (1971) 4592.
129 (a) A. Buraway and C. S. Gibson, *J. Chem. Soc.*, (1934) 360; (1936) 324. (b) S. Komiya, J. C. Huffman and J. K. Kochi, *Inorg. Chem.*, 16 (1977) 1253.
130 R. J. Puddephatt and P. J. Thompson, *J. Chem. Soc., Dalton Trans.*, (1976) 2091.
131 J. E. Parks and A. L. Balch, *J. Organomet. Chem.*, 71 (1974) 453.
132 J. A. McCleverty and M. M. M. da Mota, *J. Chem. Soc., Dalton Trans.*, (1973) 2571.
133 F. Bonati and G. Minghetti, *J. Organomet. Chem.*, 59 (1973) 403.
134 G. Minghetti, L. Baratto and F. Bonati, *J. Organomet. Chem.*, 102 (1975) 397.
135 F. Bonati and G. Minghetti, *Gazz. Chim. Ital.*, 103 (1973) 373.
136 B. Cetinkaya, P. Dixneuf and M. F. Lappert, *J. Chem. Soc., Chem. Commun.*, (1973) 206.
137 G. Minghetti, F. Bonati and G. Banditelli, *Inorg. Chem.*, 15 (1976) 1718.
138 G. Minghetti and F. Bonati, *Gazz. Chim. Ital.*, 102 (1972) 205; *J. Organomet. Chem.*, 60 (1973) C43; G. Minghetti, F. Bonati and M. Massobrio, *Inorg. Chem.*, 14 (1975) 1974.
139 K. Bartel and W. P. Fehlhammer, *Angew. Chem.*, 86 (1974) 588.
140 L. Manojlovic-Muir, *J. Organomet. Chem.*, 73 (1974) C45.
141 B. Cetinkaya, M. F. Lappert and H. Reinheimer, *J. Chem. Soc., Dalton Trans.*, (1974) 182.

142 R. Hüttel, H. Reinheimer and H. Dietl, *Chem. Ber.*, 99 (1966) 462.
143 R. Hüttel and H. Reinheimer, *Chem. Ber.*, 99 (1966) 2778.
144 P. Tauchner and R. Hüttel, *Chem. Ber.*, 107 (1974) 3761.
145 R. Hüttel, P. Tauchner and H. Forkl, *Chem. Ber.*, 105 (1972) 1; P. Tauchner and R. Hüttel, *Tetrahedron Lett.*, 46 (1972) 4733.
146 R. Hüttel and H. Forkl, *Chem. Ber.*, 105 (1972) 1664, 2913.
147 G. Wittig and S. Fischer, *Chem. Ber.*, 105 (1972) 3542.
148 A. J. Chalk, *J. Am. Chem. Soc.*, 86 (1964) 4733.
149 T. J. Leedham, D. B. Powell and J. G. V. Scott, *Spectrochim. Acta, Part A*, 29 (1973) 559.
150 R. Hüttel, H. Reinheimer and K. Nowak, *Chem. Ber.*, 101 (1968) 3761.

CHAPTER 8

COMPOUNDS WITH GOLD—METAL BONDS

Gold has a strong tendency to form bonds to itself as is testified by the facile decomposition of many gold compounds to metallic gold. However, metal—metal bonds are also present in many gold complexes, and also at high temperature in the gas phase in the Au_2 molecule.

GOLD—GOLD BONDED COMPLEXES

The structure of $[Au(dmg)_2]^+[AuCl_2]^-$, where dmgH = dimethylglyoxime, contains square-planar gold(III) and linear gold(I) ions held together in chains by weak Au—Au interactions with r(AuAu) = 326 pm [1]. Similarly, the gold(I) complex [AuCN(CNMe)] contains linear gold(I) species, but there are also weak Au—Au bonds which lead to formation of two-dimensional polymeric layers as shown in Fig. 8.1. The nearest Au—Au neighbours are separated by 352—372 pm. Although the intermolecular Au—Au interactions are weak, they are strong enough to cause the complex to be insoluble in common solvents [2]. In [AuCl(piperidine)], the linear molecules are arranged in tetrameric units with the gold atoms forming a loose square cluster as shown in Fig. 8.2, with r(AuAu) = 330 pm [3]. Similar weak Au—Au interactions are observed in the structures of $Na_3[Au(S_2O_3)_2] \cdot 2H_2O$ with r(AuAu) = 330 pm and $[ClAuS(Ph)CH_2CH_2S(Ph)\text{-}AuCl]$ with r(AuAu) = 320 pm [4].

Stronger Au—Au bonds are found in gold(I) complexes with bridging ligands. For example, in $\{AuS_2P(O^iPr)_2\}_n$, whose structure is shown in Fig. 3.6, the gold atoms may be considered to have square-planar stereochemistry with two Au—S and two Au—Au bonds forming polymeric chains. Here r(AuAu) is on average 304 pm [5]. Even shorter bonds are found in gold(I) dithiocarbamates. Thus, in

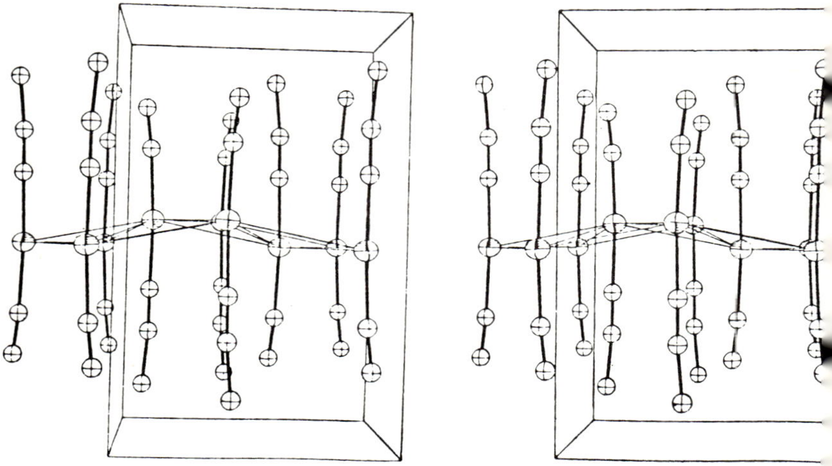

Fig. 8.1. The packing of molecules of cyano(methyl isocyanide)gold(I) in the crystal. (From S. Esperås, *Acta Chem. Scand., Ser. A*, A30 (1976) 527.)

$[(AuS_2CNPr_2)_2]$ the molecule distorts so as to allow a short Au—Au distance of 276 pm as shown in Fig. 3.5. This is shorter than the Au—Au distance in metallic gold (288.4 pm), but even so the bond order has been estimated to be only ca. 0.25 by Raman spectroscopy [6]. Significant gold—gold bonding is also expected in electron-deficient organogold complexes with bridging aromatic groups (see Chapter 7). For example, in $[(C_5H_5)FeC_5H_4(AuPPh_3)_2]^+$ the Au—Au bond distance is 277 pm [7].

The complex $[Au_2Cl_2(\mu\text{-dppm})]$ has a very short Au—Au distance and there even appears to be a gold—gold interaction in the complex $[Au_2Cl_2(\mu\text{-dppm})_2]$ which contains distorted trigonal-planar gold centres as illustrated below [8]:

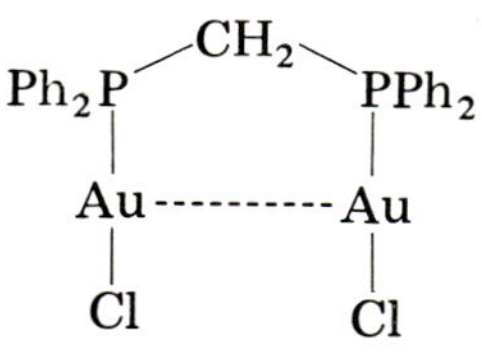

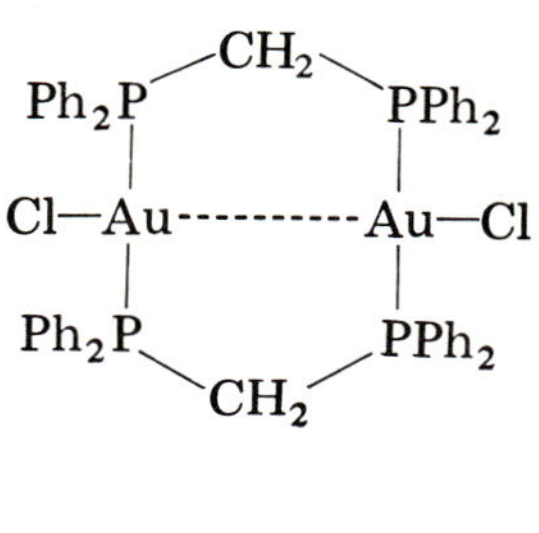

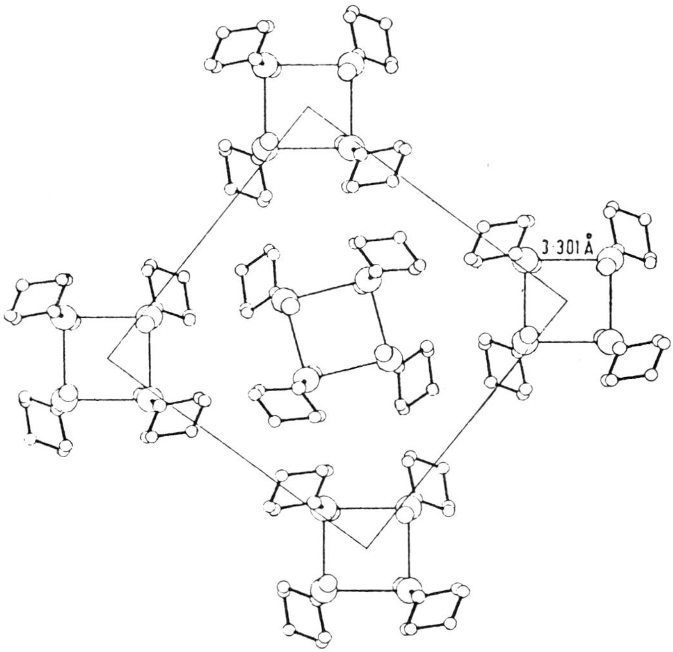

Fig. 8.2. The packing of molecules of chloro(piperidene)gold(I) in the crystal. (From J. J. Guy, P. G. Jones, M. J. Mays and G. M. Sheldrick, *J. Chem. Soc. Dalton Trans.*, (1977) 8.)

In these gold(I) complexes the nature of the gold—gold interactions is not clearly understood. Indeed, some authors have suggested that there is no bonding interaction and that short Au—Au contacts in the solid state are caused by simple packing factors. The frequency of occurrence of short Au—Au distances in different complexes, however, is not consistent with this view. Further theoretical studies are required, however, before a profitable discussion of the nature and strength of the gold—gold bonding in gold(I) complexes is possible.

A gold—gold bond is required in the gold(II) complex $[Au_2Cl_2\{(CH_2)_2PEt_2\}_2]$, whose structure is shown in Fig. 1.1.

Here the Au—Au bond distance is 260 pm [8], and a single Au—Au σ-bond is indicated. Other gold(II) compounds with Au—Au bonds were discussed in Chapter 4.

GOLD CLUSTER COMPOUNDS [9]

Reduction of complexes of the general type [$AuX(PR_3)$] with sodium borohydride in ethanol is found to give a series of cluster compounds of gold, containing several gold atoms in discrete polyhedral arrangements. So far compounds containing six, nine and eleven gold atoms have been characterised by X-ray structure determinations, the only reliable way to characterise these complex compounds. Several wrong formulations for the clusters were originally proposed, based on analytical and conductivity data only. Some of the compounds which have been fully characterised are given in Table 8.1.

The nature of the product of reduction of [$AuX(PR_3)$] by sodium borohydride is largely dependent on the group X. If X is a good ligand for gold, such as a halide or pseudohalide, the major product is the corresponding [$Au_{11}(PR_3)_7X_3$].

TABLE 8.1

Gold cluster compounds

Complex	Colour	m.p./°C	Ref.
[$Au_6\{P(C_6H_4Me\text{-}4)_3\}_6$] [$BPh_4$]$_2$	yellow		10
[$Au_9(PPh_3)_8$][NO_3]$_3$	green	230 d	11
[$Au_9(PPh_3)_8$][BF_4]$_3$	green	240 d	11
[$Au_9(PPh_3)_8$][ClO_4]$_3$	green	238 d	11
[$Au_9\{P(C_6H_4Me\text{-}4)_3\}_8$][$PF_6$]$_3$	green	280 d	11, 12
[$Au_{11}(PPh_3)_7I_3$]	red	146 d	13
[$Au_{11}(PPh_3)_7(SCN)_3$]	red	180 d	13, 14
[$Au_{11}(PPh_3)_7(CN)_3$]	red	173 d	13
[$Au_{11}\{P(C_6H_4F\text{-}4)_3\}_7I_3$]	red	215 d	13, 15
[$Au_{11}\{P(C_6H_4Cl\text{-}4)_3\}_7I_3$]	red	192 d	13, 16

These compounds with eleven gold atoms are typically red [13]. In contrast, if X is a poor ligand for gold such as $[NO_3]^-$, then reduction of $[AuX(PR_3)]$ gives $[Au_9(PR_3)_8]X_3$. These compounds with nine gold atoms are green, and contain the $[Au_9(PR_3)_8]^{3+}$ ion. The nitrate counterion is readily displaced by other large anions such as perchlorate and hexafluorophosphate [11]. Reduction of $[Au(NO_3)\{P(C_6H_4Me\text{-}4)_3\}]$ by sodium borohydride also gives in very low yield the ion $[Au_6\{P(C_6H_4Me\text{-}4)_3\}_6]^{2+}$, which can be isolated as the yellow tetraphenylborate salt [10].

Besides these well-characterised species, several other cluster compounds are known [17, 18]. For example reduction of $[(AuCl)_2(dppe)]$, where dppe = $Ph_2PCH_2CH_2PPh_2$, gives a complex formulated as $Au_6(dppe)_2Cl_2 \cdot 2H_2O$ from which other cluster compounds can be obtained thus:

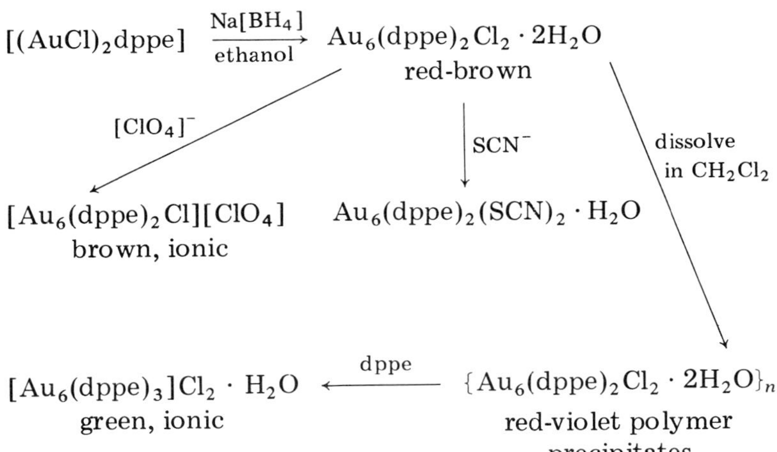

However, no X-ray structural studies on these compounds have been made, and the formulations may not be correct. For example the green $[Au_6(dppe)_3]^{2+}$ may be better formulated as $[Au_9(dppe)_4]^{3+}$ by analogy with the well-characterised green $[Au_9(PR_3)_8]^{3+}$ ions.

The structures of the three structural types $[Au_6L_6]^{2+}$, $[Au_9L_8]^{3+}$ and $[Au_{11}L_7X_3]$ are shown in Fig. 8.3. The ion

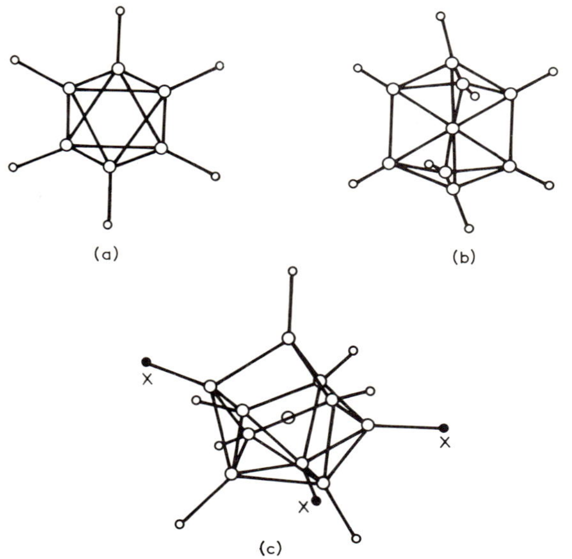

Fig. 8.3. The skeletal structures of the gold cluster complexes. (a) $[Au_6L_6]^{2+}$ in $[Au_6\{P(C_6H_4Me\text{-}4)_3\}_6][BPh_4]_2$; (b) $[Au_9L_8]^{3+}$ in $[Au_9\{P(C_6H_4Me\text{-}4)_3\}_8][PF_6]_3$; (c) $[Au_{11}L_7X_3]$ in $[Au_{11}\{P(C_6H_4F\text{-}4)_3\}_7I_3]$.

$[Au_6L_6]^{2+}$ contains a distorted centrosymmetric octahedron of gold atoms, the distortion causing the Au—Au bond lengths on the edges of two opposite faces of the octahedron to be longer (mean 307.3 pm) than the remainder (mean 296.5 pm). The $[Au_9L_8]^{3+}$ and $[Au_{11}L_7X_3]$ clusters each contain a central gold atom which is surrounded by 8 and 10 gold atoms, respectively. The structure defined by the peripheral gold atoms is not regular, but in each case approximates to an icosahedron with four or two corners missing for the Au_9 and Au_{11} clusters, respectively [12]. The bond distance between the central gold atom and the peripheral gold atoms is shorter than between pairs of peripheral gold atoms in both types of cluster. The mean respective Au—Au distances in $[Au_9L_8]^{3+}$ are 271 and 281 pm, and in $[Au_{11}L_7I_3]$ they are 268 and 298 pm.

Because of the shorter gold—gold bonds to the central gold atom in $[Au_{11}L_7I_3]$, it was proposed that there was no bonding between the peripheral gold atoms. Thus, the compounds were formulated with a central Au^{III} atom bound to ten peripheral Au° atoms [16], or with a central Au^{-VII} atom bound to 10 peripheral Au^{I} atoms [19]. In either case the central gold atom achieves the noble gas electron configuration. The structure with the peripheral gold atoms each with its ligand (L or X) protruding like quills from the central gold atom gave rise to the description "porcupine compounds" for the $[Au_{11}L_7X_3]$ clusters [19]. Recently, however, molecular orbital calculations on the gold cluster compounds have indicated that there is significant bonding between the peripheral gold atoms. It is found that overlap of the 6s orbitals on gold makes the dominant contribution to the bonding. Bonding between the peripheral gold atoms is weaker than between the central and peripheral gold atoms partly as a result of the distorted structures of the Au_9 and Au_{11} clusters. These distorted stuctures arise through the need to create a sufficiently large cavity to accommodate the central gold atom. The cavity in the Au_6 cluster is not large enough to accommodate a central gold atom, but it is suggested that incorporation of a central carbon atom into the $[Au_6L_6]^{2+}$ cluster would lead to a much more stable cluster. In the centred clusters the central gold atom carries a significantly higher positive charge than the peripheral gold atoms [20].

Another complex formulated as a gold cluster complex is formed by the following reaction:

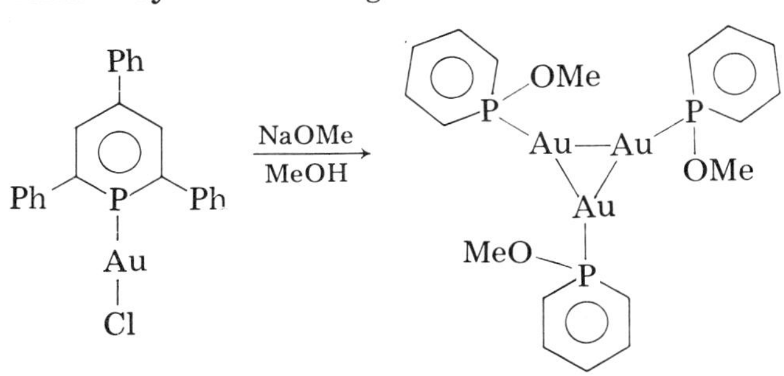

The product is a trimer, and the proposed structure is based on spectroscopic evidence [20b].

ALLOYS AND LIGAND-FREE GOLD—METAL BONDED COMPOUNDS

The simplest gold—metal bonded compounds are the dimeric molecules Au—M which are present in the gas phase above the corresponding Au—M alloy at high temperature and can be identified by mass spectrometry [21]. Some bond dissociation energies for such compounds are given in Table 8.2. More complex species are also formed. For example the compounds Sn_2Au, $SnAu_2$, Sn_2Au_2 and Sn_3Au have been identified. They are thought to contain clusters rather than chains of metal atoms [22].

TABLE 8.2

Bond dissociation energies for some gold—metal bonded molecules [21]

Compound	D(MAu) kJ mol^{-1}	Compound	D(MAu) kJ mol^{-1}
CrAu	211	PdAu	139
MnAu	182	AgAu	199
FeAu	207	AuAu	222
CoAu	218	AlAu	322
NiAu	248	SnAu	241
CuAu	232	CeAu	297

The solubility of other metals in gold to give solid solutions is found to be dependent on the size and number of valence electrons of the metal. The limit of solubility of the B metals is reached when the mean number of valence electrons per atom reaches 1.4. Thus copper with one valence electron is infinitely soluble in gold, but zinc, gallium and germanium with two, three and four valence electrons have maximum solubilities of 31, 12.5 and 3.2%, respectively. Lower solubilities are found if the second metal is significantly different in size from gold [23].

In addition to these solid solutions, phases of distinct chemical composition are also formed in alloys of gold. These frequently occur when the number of electrons per atom (e/a) have the ratio 3/2, 21/13 or 7/4. Examples of 3/2 e/a compounds are AuZn, Au_3Al and Au_5Sn; and examples of 7/4 e/a compounds are $AuZn_3$ Au_5Al_3 and Au_3Sn which all have hexagonal close-packed structures [24].

Many other phases of distinct chemical composition are formed which do not fall into the above pattern. The most common structural types are the CsCl structure found in AuRb, AuSc, AuCe and other rare-earth alloys of gold, and the $MoSi_2$ structure found in $ScAu_2$, $TiAu_2$ and rare-earth alloys such as $GdAu_2$. The alloys formed by copper and silver are often of the same composition as those of gold, and the principles governing their formation are reasonably well understood [23, 24], though a discussion of these aspects is beyond the scope of this book.

The alkali metal compounds MAu become more ionic for the heavier alkali metals M, and this leads to an interesting change in alloy properties from metallic properties when M = Li, Na or K to semiconductor properties when M = Rb or Cs.

COMPOUNDS WITH GOLD-MAIN GROUP METAL BONDS

No compounds with σ-bonds between gold and the elements of Groups I and II of the Periodic Table are known. However, Au—M interactions may be present in the complexes $[(2\text{-}Me_2NCH_2C_6H_4)_4Au_2M_2]$, where M = Li or Cu, and in $[Au_2Zn_2Ph_6]$ whose proposed structures are given in Fig. 7.5. Gold—mercury bonded species have been proposed as short-lived reaction intermediates, but no isolable compounds are known.

A compound with a gold—boron σ-bond is formed by the reaction (P—P = chelate phosphine ligand) [25]:

$$[Co(BPh_2)_2(P\text{—}P)_2] + 2[AuCl(PPh_3)] \rightarrow 2[AuBPh_2(PPh_3)] + [CoCl_2(P\text{—}P)_2]$$

Several more complex derivatives of carboranes are known. A compound with gold bridging between two boron atoms can be prepared from $Na[C_2B_4H_7]$ by the reaction [26]:

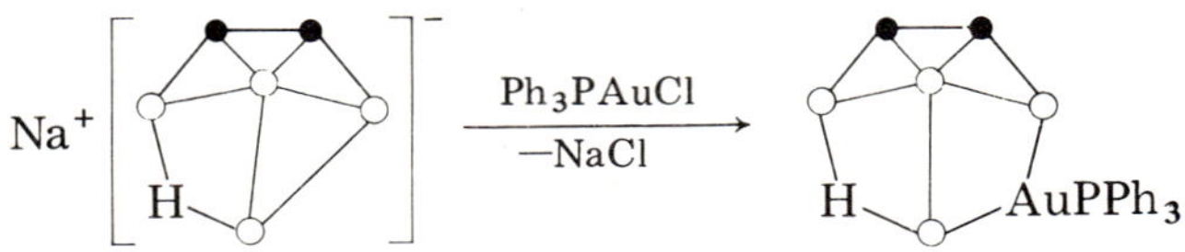

(closed circles represent CH groups, open circles BH groups)

A gold(III)—carborane complex can be prepared by the reaction [27]:

$$Tl_2[7,8\text{-}C_2B_9H_{11}] + [AuBr_2(S_2CNEt_2)] \rightarrow 2TlBr + [Au(7,8\text{-}C_2B_9H_{11})(S_2CNEt_2)]$$

Compounds containing the red anion $[Au(7,8\text{-}C_2B_9H_{11})_2]^-$ can also be prepared and can be reduced electrolytically to

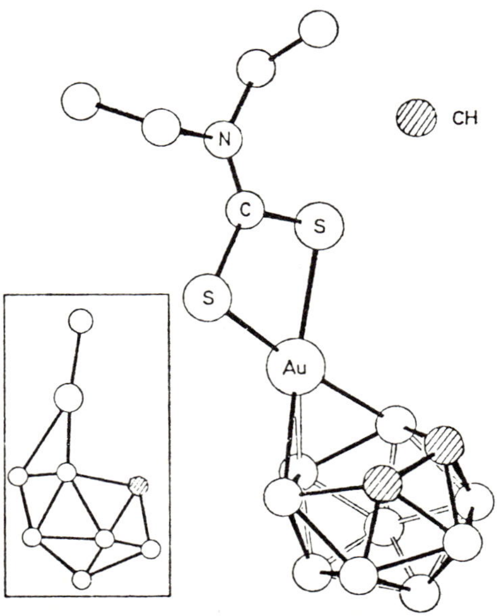

Fig. 8.4. The molecular structure of $3\text{-}(Et_2NCS_2)\text{-}3,1,2\text{-}AuC_2B_9H_{11}$. Hydrogen atoms are omitted. (From H. M. Colqhoun, T. J. Greenhough and M. G. H. Wallbridge, *J. Chem. Soc. Chem. Commun.*, (1976) 1019.)

the blue paramagnetic ion $[Au(7,8\text{-}C_2B_9H_{11})_2]^{2-}$ which is formally a gold(II) complex [28]. These compounds all contain gold atoms unsymmetrically placed above the C_2B_3 face of the carborane, a feature which is thought to be typical of d^8—d^{10} metal ions. The structure of the complex $[Au(7,8\text{-}C_2B_9H_{11})(S_2CNEt_2)]$ is shown in Fig. 8.4.

Some of the compounds containing a bond from gold to a Group IVB element are given in Table 8.3. The silyl and germyl derivatives are formed by metal for metal-exchange reactions (M = Si or Ge):

$$[AuCl(PPh_3)] + Ph_3MLi \rightarrow [Au(MPh_3)(PPh_3)] + LiCl$$

$$2[AuCl(PPh_3)] + (Me_3Ge)_2Hg \rightarrow 2[Au(GeMe_3)(PPh_3)] + HgCl_2$$

Many of these compounds are thermally and photochemically unstable. The best characterised is $[Au(GePh_3)(PPh_3)]$. This reacts with more Ph_3GeLi with displacement of the phosphine ligand to give $[Au(GePh_3)_2]^-$, isolable as the $[Et_4N]^+$ salt [30, 31]. Many reactions which involve cleavage of the Au—Ge bond have been studied, as illustrated in Fig. 8.5.

TABLE 8.3

Compounds with gold main-group-metal σ-bonds

Compound	Colour	m.p./°C	Ref.
Group III B			
$[Ph_3PAuBPh_2]$	white	100	25
Group IV B			
$[Ph_3PAuSiPh_3]$	pale tan	155—160 dec.	29
$[Ph_3PAuGePh_3]$	yellow	185 dec.	30, 31
$[Ph_3PAuGeMe_3]$	buff	126 dec.	31
$[Me_3PAuGePh_3]$		125—130 dec.	30
$[Et_3PAuGePh_3]$		159	31
$Et_4N[Au(GePh_3)_2]$	pale yellow	195—200 dec.	30
$[(Ph_3P)_2AuSnCl_3]$		189	31
$[\{(Me_3SiCH_2)_3P\}_2AuSnCl_3]$	white	217—218	32
$[(Ph_3P)_3AuSnCl_3]$	white	193—195	33

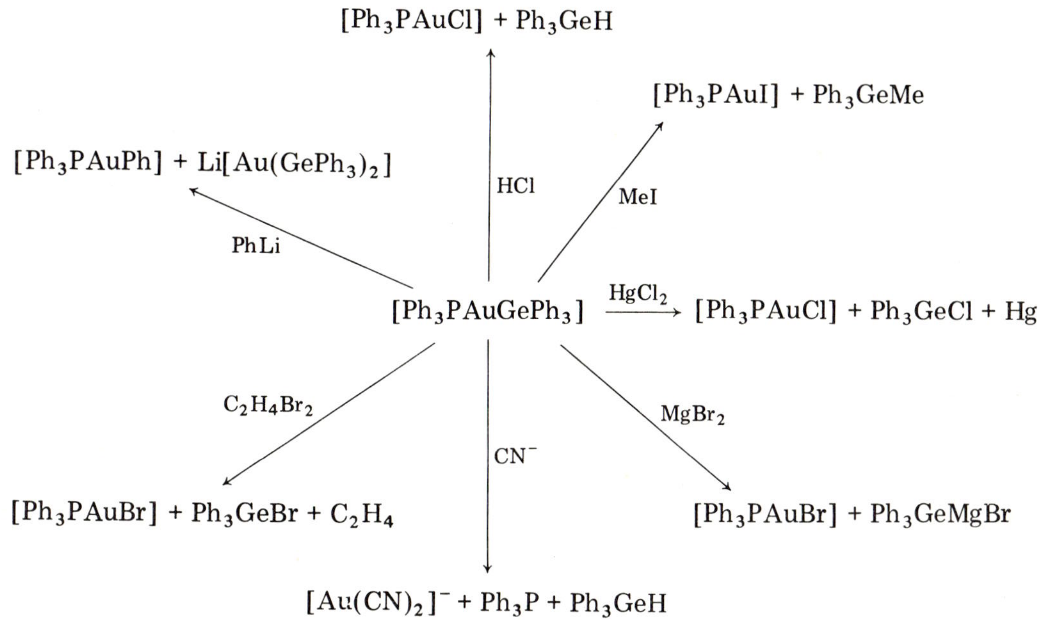

Fig. 8.5. Some reactions of triphenylgermyl(triphenylphosphine)gold(I).

The Au—Sn bonded compounds are usually prepared by insertion of $SnCl_2$ into the AuCl bond [33].

$$[AuCl(PPh_3)] + SnCl_2 + 2PPh_3 \rightarrow [Au(SnCl_3)(PPh_3)_3]$$

The $[SnCl_3]^-$ evidently stabilises higher co-ordination numbers for gold(I) [32, 33], and the linear $[Au(SnCl_3)(PPh_3)]$ is unstable [31]. However, phosphine dissociation from the tetrahedral $[Au(SnCl_3)(PPh_3)_3]$ occurs more readily than from the similar copper(I) and silver(I) analogues [33].

$$[Au(SnCl_3)(PPh_3)_3] \rightleftharpoons [Au(SnCl_3)(PPh_3)_2] + PPh_3$$

COMPOUNDS WITH GOLD-TRANSITION METAL BONDS

A selection of compounds containing a direct bond between gold and a transition element are given in Table 8.4. Some general aspects are discussed below.

TABLE 8.4

Compounds with gold-transition metal σ-bonds

Compound	Colour	m.p./°C	Ref.
Group V A			
$[Ph_3PAuV(CO)_6]$	orange-yellow	134—136 dec.	34, 35
$[Ph_3PAuV(CO)_5(PPh_3)]$	orange	145—150 dec.	35
$[Ph_3PAuNb(CO)_6]$	red-orange	125 dec.	35
$[Ph_3PAuNb(CO)_5(PPh_3)]$	red-orange	150 dec.	35
$[Ph_3PAuTa(CO)_6]$	orange-red	120 dec.	35
$[Ph_3PAuTa(CO)_5(PPh_3)]$	orange-red	160 dec.	35
$[(Ph_3PAu)_3V(CO)_5]$	bright-purple	—	35b
Group VI A			
$[Ph_3PAuCr(CO)_3(\eta\text{-}C_5H_5)]$	pale yellow	155—160 dec.	36
$[Ph_3PAuMo(CO)_3(\eta\text{-}C_5H_5)]$	pale yellow	168—170 dec.	36
$[(PhO)_3PAuMo(CO)_3(\eta\text{-}C_5H_5)]$	pale yellow	75—78	36
$Et_4N[Au\{Mo(CO)_3(\eta\text{-}C_5H_5)\}_2]$			37
$[Ph_3PAuW(CO)_3(\eta\text{-}C_5H_5)]$	pale yellow	201—203 dec.	36, 38

TABLE 8.4—*continued*

Compound	Colour	m.p./°C	Ref.
Group VII A			
$[Ph_3PAuMn(CO)_5]$	pale brown	151 dec.	39—42
$[Ph_3AsAuMn(CO)_5]$	brown	64 dec.	43
$[Ph_3SbAuMn(CO)_5]$	brown		43
$[Ph_3PAuMn(CO)_4(PPh_3)]$	yellow-brown	140 dec.	43
$[Ph_3PAuMn(CO)_4(AsPh_3)]$	light brown	135 dec.	43
$[Ph_3PAuMn(CO)_4(py)]$	light brown	121 dec.	43
$Et_4N[Au\{Mn(CO)_5\}_2]$	—	—	37
$[(Ph_3PAu)_3Mn(CO)_4]$	bright yellow	174 dec.	44
$[Ph_3PAuRe(CO)_5]$	—	—	42
$[(Ph_3PAu)_3Re(CO)_4]$	gold	126 dec.	44
Group VIII			
$[Me_3PAuFe(CO)_3(NO)]$	yellow-red	86—88 dec.	45
$[(MeO)_3PAuFe(CO)_3(NO)]$	yellow-red	50—51 dec.	45
$[Ph_3PAuFe(CO)_3(NO)]$	yellow-red	98—99 dec.	45
$[Ph_3PAuFe(CO)_2(NO)(PPh_3)]$	yellow-red	80 dec.	45
$[Bu_4N][Au\{Fe(CO)_2(\eta\text{-}C_5H_5)\}_2]$			37
cis-$[(Ph_3PAu)_2Fe(CO)_4]$	pale yellow	150 dec.	39
cis-$[(Ph_3PAu)_2Ru(CO)_4]$			46
cis-$[Ph_3PAuRu(CO)_4(SiMe_3)]$	bronze	106—107	46
cis-$[Ph_3PAuRu(CO)_4(GeMe_3)]$	cream	95—105 dec.	47
cis-$[(Ph_3PAu)_2Os(CO)_4]$	cream	155—165 dec.	48
$[Ph_3PAuCo(CO)_4]$	pale brown	112—114	39, 40, 42, 45, 49—51
$[Ph_3PAuCo(CO)_3(PPh_3)]$			50
$[Et_4N][Au\{Co(CO)_4\}_2]$			37
$[Ph_3PAuRh(PF_3)_4]$	white	158—159	52
$[Ph_3PAuIr(PF_3)_4]$	white	161—162	52
$[Ph_3PAuIr(CO)_3(PPh_3)]$	white		53
$[Ph_3PAuPt(PPh_3)_2Cl]$	yellow	115	54
$[Ph_3PAuPt(PPh_3)_2Br]$	orange	130 dec.	54

(a) Preparation from metal carbonyl anions

This is by far the most common method for preparing the transition metal carbonyl derivatives, and its generality is illustrated by the following equations.

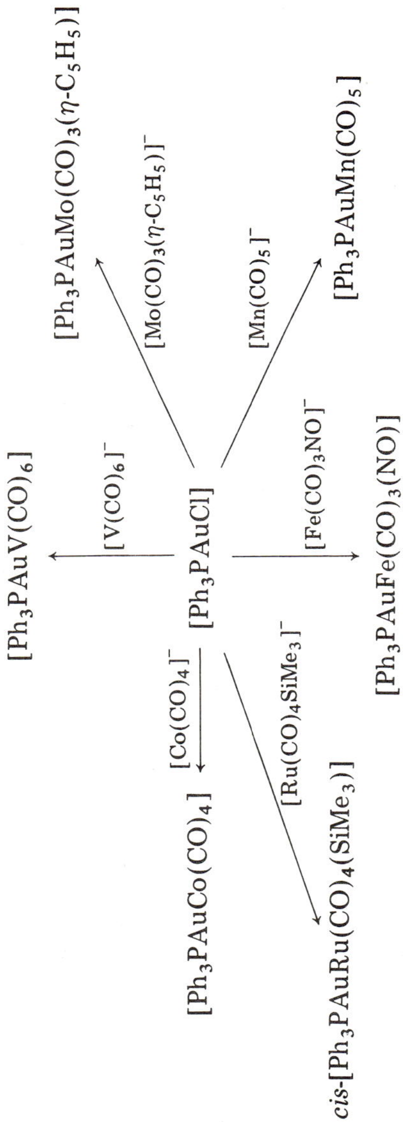
[Ph3PAuCl]
[V(CO)6]⁻
[Ph3PAuV(CO)6]
[Mo(CO)3(η-C5H5)]⁻
[Ph3PAuMo(CO)3(η-C5H5)]
[Mn(CO)5]⁻
[Ph3PAuMn(CO)5]
[Fe(CO)3NO]⁻
[Ph3PAuFe(CO)3(NO)]
[Co(CO)4]⁻
[Ph3PAuCo(CO)4]
[Ru(CO)4SiMe3]⁻
cis-[Ph3PAuRu(CO)4(SiMe3)]

Complexes containing more than one Au—M bond can be formed by variations of this method [55].

$$MeC\{CH_2As(Me_2)AuCl\}_3 + 3Na[Mn(CO)_5] \downarrow MeC\{CH_2As(Me_2)AuMn(CO)_5\}_3 + 3NaCl$$

From $[AuCl_2]^-$, compounds with linear $[AuM_2]^-$ groupings can be prepared [37].

$$[Et_4N][AuCl_2] + 2Na[Mn(CO)_5] \rightarrow [Et_4N][Au\{Mn(CO)_5\}_2] + 2NaCl$$

Compounds with Au_2M and Au_3M groupings can be prepared similarly.

$$2[Ph_3PAuCl] + [Fe(CO)_4]^{2-} \rightarrow cis\text{-}[(Ph_3PAu)_2Fe(CO)_4] + 2Cl^-$$

$$3[Ph_3PAuCl] + [Re(CO)_4]^{3-} \rightarrow [(Ph_3PAu)_3Re(CO)_4] + 3Cl^-$$

Some rhodium and iridium trifluorophosphine derivatives can be prepared by the analogous method [52].

$$[Ph_3PAuCl] + [Rh(PF_3)_4]^- \rightarrow [Ph_3PAuRh(PF_3)_4] + Cl^-$$

(b) Preparation from metal carbonyl hydrides

This method is related to the previous one, and sometimes gives better yields [36]. Thus, the preparation of $[Ph_3PAuMo(CO)_3(\eta\text{-}C_5H_5)]$ from $[Ph_3PAuCl]$ and $[(\eta\text{-}C_5H_5)(CO)_3MoH]$ is more satisfactory than the preparation from the anion $[(\eta\text{-}C_5H_5)(CO)_3Mo]^-$. Reaction of $[Ph_3PAuCl]$ with a mixture of $[Os(CO)_4]^{2-}$ and $[Os(CO)_4H]^-$ in liquid ammonia gives *cis*-$[(Ph_3PAu)_2Os(CO)_4]$ [48]. A complex with a chelating gold ligand can be prepared from $[H_2Fe(CO)_4]$ [56].

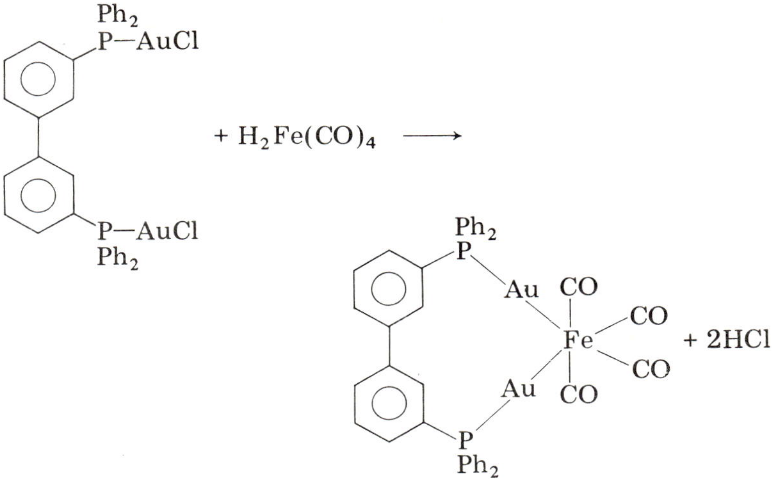

(c) Preparation by metal—metal exchange

Gold-transition metal bonds can be formed by reaction of a gold halide with a trialkyltin derivative of a metal carbonyl.

$$[Ph_3PAuCl] + [Me_3SnCo(CO)_4] \rightarrow [Ph_3PAuCo(CO)_4] + Me_3SnCl$$

$$[Ph_3PAuCl] + [Ph_3SnTa(CO)_6] \rightarrow [Ph_3PAuTa(CO)_6] + Ph_3SnCl$$

The reactions can be carried out under mild conditions and give excellent yields [35, 40].

(d) Preparation by oxidative addition

Reports that the AuCl bond of $[Ph_3PAuCl]$ would undergo oxidative addition to platinum(0) [54] and iridium(I) [57] complexes to give compounds with gold—platinum and gold—iridium bonds have not been confirmed [58]. However, interesting compounds are formed by oxidative addition to $[Os_3(CO)_{12}]$.

$$[Ph_3PAuX] + [Os_3(CO)_{12}] \xrightarrow[\text{or SCN}]{\text{X = Cl, Br, I,}} [Os_3(CO)_{10}(AuPPh_3)X] + 2CO$$

The products have the structure shown in Fig. 8.6 with gold bridging between two osmium atoms, as shown by X-ray and spectroscopic structural studies. It is suggested that oxidative addition to give a chain compound, $Ph_3PAuOs(CO)_4Os(CO)_4$-$Os(CO)_4X$, occurs followed by cyclisation with loss of two carbonyl ligands to give the final product [59].

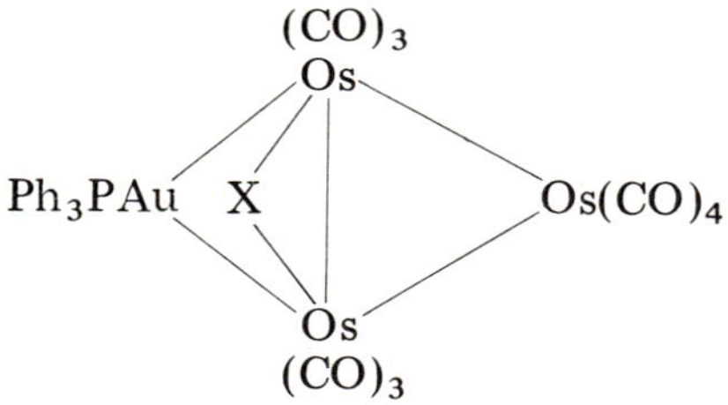

Fig. 8.6. The structure of the complexes $[Os_3(CO)_{10}(AuPPh_3)X]$, X = Cl, Br, I or NCS.

(e) Chemical properties

The gold—metal bond is readily cleaved on reaction with acids or halogens [34, 39, 49].

$$[Ph_3PAuMn(CO)_5] + I_2 \rightarrow [Ph_3PAuI] + [Mn(CO)_5I]$$

$$2[Ph_3PAuV(CO)_6] + I_2 \rightarrow 2[Ph_3PAuI] + [V(CO)_6]$$

$$[Ph_3PAuCo(CO)_4] + HCl \rightarrow [Ph_3PAuCl] + [HCo(CO)_4]$$

The conclusion from the last reaction that the gold—metal bond is polar in the sense $Au^{\delta+}$—$Co^{\delta-}$ is not justified since the mechanism is not known. Thus, reaction might occur by an oxidative-addition reductive-elimination sequence at the gold centre, when factors other than the polarity of the Au—Co bond would influence the course of reaction.

In some cases, the gold—metal bond can be cleaved heterolytically by addition of a ligand [35, 45, 50].

$$[Ph_3PAuTa(CO)_6] + 2Ph_3P \longrightarrow [Au(PPh_3)_3]^+[Ta(CO)_6]^-$$

$$[Ph_3PAuCo(CO)_4] + Ph_2PCH_2CH_2PPh_2 \longrightarrow$$

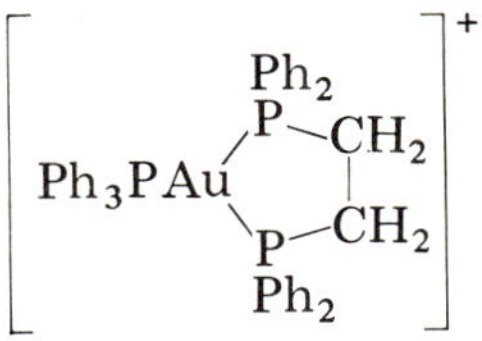

$[Co(CO)_4]^-$

The ease with which this reaction occurs is dependent on the particular complex, particularly on the stability of the metal carbonyl anion. In some cases dissociation occurs on dissolution of the gold—metal bonded complex in methyl cyanide [45].

$$[Ph_3PAuFe(CO)_3(NO)] \xrightarrow{MeCN} [Ph_3PAu(MeCN)_n]^+ [Fe(CO)_3(NO)]^-$$

In other compounds, reaction with a ligand may lead to substitution at the metal carbonyl [42, 43, 45, 60].

$$[Ph_2PAuMn(CO)_5] + Ph_3As \rightarrow cis\text{-}[Ph_3PAuMn(CO)_4(AsPh_3)] + CO$$

At least in some cases, it is thought that the ligand first adds to the gold centre, for example in the reaction:

$$[Ph_3PAuCo(CO)_4] + Ph_3P \rightarrow [Ph_3PAuCo(CO)_3(PPh_3)] + CO$$

There is an induction period before the substitution reaction takes place, and it is suggested that the phosphine first adds to gold perhaps with formation of $[Au(PPh_3)_n][Co(CO)_4]$ followed by rearrangement to the final product [50].

(f) Structure and bonding

The structures of several complexes containing gold-transition metal bonds have been determined by X-ray methods [38, 51, 61, 62]. The complexes derived from $[Ph_3PAuCl]$ contain linear P—Au—M groupings, though significant distortions may occur [62] as in *trans*-[$Ph_3PAuMn(CO)_4\{P(OPh)_3\}$] for which the ∠PAuMn is 165° rather than 180°. The Au—M bond distances are close to or slightly less than the sum of the covalent bond radii perhaps due to polarity or π-bonding in the Au—M bond. For example, it has been argued that the AuW bond length of 270 pm in [$Ph_3PAuW(CO)_3(\eta$-$C_5H_5)$] is shorter than expected for a single bond [38].

In the complexes [$XMn(CO)_5$], it is found that there is a progressive change in the carbonyl stretching frequencies in the infrared spectrum along the series X = Cl, Br, I and $AuPPh_3$ [39, 41—43]. On this basis, it was suggested that gold acts as a pseudohalogen, with electronegativity less than that of iodine, in the transition metal derivatives [39]. However, a more recent analysis indicates that in [$Ph_3PAuMn(CO)_5$] the Ph_3PAu unit acts as a fairly good π-acceptor and a strong σ-donor ligand to manganese, which is the opposite of the halogen—manganese interaction [41].

Finally, the low steric requirement of the linear Ph_3PAu group allows formation of complexes with co-ordination number seven such as [$Ph_3PAuV(CO)_6$] and [$(Ph_3PAu)_3$-$Mn(CO)_4$] and even co-ordination number eight in [$(Ph_3PAu)_3$-$V(CO)_5$] [35, 35b, 44].

REFERENCES

1 R. E. Rundle, *J. Am. Chem. Soc.*, 76 (1954) 3101.
2 S. Esperas, *Acta. Chem. Scand., Ser. A*, A30 (1976) 527.
3 J. J. Guy, P. G. Jones, M. J. Mays and G. M. Sheldrick, *J. Chem. Soc., Dalton Trans.*, (1977) 8.
4 H. Ruben, A. Zalkin, M. O. Faltens and D. H. Templeton, *Inorg. Chem.*, 13 (1974) 1836; M. G. B. Drew and M. J. Riedl, *J. Chem. Soc., Dalton Trans.*, (1973) 52.
5 S. L. Lawton, W. J. Rohrbaugh and G. T. Kokotailo, *Inorg. Chem.*, 11 (1972) 2227.
6 F. J. Farrell and T. G. Spiro, *Inorg. Chem.*, 10 (1971) 1606.
7 V. G. Andrianov, Yu. T. Struchkov and E. R. Rossinskaya, *J. Chem. Soc., Chem. Commun.*, (1973) 338.
8 H. Schmidbaur, J. R. Mandl, A. Frank and G. Huttner, *Chem. Ber.*, 109 (1976) 466; 110 (1977) 1748, 2751.
9 L. Malatesta, *Gold Bull.*, 8 (1975) 48.
10 P. L. Bellon, M. Manassero, L. Naldini and M. Sansoni, *J. Chem. Soc., Chem. Commun.*, (1972) 1035; P. Bellon, M. Manassero and M. Sansoni, *J. Chem. Soc., Dalton Trans.*, (1973) 2423.

11 F. Cariati and L. Naldini, *J. Chem. Soc., Dalton Trans.*, (1972) 2286.
12 P. L. Bellon, F. Cariati, M. Manassero, L. Naldini and M. Sansoni, *J. Chem. Soc., Chem. Commun.*, (1971) 1423.
13 F. Cariati and L. Naldini, *Inorg. Chim. Acta*, 5 (1971) 172.
14 M. McPartlin, R. Mason and L. Malatesta, *J. Chem. Soc., Chem. Commun.*, (1969) 334.
15 P. Bellon, M. Manassero and M. Sansoni, *J. Chem. Soc., Dalton Trans.*, (1972) 1481.
16 V. G. Albano, P. L. Bellon, M. Manassero and M. Sansoni, *J. Chem. Soc., Chem. Commun.*, (1970) 1210.
17 L. Malatesta, L. Naldini, G. Simonetta and F. Cariati, *J. Chem. Soc., Chem. Commun.*, (1965) 212, and (1966) 647; *Coord. Chem. Rev.*, 1 (1966) 255.
18 F. Cariati, L. Naldini, G. Simonetta and L. Malatesta, *Inorg. Chim.-Acta*, 1 (1967) 34, and 315.
19 R. B. King, *Prog. Inorg. Chem.*, 15 (1972) 287.
20 (a) D. M. P. Mingos, *J. Chem. Soc., Dalton Trans.*, (1976) 1163.
(b) H. Kanter and K. Dimroth, *Tetrahedron Lett.*, (1975) 545.
21 K. A. Gingerich and G. D. Blue, *J. Chem. Phys.*, 59 (1973) 185.
22 K. A. Gingerich, D. L. Cocke and U. V. Choudary, *Inorg. Chim. Acta*, 14 (1975) L47.
23 G. V. Raynor, *Gold Bull.*, 9 (1976) 12.
24 G. V. Raynor, *Gold Bull.*, 9 (1976) 50.
25 G. Schmid and H. Nöth, *Chem. Ber.*, 100 (1967) 2899.
26 C. P. Magee, L. G. Sneddon, D. C. Beer and R. N. Grimes, *J. Organomet. Chem.*, 86 (1975) 159.
27 H. M. Colquhoun, T. J. Greenhough and M. G. H. Wallbridge, *J. Chem. Soc., Chem. Commun.*, (1976) 1019.
28 L. F. Warren and M. F. Hawthorne, *J. Am. Chem. Soc.*, 90 (1968) 4823.
29 M. C. Baird, *J. Inorg. Nucl. Chem.*, 29 (1967) 367.
30 F. Glockling and K. A. Hooton, *J. Chem. Soc.*, (1962) 2658.
31 F. Glockling and M. D. Wilbey, *J. Chem. Soc. A*, (1968) 2168.
32 A. T. T. Hsieh, J. D. Ruddick and G. Wilkinson, *J. Chem. Soc., Dalton Trans.*, (1972) 1966.
33 J. A. Dilts and M. P. Johnson, *Inorg. Chem.*, 5 (1966) 2079; R. V. Parish and P. J. Rowbotham, *J. Chem. Soc., Dalton Trans.*, (1973) 37.
34 A. S. Kasenally, R. S. Nyholm, R. J. O'Brien and M. H. B. Stiddard, *Nature (London)*, 204 (1964) 871.
35 (a) A. Davison and J. E. Ellis, *J. Organomet. Chem.*, 23 (1970) Cl; 36 (1972) 113.
(b) J. R. Ellis and M. C. Palazzotto, *J. Am. Chem. Soc.*, 98 (1976) 8264.
36 R. J. Haines, R. S. Nyholm and M. H. B. Stiddard, *J. Chem. Soc. A*, (1968) 46.
37 P. Braunstein and J. Dehand, *J. Organomet. Chem.*, 88 (1975) C24.
38 J. B. Wilford and H. M. Powell, *J. Chem. Soc. A*, (1969) 8.
39 E. C. Coffey, J. Lewis and R. S. Nyholm, *J. Chem. Soc.*, (1964) 1741.
40 E. W. Abel and G. V. Hutson, *J. Inorg. Nucl. Chem.*, 30 (1968) 2339.
41 W. A. G. Graham, *Inorg. Chem.*, 7 (1968) 315.
42 L. M. Bower and M. H. B. Stiddard, *J. Chem. Soc. A*, (1968) 706.
43 A. S. Kasenally, J. Lewis, A. R. Manning, J. R. Miller, R. S. Nyholm and M. H. B. Stiddard, *J. Chem. Soc.*, (1965) 3407.
44 J. Ellis and R. A. Faltynek, *J. Chem. Soc., Chem. Commun.*, (1975) 966; *J. Am. Chem. Soc.*, 99 (1977) 1801.
45 M. Casey and A. R. Manning, *J. Chem. Soc. A*, (1971) 2989.
46 S. A. R. Knox and F. G. A. Stone, *J. Chem. Soc. A*, (1969) 2559.
47 S. A. R. Knox and F. G. A. Stone, *J. Chem. Soc. A*, (1971) 2874.
48 R. D. George, S. A. R. Knox and F. G. A. Stone, *J. Chem. Soc., Dalton Trans.*, (1973) 972.
49 A. Davison, W. McFarlane, L. Pratt and G. Wilkinson, *J. Chem. Soc.*, (1962) 3653.

50 D. A. Brown and R. T. Sane, *Curr. Sci.*, 41 (1972) 877; R. T. Sane and P. R. Kulkarni, *Curr. Sci.*, 43 (1974) 42.
51 T. L. Blundell and H. M. Powell, *J. Chem. Soc. A*, (1971) 1685.
52 M. A. Bennett and D. J. Patmore, *Inorg. Chem.*, 10 (1971) 2387.
53 J. P. Collman, F. D. Vastine and W. R. Roper, *J. Am. Chem. Soc.*, 88 (1966) 5035; 90 (1968) 2282.
54 A. J. Layton, R. S. Nyholm, G. A. Pneumaticakis and M. L. Tobe, *Chem. Ind.*, (1967) 465.
55 A. S. Kasenally, R. S. Nyholm and M. H. B. Stiddard, *J. Am. Chem. Soc.*, 86 (1964) 1884.
56 B. Chiswell and L. M. Venanzi, *J. Chem. Soc. A*, (1966) 901.
57 G. Dolcetti, M. Nicolini, M. Giustiniani and U. Belluco, *J. Chem. Soc. A*, (1969) 1389.
58 C. A. Reed and W. R. Roper, *J. Chem. Soc. A* (1970) 506.
59 R. J. H. Clark, C. W. Bradford, W. van Bronswijk and R. S. Nyholm, *J. Chem. Soc. A*, (1970) 2889.
60 S. Breitschaft and F. Basolo, *J. Am. Chem. Soc.*, 88 (1966) 2702.
61 H. M. Powell, K. Mannan, B. T. Kilbourn and P. Porta, *Proc. 8th Int. Conf. Coord. Chem.*, Vienna, 1964, p. 155.
62 K. A. I. F. M. Mannan, *Acta Crystallogr.*, 23 (1967) 649.

CHAPTER 9

REACTION MECHANISMS IN GOLD CHEMISTRY

The mechanisms of only a few reactions of gold complexes have been investigated in detail. Thus, the reaction mechanisms discussed in this chapter are necessarily rather few and are confined to ligand substitution reactions very largely in gold(III) complexes and to redox reactions.

LIGAND SUBSTITUTION REACTIONS IN GOLD(III) COMPLEXES

(a) The general mechanism of reaction [1]

The displacement of a ligand X, co-ordinated to a square-planar d^8 metal ion, by a second ligand Y is shown for the general case in the equation:

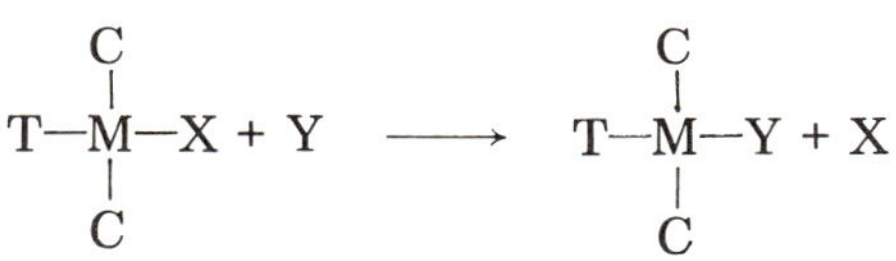

Here T is the *trans*-ligand and C are the *cis*-ligands with respect to X and, as indicated, the reactions take place with retention of stereochemistry at the metal centre. These substitution reactions almost always take place by an associative mechanism, but the observed rate law is usually expressed by the two-term equation:

$$\text{Rate} = -\mathrm{d}(\text{substrate})/\mathrm{d}t = (k_1 + k_2[\mathrm{Y}])[\text{substrate}]$$

The k_2 term is expected and is due to bimolecular attack by Y on the substrate. The k_1 term does not indicate a unimolecular mechanism, but is due to rate-determining displacement of X by solvent, S, by the associative mechanism followed by rapid displacement of solvent by Y.

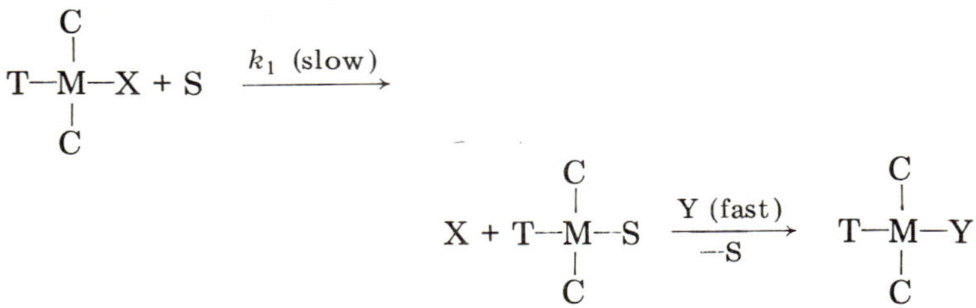

Since the concentration of solvent is large and constant, an apparent first-order term k_1 results. Of course, the relative magnitudes of the k_1 and $k_2[Y]$ terms will be expected to depend on the substrate, the entering group Y and the solvent S.

The stereochemical course of a ligand displacement reaction at a planar complex is illustrated by the equation:

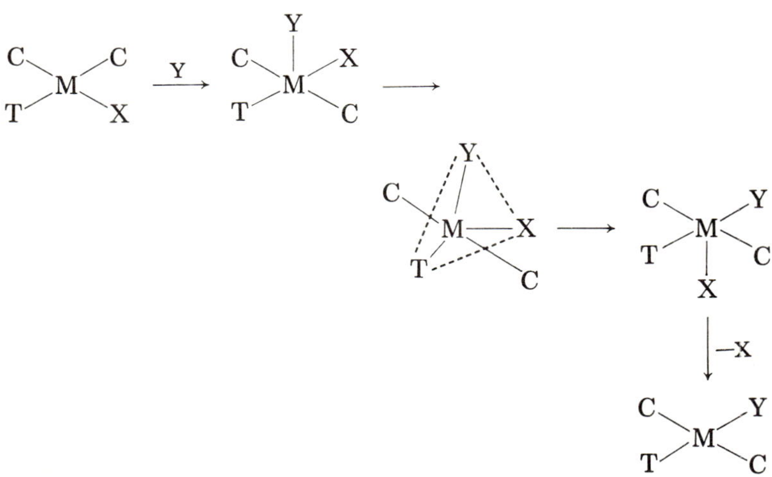

This reaction sequence involves initial interaction with Y to give a square-pyramidal species, rearrangement to an approximately trigonal-bipyramidal species with the *trans*-ligand T, the entering ligand Y and leaving ligand X in the trigonal plane. The ligand X is then lost probably from the axial site of another square-pyramidal species as shown. The retention of stereochemistry at the metal centre follows naturally from this mechanism.

The precise nature of the transition state is very difficult to determine. It is generally thought to be close to the trigonal bipyramidal species, but the structure may vary from case to case. Thus in some rhodium(I) complexes the trigonal bipyramidal complex is formed as a stable intermediate which then slowly loses a ligand.

(b) Evidence for five co-ordinate gold(III) intermediates

The evidence for five-co-ordinate intermediates in ligand substitution reactions of gold(III) complexes is usually obtained from the kinetic order, but in some cases direct evidence for association has been obtained. In the stable complexes such as $[AuCl(diphos)_2]^{2+}$,

diphos = 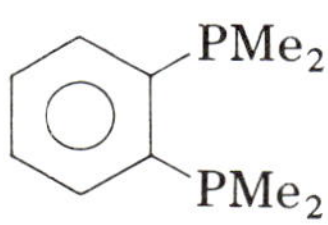

,

or $[AuCl_3(2,2'\text{-biquinolyl})]$, whose structure is given in Fig. 5.1, gold(III) shows a preference for square-pyramidal stereochemistry [2, 3]. Six-co-ordinate tetragonal complexes such as $[AuCl_2(diphos)_2]^+$ are also known [2].

In a few cases, stable five-co-ordinate intermediates in substitution reactions have been identified. For example, the reaction of thiocyanate with $[AuCl_4]^-$ rapidly gives an intermediate, C, which subsequently breaks down to give $[AuCl_3(SCN)]^-$, which in turn reacts rapidly with more thiocyanate to give $[Au(SCN)_4]^-$.

$$[AuCl_4]^- + SCN^- \underset{k_{-1}}{\overset{k_1}{\rightleftharpoons}} C \xrightarrow{k_2} [AuCl_3(SCN)]^- + Cl^-$$

$$[AuCl_3(SCN)]^- + 3SCN^- \xrightarrow{\text{fast}} [Au(SCN)_4]^-$$

Formation of C takes place therefore as a fast pre-equilibrium step, and the equilibrium constant $K = k_1/k_{-1} = 350$ l mol^{-1} has been determined. Since $k_1 = 3.33 \times 10^3$ l mol^{-1} s^{-1} and $k_2 = 10.4$ s^{-1} under the experimental conditions used, decomposition of C, which is presumed to be the five-co-ordinate $[AuCl_4(SCN)]^-$, is the slow step in the reaction sequence [4]. Similarly a long-lived five-co-ordinate intermediate has been detected in the reaction of PhC(=S)NHR with $[AuCl_4]^-$ to give finally $[AuCl_3\{S{=}C(Ph)NHR\}]$ and chloride ion [5]. It is difficult to determine the stereochemistry of these five-co-ordinate intermediates since they cannot be isolated, but they are expected to have square-pyramidal rather than trigonal-bipyramidal stereochemistry.

(c) The role of the entering group in determining the rate of substitution

In general, the dependence of the rate of ligand substitution on the nature of the entering group is related to the softness or polarisability of the group rather than on its basicity. This is illustrated by the substitution reactions of $[Au(dien\text{-}H)Cl]^+$.

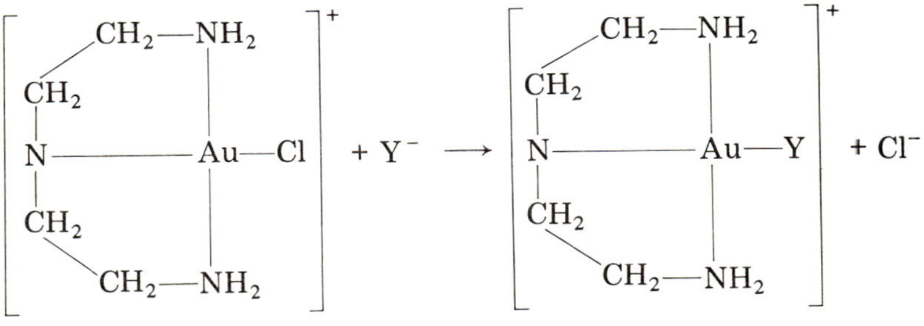

The rate varies as $Y^- = I^- > SCN^- > Br^- > N_3^- > OH^-$. The reactions followed the usual rate law, rate = $\{k_1 + k_2[Y^-]\}$ $[Au(dien\text{-}H)Cl]^+$, but the k_1 term was negligibly small for many of the reactions and the values of k_2 showed a strong dependence on the nature of Y^-. Thus the gold(III) centre has a high discriminating power for the softer entering groups. This discriminating power of gold(III) is considerably greater than that of platinum(II) and suggests a higher degree of bond-making in the transition state in substitution reactions of square-planar gold(III) compared with platinum(II) complexes [6]. This effect was first found in the exchange reaction of radioactively labelled chloride with $[AuCl_4]^-$ in aqueous solution, when it was found that the ratio of k_2/k_1 was much greater for $[AuCl_4]^-$ than for $[PtCl_4]^{2-}$. Thus gold(III) discriminates more than platinum(II) between chloride and solvent as entering groups [7].

Similar conclusions have been reached in several more recent studies, Thus, the rate of displacement of chloride in $[AuCl_4]^-$ by Y follows the series Y = $I^- > SCN^- >$ pyridine $> N_3^- > Br^- > NO_2^-$. In these reactions the ratio $k_2(I^-)/k_2(Br^-)$ was 600 for substitution in $[AuCl_4]^-$ but only 3.6 for $[PtCl_4]^{2-}$, again indicating the greater discriminating power of gold(III) for soft entering ligands [8]. A similar effect is seen in the displacement of pyridine derivatives, (am), from $[AuCl_3(am)]$ by the ligands dibenzylsulphide, bromide and chloride. Here, it is also found that, for a given entering group, the second-order rate constant k_2 is dependent on the basicity of the leaving group, (am), as measured by the pK_a of the conjugate acid $(amH)^+$. Thus, provided there are no differences in steric effects of the amine, a graph of log k_2 *vs.* pK_a gives a straight line whose slope depends on the entering group. The slopes are −0.68, −0.37 and −0.2 for entering groups Cl^-, Br^- and $(PhCH_2)_2S$, respectively. The slope is expected to depend on the extent of bond-breaking in the transition state, and hence it can be argued that the extent of bond-making rather than bond-breaking in the transition state follows the series $(PhCH_2)_2S > Br^- > Cl^-$. This is the order of decreasing softness of the entering groups [9, 10].

Although in general, the softness of the entering group is the dominant factor in determining the rate of ligand substitution in gold(III) complexes, if this factor is kept constant the ligand basicity can be shown to influence the rate. This has been demonstrated elegantly by study of the rates of the following reactions in methanol where (am) is pyridine or a substituted derivative of pyridine.

$[AuCl_4]^- + (am) \rightarrow [AuCl_3(am)] + Cl^-$

Provided that the amines have similar steric effects, then a linear relationship is found between log k_2, where k_2 is the second-order rate constant, and the pK_a of the conjugate acid of the amine, the rate increasing with the basicity of the amine. The slope of the graph, α, is a measure of the discriminating power of $[AuCl_4]^-$ for the amines, and is expected to increase as the extent of bond formation in the transition state increases [11]. Similar relationships are obtained for substitution in cationic gold(III) complexes:

$[Au(bipy)Cl_2]^+ + (am) \rightarrow [Au(bipy)(am)Cl]^{2+} + Cl^-$

The values of α for some of these reactions are given in Table 9.1. It can be seen that gold(III) discriminates more than does platinum(II) between the amine ligands, indicating again a greater degree of bond formation in the transition state for gold(III). For the reactions with cationic gold(III) complexes $[Au(N\frown N)Cl_2]^+$ the discrimination increases markedly in the series N—N = $Me_2NCH_2CH_2NMe_2$ < 1,10-phenanthroline < 2,2′-bipyridine < 5-nitro-1,10-phenanthroline which is also the order of decreasing basicity of the bidentate ligands. Thus, it seems that the complexes with the

TABLE 9.1

Discrimination factors (α) for reactions of amines with square-planar complexes

Complex	Solvent	Leaving group	α
$[Pt(bipy)Cl_2]$	MeOH	Cl^-	0.06
$[Au(tmen)Cl_2]^+$	Acetone	Cl^-	0.14
$[AuCl_4]^-$	Acetone or MeOH	Cl^-	0.15
$[Au(phen)Cl_2]^+$	Acetone	Cl^-	0.22
$[Au(bipy)Cl_2]^+$	Acetone	Cl^-	0.46
$[Au(5\text{-}NO_2phen)Cl_2]^+$	Acetone	Cl^-	0.89

less basic ligands have a higher effective positive charge on the gold atom, and hence bond-making takes place to a greater degree in the transition state [12].

(d) The role of the leaving group in determining the rate of reaction

This aspect has been investigated by studies of the kinetics of the reactions:

$[AuCl_3(am)] + Y^- \rightarrow [AuCl_3(Y)]^- + am$

trans-$[Au(CN)_2Cl(am)] + Y^- \rightarrow$ *trans*-$[Au(CN)_2ClY]^- + am$

The ratio of the observed second-order and first-order rate constants measures the discrimination of gold(III) for Y^- rather than the solvent, and it was found that a graph of $\log_{10} k_2/k_1$ *vs.* pK_a of the conjugate acid of the amine, (am), gave linear correlations when Y^- was Br^- or Cl^- as shown in Fig. 9.1. The more basic amines give stronger Au—N bonds and the rates are then lower. Thus it seems that bond-breaking occurs to a significant extent in the transition state, this being less for bromide as entering group (slope −0.35) than with chloride (slope −0.61) in the reactions with $[AuCl_3(am)]$.

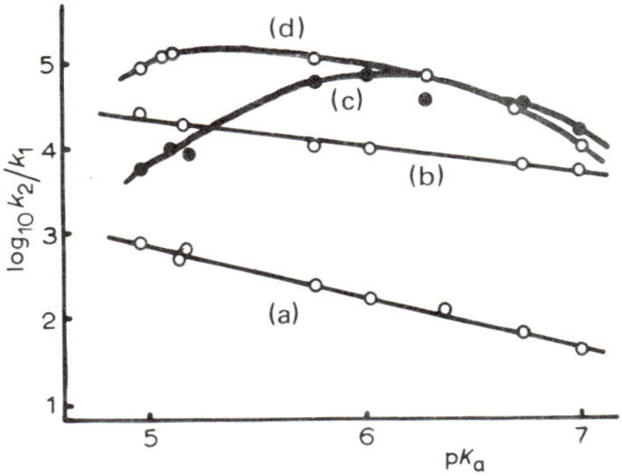

Fig. 9.1. The dependence of relative reactivity on the pK_a of derivatives of pyridine (am) displaced from $[AuCl_3(am)]$ by various entering groups. (a) Cl^-; (b) Br^-; (c) N_3^-; (d) NO_2^-.

The non-linear relationship for NO_2^- and N_3^- entering groups is thought to arise because these are "biphilic" groups, being σ-donor and π-acceptor ligands. Thus the more basic amines enhance the π-bonding between gold(III) and the entering ligands, thus leading to faster reaction rate, but this effect is counteracted by the increasing difficulty in cleaving the stronger Au—N bond. The result is that the rate reaches a maximum with the basicity of the amine and then decreases [9, 13].

(e) Steric effects on the reaction rates

The importance of steric effects on the rate of ligand substitution reactions in gold(III) complexes has been demonstrated by studying the rate of displacement of chloride by bromide in $[Au(dien\text{-}H)Cl]^+$, and in derivatives obtained by substitution of terminal NH groups in the ligand, dien = $H_2NCH_2CH_2NHCH_2CH_2NH_2$, by alkyl groups. Some results are given in Table 9.2 and show very clearly that, as the bulk of the substituents on the ligand increases, both the k_1 and k_2 terms decrease markedly. This is clearly due to the bulky substituents blocking access of the entering ligand (either bromide or water) to the gold centre [14, 15].

Steric effects can also be observed in reactions of pyridine derivatives (am) with $[AuCl_4]^-$ or $[Au(bipy)Cl_2]^+$.

$$[Au(bipy)Cl_2]^+ + (am) \rightarrow [Au(bipy)(am)Cl]^{2+} + Cl^-.$$

TABLE 9.2

First- and second-order rate constants for bromide substitution in $[Au(am\text{-}H)Cl]^+$ in water (pH 6.8) at 25 °C

Complex	k_1/s^{-1}	$k_2/M^{-1}\ s^{-1}$
$[Au(dien\text{-}H)Cl]^+$	0.6	90
$[Au(Me_2dien\text{-}H)Cl]^+$	0.20	44
$[Au(Me_4dien\text{-}H)Cl]^+$	0.062	0.76
$[Au(Et_2dien\text{-}H)Cl]^+$	0.023	3.8
$[Au(Et_4dien\text{-}H)Cl]^+$	0.00012	0.0085

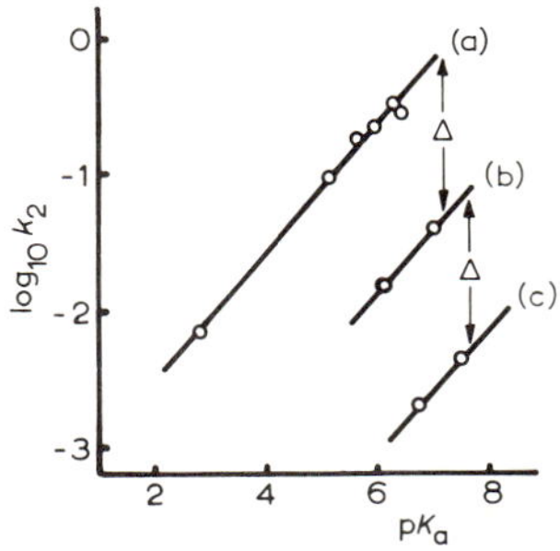

Fig. 9.2. Linear free energy relationships for displacement of chloride from $[Au(bipy)Cl_2]^+$ by derivatives of (a) pyridine; (b) 2-methylpyridine and (c) 2,6-dimethylpyridine.

Thus a graph of $\log_{10} k_2$ *vs.* pK_a of $(amH)^+$ gives three straight lines as illustrated in Fig. 9.2. The upper line refers to 3- or 4-substituted pyridine derivatives, the centre line to similar derivatives of 2-methylpyridine and the bottom line to derivatives of 2,6-dimethylpyridine. The significance of the slope α was discussed in Section (c), and the separation Δ between the lines gives a measure of the steric retardation due to the *ortho*-methyl substituents [11, 12]. Interestingly, the rate of the reverse reaction involving displacement of amine from $[AuCl_3(am)]$ by Cl^- is not affected by substituents in the *ortho*-position. This can be understood if bond-formation is essentially complete in the transition state since steric effects in $[AuCl_3(am)]$ and in the trigonal bipyramidal intermediate will then be similar but considerably greater than in $[AuCl_4]^-$ as illustrated:

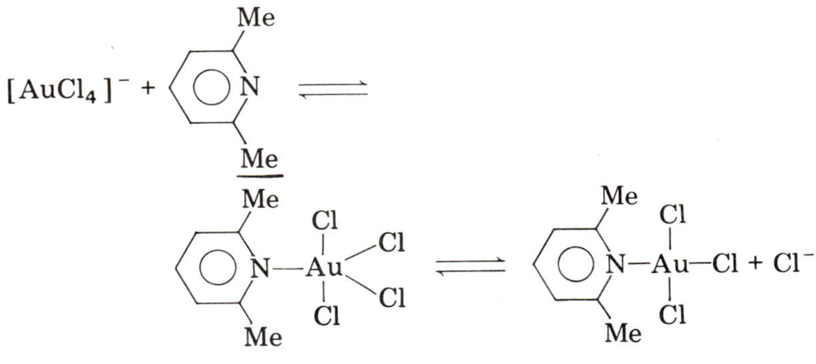

With the bulkier dibenzylsulphide entering group, steric effects due to the amine were however observed [9—12].

$$[AuCl_3(am)] + (PhCH_2)_2S \rightarrow [AuCl_3\{S(CH_2Ph)_2\}] + am$$

(f) Solvent effects and the trans-effect

The effect of the solvent on the rate of reaction arises through differences in the relative solvation of the ground and the transition state. The rate of the following reaction in both forward and reverse directions was studied in various alcohols (am = pyridine derivative).

$$[AuCl_3(am)] + Cl^- \rightleftharpoons [AuCl_4]^- + (am)$$

In the forward direction the rate follows the series $^{i}PrOH > {}^{n}PrOH > EtOH > MeOH > PhCH_2OH$, that is the least polar alcohols give the highest rates. This is clearly due to the five-co-ordinate transition state being more weakly solvated than the reagents, probably as a result of the partial loss of the high solvation energy of the chloride ion. The opposite solvent effect was observed from the reverse reaction indicating an increase in solvation on going from $[AuCl_4]^-$ and (am) to the transition state, perhaps arising from incipient formation of the strongly solvated chloride ion [16].

The *trans*-effect may be defined as the ability of a ligand to labilise the ligand *trans* to itself. Rather like the solvent effect, the *trans*-effect may arise through the ligand weakening the bond *trans* to itself in the ground state of the molecule and hence making it easier to displace, or by stabilising the five-co-ordinate transition state and hence lowering the activation energy. It is the latter effect which necessitates a distinction between the *trans*-effect and *trans*-influence. The *trans*-effect has proved to be a very valuable concept in studying substitution reactions in square-planar platinum(II) complexes, but similar applications for gold(III) complexes have not been developed as yet. For this reason, detailed discussion will not be given here, but only a single example.

The substitution of chloride by pyridine in $[AuCl_3(PPh_3)]$ is too fast to be studied by stopped-flow techniques, whereas similar substitution reactions in $[AuCl_4]^-$, $[AuCl_3(py)]$ or $[AuCl_3(SR_2)]$ are slow enough to be studied without difficulty. This high reactivity is then due to the high *trans*-effect exerted by the triphenylphosphine ligand. The structure of the complex has been determined and shows that the Au—Cl bond *trans* to phosphorus is considerably longer than the other two, indicating a high *trans*-influence of Ph_3P. In addition, the molecule is considerably distorted from square-planar stereochemistry, so that the angle between the *trans*-ligands $<$PAuCl is 170° rather than 180° expected for a regular square-planar complex. Formation of the five-co-ordinate transition state may be expected to be relatively easy therefore. In this case then, the high *trans*-effect of Ph_3P may operate partly through the ground state bond-weakening effect and partly by stabilisation of the transition state [17].

(g) The intimate mechanism of reaction

In Fig. 9.3 some possible reaction profiles for substitution in square-planar complexes are shown. They vary from (a) where the trigonal bipyramidal species is the transition state and a concerted mechanism operates, to (e) where the trigonal bipyramidal intermediate is stable enough to be isolated and the bond-making and bond-breaking take place as separate steps. In (b), the trigonal bipyramidal species is a short-lived intermediate and the transition states are distorted forms of the trigonal bipyramid, while (c) and (d) represent similar situations in which bond-making or bond-breaking, respectively, are rate-determining.

Substitution reactions in gold(III) complexes are always faster than in platinum(II) complexes, and the gold(III) centre discriminates more than platinum(II) between both entering and leaving groups, as discussed in earlier sections. Thus, both bond-making and bond-breaking take place to a greater extent in the transition state for gold(III) compared to platinum(II), suggesting that the reaction profile is close to curve (a) in Fig. 9.3. and that the substitution is a concerted process.

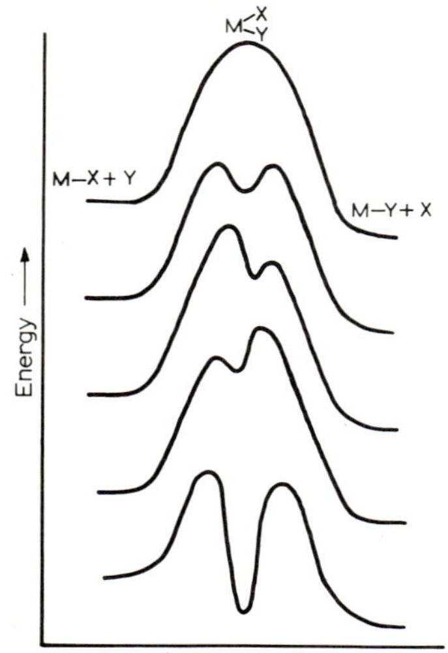

Fig. 9.3. Reaction profiles for ligand substitution reactions of square-planar complexes.

This conclusion is not invalidated by the observation of stable five-co-ordinate intermediate species in some substitution reactions [4, 5], since such intermediates are most likely to have square-pyramidal rather than trigonal-bipyramidal stereochemistry.

Some intramolecular reactions probably occur by the same type of mechanism, though the reactions may be much faster. For example, in the complex $[AuCl_2(pz_3BH)]$, where pz_3BH is tripyrazolylborate and whose structure contains only two-co-ordinated pyrazolyl groups as shown in Fig. 5.2, the co-ordinated and free pyrazolyl groups exchange rapidly on the n.m.r. time scale at room temperature. An intermediate or transition state with all three pyrazolyl groups co-ordinated to a trigonal-bipyramidal gold(III) centre is considered likely [18]. Similarly, *cis*-$[AuMe_2Cl(1,8\text{-naphthyridine})]$ has only one nitrogen atom of the naphthyridine co-ordinated in the ground state, but the rapid exchange of the gold atom between the two nitrogen donors of the ligand probably occurs through a five-co-ordinate intermediate in which the naphthyridine ligand is bidentate [19].

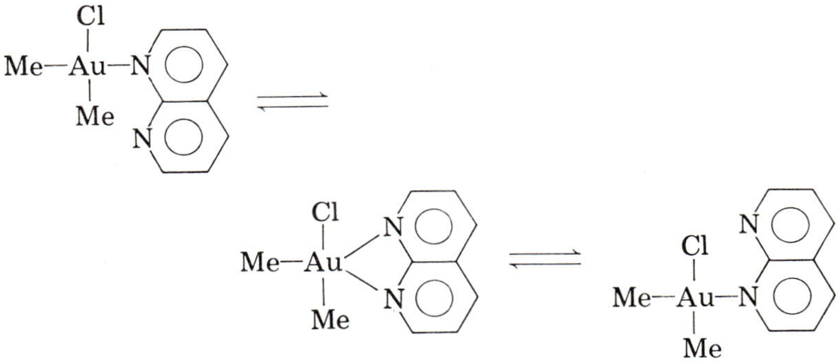

(h) Displacement of chelate ligands from gold(III)

A number of studies indicate that bidentate nitrogen ligands can be displaced from gold(III) by halide in acidic solution. The mechanism may be similar to that for displacement of unidentate ligands, but aided by protonation of the intermediate complex with a unidentate ligand. This is found for N,N,N′,N′-tetramethylethylenediamine complexes (N—N = $Me_2NCH_2CH_2NMe_2$) [20].

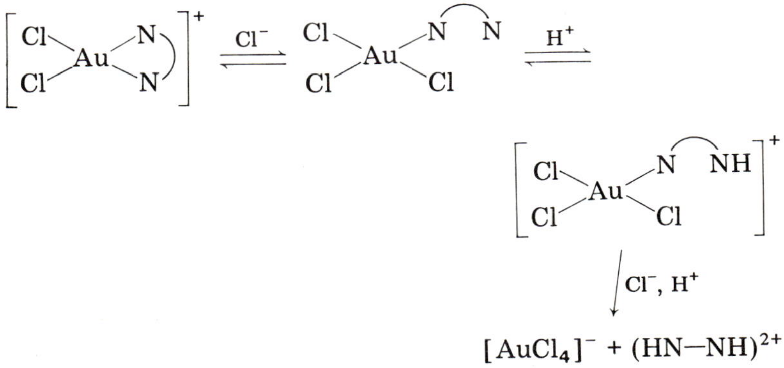

The similar reaction of $[AuCl_2(bipy)]^+$ gives a kinetic expression with a higher order in $[Cl^-]$, and it is thought that part of the reaction proceeds by way of a six-co-ordinate gold(III) intermediate (N—N = 2,2′-bipyridine) [21].

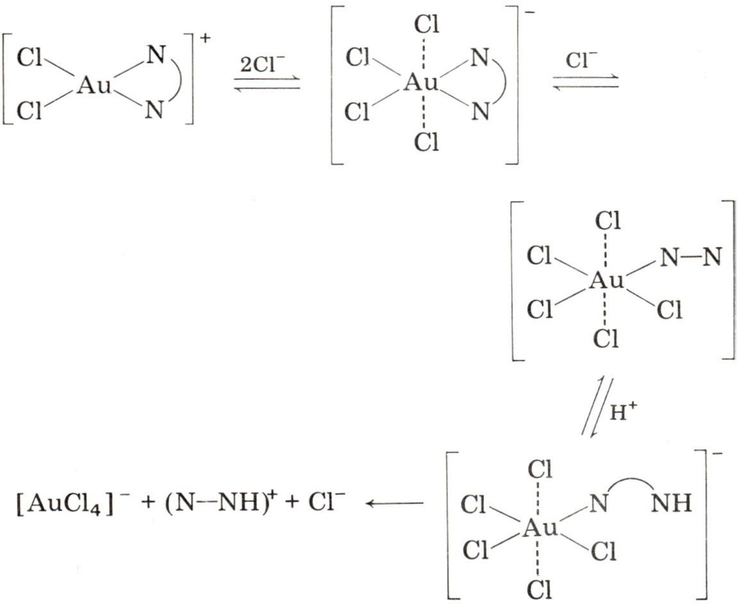

A similar six-co-ordinate species is implicated in the displacement of the chelate ligand by chloride from $[AuCl_2(5\text{-nitro-}1,10\text{-phenanthroline})]^+$. However, in this case the reaction is considerably slower since the rigid phenanthroline ligand cannot readily act as a unidentate ligand, and so its displacement must occur in one step [22].

The tridentate ligand, dien = $H_2NCH_2CH_2NHCH_2CH_2NH_2$, is displaced from the complex $[AuCl(dien)]^{2+}$ in a stepwise manner, with intermediates in which the ligand is bidentate and then unidentate being implicated [23].

The complex $[AuMe_2(acac)]$ has been shown to exchange the acetylacetonato ligand with free acetylacetone, the reaction occurring very readily. This may be due to the high *trans*-effect of the methyl groups but the detailed mechanism is not known [24].

SUBSTITUTION REACTIONS OF GOLD(I) COMPLEXES

Ligand substitution reactions of linear gold(I) complexes generally occur rapidly, but there has been little detailed study of the mechanism. However, the exchange of tertiary phosphine ligands in methyl(tertiary phosphine)gold(I) complexes has been studied by the n.m.r. technique, and an associative mechanism is proposed (L,L′ = tertiary phosphine):

$$\text{Me—Au—L} + \text{L}' \rightleftharpoons \text{Me—Au}\begin{matrix}\diagup \text{L} \\ \diagdown \text{L}'\end{matrix} \rightleftharpoons \text{Me—Au—L}' + \text{L}$$

The intermediate is presumably a trigonal-planar species analogous to the stable trigonal-planar $[AuCl(PPh_3)_2]$. It is interesting that trimethyl(tertiary phosphine) gold(III) complexes undergo a similar exchange with free phosphine ligands, but these reactions are much slower than for the gold(I) complexes. Evidently, the activation energy for formation of a five-co-ordinate gold(III) intermediate is considerably higher than for the trigonal gold(I) species [25].

Before finishing this discussion of nucleophilic substitution reactions at gold(I) and gold(III) centres, it should be pointed out that the ease with which these reactions occur can be important in some reactions which are catalysed by gold complexes. The catalysed reaction of alcohols with isocyanides was discussed in Chapter 7, and involves substitution in carbene complexes of gold(I). The hydrolysis of N-cyclohexyl methylthiobenzimidate, $PhC(SMe) = NC_6H_{11}$ to $PhCONHC_6H_{11}$ is catalysed by $[AuCl_4]^-$, and the reaction has been shown to involve rapid substitution at gold(III) to give

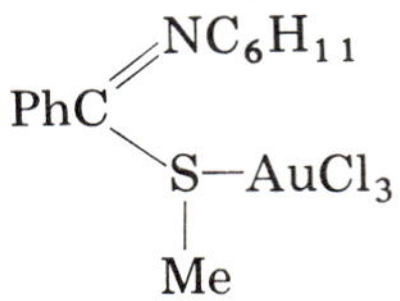

followed by decomposition of this complex to give the final product [26]. Oxidation of alkenes by gold(III) complexes also involves substitution and formation of reactive alkene—gold(III) intermediates.

ELECTROPHILIC SUBSTITUTION REACTIONS

Substitution reactions of inorganic gold complexes generally involve attack of a nucleophile at the gold centre, but substitution in organogold compounds is dominated by electrophilic attack which may occur initially, either at the metal or carbon centre. It is useful to compare some of these reactions whose mechanisms have been studied with the nucleophilic substitutions described earlier.

If we consider briefly the general case in which an alkyl-metal bond, R—M, is cleaved by an electrophilic reagent $E^{\delta+}$—$X^{\delta-}$(e.g., H—Cl), then a number of reaction mechanisms are possible [27]. Most commonly, the reactions are bimolecular and proceed with retention of configuration at carbon if the alkyl group R is optically active. There are three limiting mechanisms which are consistent with these conditions. In two of these the electrophile attacks the carbon atom of the alkyl group, and the nucleophile may simultaneously attack the metal centre giving a four-centre transition state [S_E2 (cyclic) mechanism] or not [S_E2 (open) mechanism]. These are illustrated respectively in the equations:

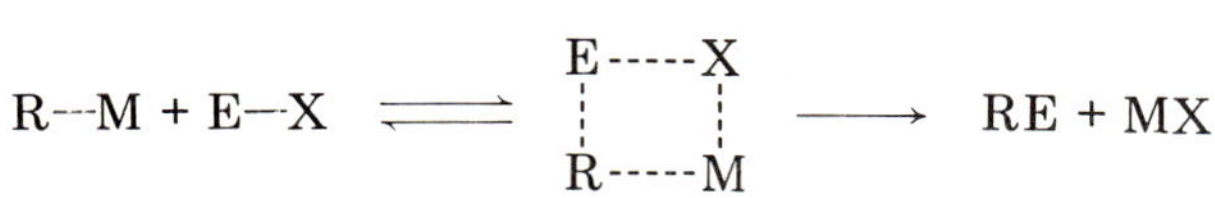

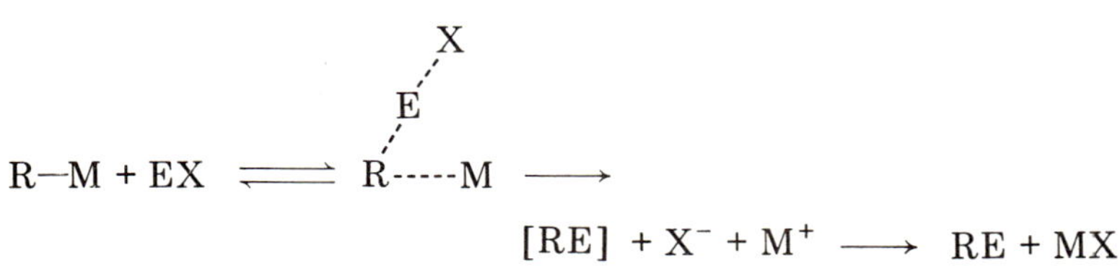

In the case of organometallic compounds which readily undergo oxidative addition reactions, EX may first attack the metal centre followed by reductive elimination to give the final products.

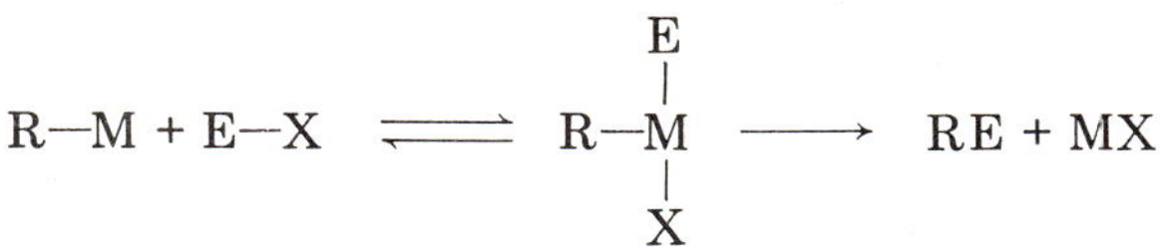

A priori, we might expect this mechanism in reactions of alkylgold(I) complexes which readily undergo oxidative addition, but not for alkylgold(III) complexes which are not oxidised readily. Obviously, therefore, comparison of the reactions of alkylgold(I) and alkylgold(III) complexes with electrophilic reagents should be of particular interest.

(a) Reactions with protic reagents

One of the gold—carbon bonds of trialkylgold complexes $[R_3AuL]$ is cleaved by acids, HX, to give the alkane RH and *cis*-$[R_2AuXL]$. One of the mutually *trans* methyl groups is displaced selectively from $[AuMe_3(PPh_3)]$ as shown by the labelling experiments ($L = PPh_3$).

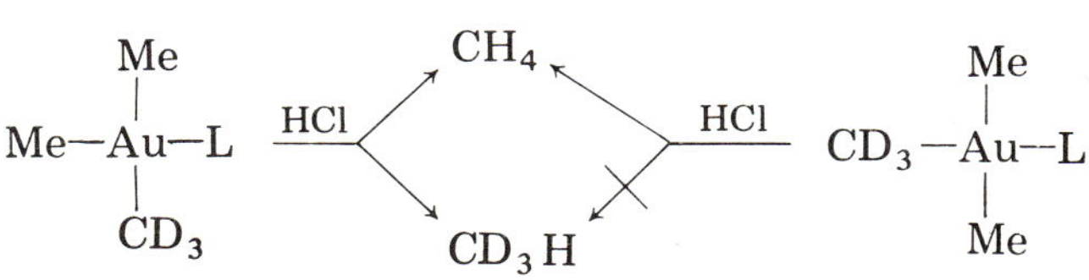

The mechanism is thought to correspond most closely to the S_E2 (open) mechanism since the rates of reaction are determined largely by the acid strength in the order trifluoromethanesulphonic acid > nitric acid > trifluoroacetic acid > acetic acid > benzenethiol and not by the nucleophilicity of the conjugate base X^-. By study of reactions of *cis*-$[AuMe_2R(PPh_3)]$ with acids, the order of reactivity of alkyl groups R to cleavage by acids is found to be Me $\gg$ Et > ^{n}Pr >

${}^{i}Pr$ which reflects increasing steric hindrance at the site of protonation. The cleavage of one of the mutually *trans*-alkyl groups, rather than the alkyl group *trans* to phosphine, in these reactions may be ascribed to the greater *trans*-effect of the methyl group over the phosphine ligand. However, in these reactions the effect must operate through electron release to enhance the carbanionic character of the *trans* methyl group either in the ground state or transition state. Thus, the *trans*-effect series deduced from nucleophilic substitution reactions is not expected to operate in these reactions with electrophiles [28].

In contrast to these reactions, cleavage of a gold—carbon bond in $[MeAuR]^-$ by HCl showed remarkably little selectivity. Thus $[MeAu^tBu]^-$ gave equal amounts of methane and isobutane, and $[MeAuEt]^-$ gave slightly more ethane than methane. These results cannot readily be explained by a mechanism involving electrophilic attack at carbon, and it is suggested that the proton attacks at the gold centre to give a hydridogold(III) intermediate which then undergoes reductive elimination of alkane ($L = PPh_3$) [29].

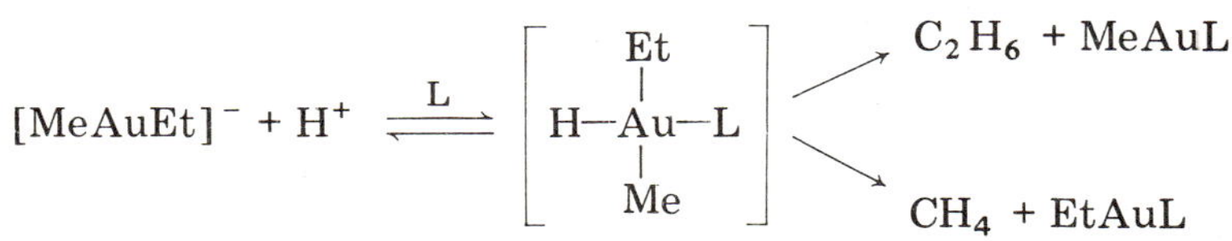

A similar mechanism is thought to operate in the reactions of [MeAuL] with benzenethiol and benzeneselenol, which occur very much more rapidly than similar reactions of $[Me_3AuL]$. In these cases, the oxidative addition step has been shown to occur by a free-radical chain mechanism however [30, 31].

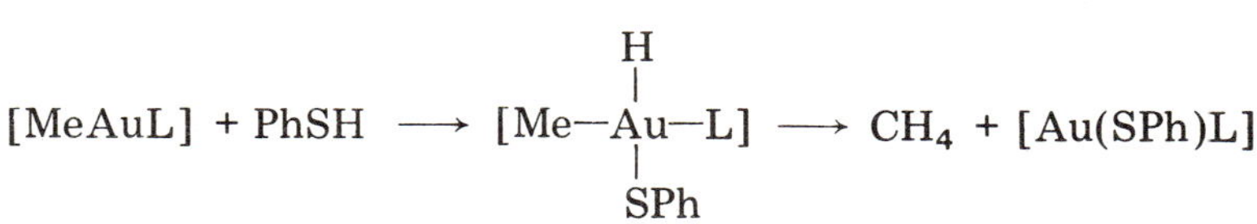

In all of these examples, the evidence for hydridogold(III) intermediates is circumstantial since such compounds are too short-lived to be identified directly. The different reactivity patterns shown by the gold(I) and gold(III) complexes do, however, strongly indicate that quite different mechanisms are involved.

(b) Reactions with metal halides

Cleavage of gold—carbon bonds in $[AuMe(PPh_3)]$, $[AuEt(PPh_3)]$ and $[AuMe_3(PPh_3)]$ by mercury(II) salts is thought to occur by the S_E2 (open) mechanism. Thus, the reactions follow second-order kinetics, first order in both alkylgold complex and mercury(II) compound. The reactivity series $Hg(NO_3)_2 \gg Hg(OAc)_2 \gg HgI_2 > HgBr_2 > HgCl_2$ and $[AuEt(PPh_3)] > [AuMe(PPh_3)] \sim [AuMe_3(PPh_3)]$ was observed. The similar rates of reaction of the methylgold(I) and trimethylgold(III) complexes also suggest a common mechanism, in contrast with the reactions of methylgold complexes with acids. With the trimethylgold complex, cleavage of one of the mutually *trans* methyl groups took place as expected [32].

$$[AuMe_3(PPh_3)] + HgCl_2 \rightarrow cis\text{-}[AuMe_2Cl(PPh_3)] + MeHgCl$$

Cleavage of methyl—gold bonds by methylmercury(II) halides occurs by the same mechanism but more slowly.

$$[AuMe(PPh_3)] + MeHgCl \rightarrow [AuCl(PPh_3)] + Me_2Hg$$

Cleavage of the gold—carbon bond in 1-cyano-1-ethoxycarbonylpentyl(triphenylphosphine)gold(I) by methylmercury(II) halides in 1,4-dioxane again occurs by the S_E2 mechanism. However, in dioxane—dimethylsulphoxide mixtures the reaction is first order in gold complex but zero'th order in mercury(II) and an S_E1 mechanism is implicated [32].

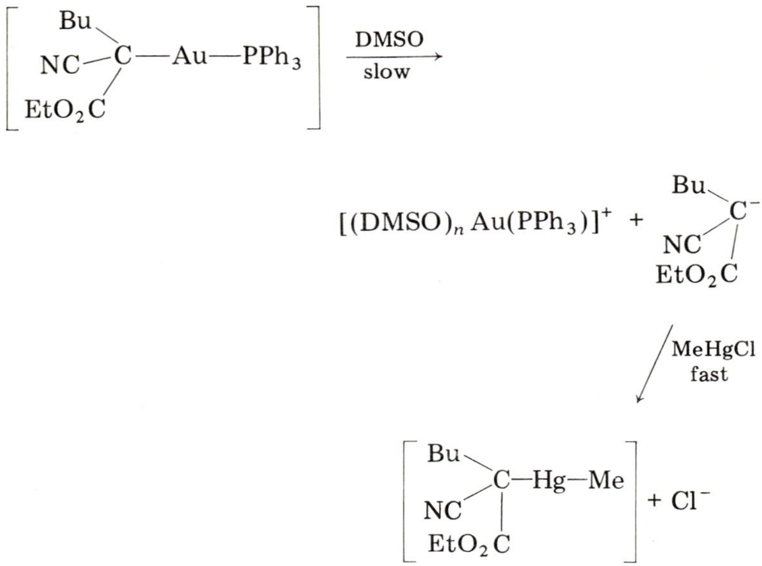

The rate-determining heterolysis of the gold—carbon bond is no doubt aided by solvation of gold(I) by dimethylsulphoxide, and by the high electronegativity of the alkyl group.

Methyl—gold bonds are also cleaved by other metal halides, for example by palladium and platinum halides. However, with these reagents methylgold(I) complexes are found to react much faster than trimethylgold(III) complexes, and an S_E2 (cyclic) mechanism is thought likely. Thus, since gold(I) complexes undergo nucleophilic substitution more rapidly than do gold(III) complexes, the S_E2 (cyclic) mechanism should lead to the observed reactivity series, whereas the S_E2 (open) mechanism should lead to similar rates of reaction for the gold(I) and gold(III) derivatives. It is also possible that the gold(I) complexes react by the oxidative-addition reductive-elimination mechanism [33].

The methylgold(I) bond can be cleaved by gold(III) halides, and the cleavage by *cis*-[$AuMe_2I(PPh_3)$] has been examined in some detail [34]. Labelling studies show that the reaction proceeds in the following way ($Au^* = {}^{198}Au$):

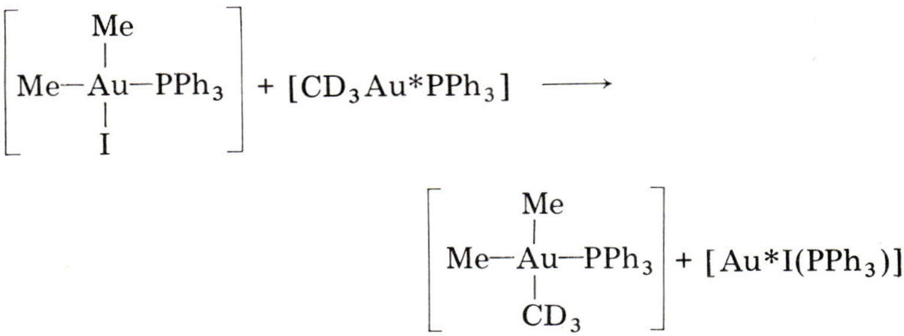

Thus, the reaction proceeds to a large extent with retention of stereochemistry at the gold(III) centre, as shown by deuterium labelling, and takes place by simple methyl for iodo exchange rather than by a redox mechanism as shown by the gold-labelling experiment. Again an S_E2 (cyclic) mechanism is proposed, and the transition state might be expected to contain trigonal-bipyramidal gold(III) and trigonal-planar gold(I) centres to explain the stereochemistry of reaction ($L = PPh_3$) [33—37].

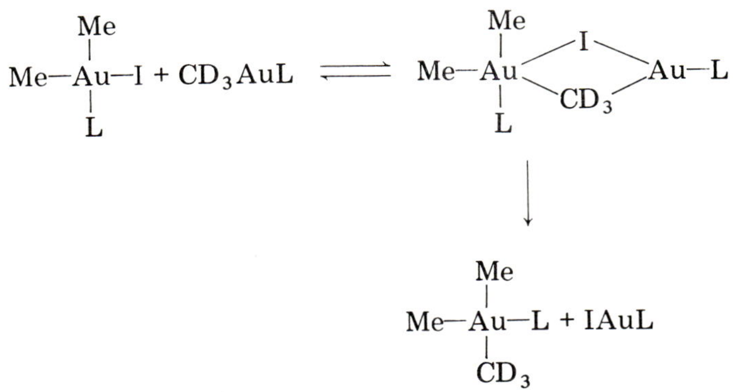

It is likely that other electrophilic reagents, such as halogens and acyl halides, react by similar reaction mechanisms as acids and metal halides with alkylgold complexes but detailed mechanistic investigations have not been carried out as yet.

HOMOLYTIC SUBSTITUTION REACTIONS

Homolytic substitution reactions in methylgold(I) complexes have been demonstrated by application of the CIDNP (Chemically Induced Dynamic Nuclear Polarisation) effect. This effect may lead to anomalous intensities of peaks in the n.m.r. spectra of reacting mixtures for products which are formed directly from free radicals. The effect arises since such products are formed with a population of the nuclear-spin energy levels which does not correspond with the Boltzmann distribution.

Photolysis of dibenzoylperoxide gives $PhCO_2^{\cdot}$ radicals, which displace a methyl radical from $[AuMe(PPh_3)]$ by the bimolecular homolytic substitution mechanism, S_H2.

$$PhCO_2^{\cdot} + [AuMe(PPh_3)] \rightarrow [Au(O_2CPh)(PPh_3)] + Me^{\cdot}$$

Methyl radicals then combine with $PhCO_2^{\cdot}$ radicals or the phenyl radicals, which are formed by loss of CO_2, to give $PhCO_2Me$ and PhMe, respectively. The n.m.r. spectrum of a reaction mixture shows CIDNP effects for both these products, thus confirming that they are formed by a free radical mechanism.

Photolysis of $[AuMe(PPh_3)]$ in chloroform solution gives methyl radicals, which have been shown to exchange methyl groups with $[AuMe(PPh_3)]$ by the S_H2 mechanism. Again, the evidence is based on observation of CIDNP effects [38, 39].

$$Me^* + [AuMe(PPh_3)] \rightarrow [AuMe^*(PPh_3)] + Me$$

Similarly, photolysis of PhSSPh gives $PhS^{\cdot}$ radicals which displace a methyl radical from $[AuMe(PPh_3)]$. In this case, the methyl radicals were trapped by ^tBuNO to give the stable $^tBuMeNO^{\cdot}$ radical, which was identified by e.s.r. spectroscopy. The same radical could be trapped on reaction of $[AuMe(PMe_3)]$ with benzenethiol in the presence of ^tBuNO, and homolytic substitution as part of a free-radical chain mechanism was proposed. Further evidence of a radical chain mechanism was obtained by addition of radical scavengers such as galvinoxyl which were found to inhibit the reaction. The propagation steps are shown below:

$$PhS^{\cdot} + [AuMe(PMe_3)] \rightarrow [Au(SPh)(PMe_3)] + Me^{\cdot}$$

$$Me^{\cdot} + PhSH \rightarrow CH_4 + PhS^{\cdot}$$

In this case, it is suggested that the S_H2 reaction proceeds in a stepwise manner, with formation of a gold(II) intermediate which can react with benzenethiol.

$$PhS^{\cdot} + [AuMe(PMe_3)] \longrightarrow [AuMe(SPh)(PMe_3)]^{\cdot}$$

$$[AuMe(SPh)(PMe_3)]^{\cdot} \xrightarrow{-Me^{\cdot}} [Au(SPh)(PMe_3)]$$

$$[AuMe(SPh)(PMe_3)]^{\cdot} \xrightarrow{PhSH} [AuMe(H)(SPh)(PMe_3)] + PhS^{\cdot}$$

$$[AuMe(H)(SPh)(PMe_3)] \xrightarrow{-CH_4} [Au(SPh)(PMe_3)]$$

Complexes $[AuMe_3L]$, L = tertiary phosphine, do not react with benzenethiol by a free-radical chain mechanism presumably because S_H2 substitution at the gold(III) centre does not occur readily [30]. Similar results are found in reactions with benzeneselenol [31].

A free-radical chain mechanism is also implicated in reactions of methylgold(I) complexes with CF_3I, and S_H2 substitution may be important in some cases [37], as in the propagation steps shown below:

$$CF_3^{\cdot} + [AuMe(PPh_3)] \rightarrow [Au(CF_3)(PPh_3)] + Me^{\cdot}$$

$$Me^{\cdot} + CF_3I \rightarrow MeI + CF_3^{\cdot}$$

Free-radical mechanisms, which may involve S_H2 substitution at gold(I), have been suggested for reactions of methylgold(I) complexes with unsaturated reagents [40] and, as investigators are now aware that such reactions often occur very readily, more examples are likely to be reported in the near future.

OXIDATION OF GOLD(I) COMPLEXES

The oxidation of $[Au(CN)_2]^-$ by I_2 or I_3^- to give *trans*-$[Au(CN)_2I_2]^-$ is first order in both $[Au(CN)_2]^-$ and I_2 or I_3^-, respectively. It is found that I_3^- reacts faster than does

I_2 which is unusual. To explain this effect, a single-step *trans* oxidative addition is suggested, and the higher rate for I_3^- is then due to the greater bond separation as illustrated by the proposed transition states [41].

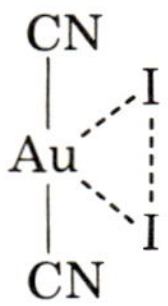

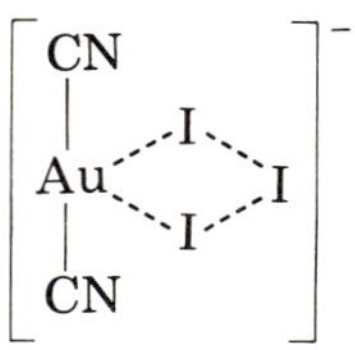

The oxidation of [AuClL], L = tertiary phosphine, by bromine is considerably more complex, apparently because the initial product [$AuClBr_2L$] is more reactive than bromine towards [AuClL]. Thus complete scrambling of halogens occurs by reactions such as:

$$[AuClBr_2L] + [AuClL] \rightleftharpoons [AuBrL] + [AuCl_2BrL]$$

The result is that all possible isomers of all possible species [$AuCl_nBr_{3-n}L$] are formed [33, 42].

Alkyl halides have been shown to add oxidatively to platinum(0) complexes by a variety of mechanisms. With benzyl halides an S_N2 mechanism is found (L = tertiary phosphine).

$$PtL_2 + PhCH_2Cl \rightarrow [PhCH_2PtL_2]^+ Cl^- \rightarrow [PhCH_2PtClL_2]$$

With secondary alkyl halides free-radical mechanisms dominate, but both chain and non-chain processes are possible.

$$PtL_2 + R^\cdot \rightarrow RPtL_2^\cdot \xrightarrow{RI} RPtIL_2 + R^\cdot$$

$$PtL_2 + RI \rightarrow [PtIL_2]^\cdot R^\cdot \rightarrow RPtIL_2$$

With methyl halides, there is still some dispute as to whether the S_N2 or free-radical non-chain mechanism operates in oxidative addition to platinum(0) complexes.

Similar results may be expected in oxidative addition reactions of the isoelectronic gold(I) complexes, but fewer investigations have been made. The reactions of [$AuMe(PMe_3)$]

with methyl iodide to give initially *cis*-$[AuIMe_2(PMe_3)]$, and of $[AuMe_2]^-$ with methyl iodide to give $[AuMe_3L]$ have been suggested to occur by the S_N2 mechanism [29, 36], but CF_3I reacts with $[AuMe(PMe_3)]$, at least in part, by a free-radical chain process which is inhibited by radical scavengers. The propagation steps are shown below [37]:

$$CF_3^{\cdot} + [AuMe(PMe_3)] \rightarrow [AuMe(CF_3)(PMe_3)]$$

$$[AuMe(CF_3)(PMe_3)] + CF_3I \rightarrow [AuI(Me)(CF_3)(PMe_3)] + CF_3^{\cdot}$$

Oxidation of gold(I) complexes may also occur by a one-electron transfer process. This has been proposed for the reaction of gold(I) complexes with acyl halides.

$$[AuPh(PPh_3)] + PhCOCl \rightleftharpoons [AuPh(PPh_3)]^{\dot{+}} [PhCOCl]^{\dot{-}}$$

This initial product could rearrange to the product of oxidative addition, but actually gives $[AuCl(PPh_3)]$ and various products arising from $Ph^{\cdot}$ and $PhCO^{\cdot}$ radicals [43]. One-electron transfer is also thought to account for the formation of $TCNE^{\dot{-}}$ from $[AuMe(PMe_3)(TCNE)]$ where TCNE is tetracyanoethylene, but the general scope of such reactions has not yet been determined [44].

REDUCTION OF GOLD(III) COMPLEXES

The reduction of gold(III) to gold(I) may proceed in one-electron or two-electron steps. A good example of the former is the reduction of $[AuCl_4]^-$ by iron(II) which was shown to give a gold(II) species, perhaps $[AuCl_4]^{2-}$, in the first step [45].

The reduction of $[AuCl_4]^-$ by phosphorus(III) and sulphur(II) ligands has been studied by several research groups. The group V derivatives YPh_3 (Y = P, As or Sb) react with $[AuCl_4]^-$ to give the transition state $[AuCl_4(YPh_3)]^-$ which breaks down directly to give $[AuCl_2]^-$ and YPh_3Cl_2, when Y = As or Sb, but to give the substitution product

$[AuCl_3(YPh_3)]$ when Y = P. This then reacts with more phosphine to give reduction to gold(I):

$$[AuCl_3(PPh_3)] + PPh_3 \rightarrow [AuCl(PPh_3)] + PPh_3Cl_2$$

Whether formation of YPh_3Cl_2 takes place in a concerted step or *via* intermediates such as $[Cl_3Au\text{-}YPh_3Cl]^-$ is not known [46]. A similar mechanism is involved in reduction by dialkylsulphides, such as thiomorpholin-3-one, T, though here hydrolysis of gold(III) intermediates and direct formation of the sulphoxide is proposed [47]:

$$[AuX_4]^- + H_2O \underset{k_2}{\overset{k_1}{\rightleftharpoons}} [AuX_3(OH)]^- + H^+ + X^-$$

$$[AuX_3(OH)]^- + T \xrightarrow{\text{fast}} [AuX_3(T)] + HO^-$$

$$[AuX_3(T)] + H_2O \xrightarrow{\text{fast}} [AuX_2(OH)T] + H^+ + X^-$$

$$[AuX_2(OH)T] + T \xrightarrow{k_3} [AuX(T)] + T=O + H^+ + X^-$$

The oxidation of S-methionine by $[AuCl_4]^-$ takes place stereospecifically to give S-methionine-S-sulphoxide by a similar mechanism [48].

Platinum(II) complexes can be oxidised by $[AuCl_4]^-$ in the presence of Cl^-.

$$[PtClPh(PEt_3)_2] + [AuCl_4]^- \rightarrow [PtCl_3Ph(PEt_3)_2] + [AuCl_2]^-$$

A major part of the reaction followed the rate law, rate = $k[PtClPh(PEt_3)_2][AuCl_4^-][Cl^-]$, and a transition state such as that shown below is suggested [49].

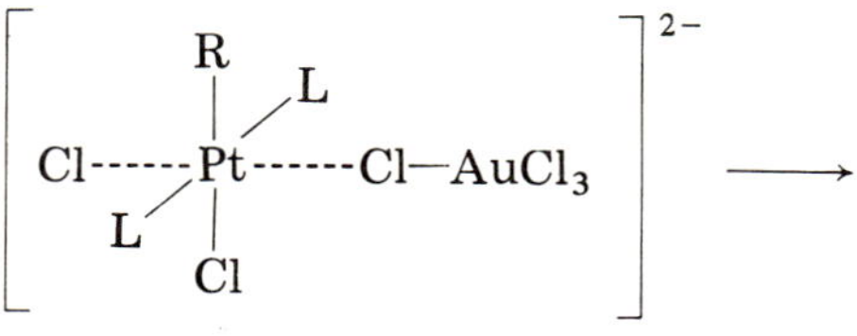

$$[PtCl_3RL_2] + [AuCl_2]^- + Cl^-$$

The rate of the reaction of $[AuCl_4]^-$ with $[PtCl_4]^{2-}$ to give $[AuCl_2]^-$ and $[PtCl_6]^{2-}$ is unaffected by the concentration of chloride indicating that, in this case, a chloride assisted process does not occur [49b].

The reduction of $[Au(NH_3)_4]^{3+}$ by thiosulphate occurs by initial substitution of thiosulphate at gold(III) followed by the reaction [50]:

$$[Au(S_2O_3)_3(H_2O)]^{3-} + S_2O_3^{2-} \rightarrow [Au(S_2O_3)_2]^{3-} + [S_4O_6]^{2-} + H_2O$$

This probably involves reductive elimination with coupling of two-co-ordinated thiosulphate groups to give $[S_4O_6]^{2-}$, and a similar process is implicated in the decomposition of $[AuCl_2\{S{=}C(NHPh)_2\}_2]^+ Cl^-$ [51].

$$[AuCl_2\{S{=}C(NHPh)_2\}_2]^+ \longrightarrow [AuCl_2]^- + 2H^+ + (PhNHC({=}NPh){-}S{-})_2$$

These reactions resemble the reductive elimination of alkanes from alkylgold(III) complexes. Such reactions were discussed in Chapter 7 and only a brief summary of the results necessary to deduce the reaction mechanism will be given here.

The reductive elimination of ethane from $[AuMe_3(PPh_3)]$ occurs after loss of the triphenylphosphine, as shown by the retardation of reductive elimination in the presence of free phosphine. In complexes $[AuMe_2R(PPh_3)]$, there is some evidence that two mutually *cis* alkyl groups are eliminated, though this issue is complicated by the occurrence of *cis*—*trans* isomerism and of intermolecular alkyl exchange reactions. The conclusion that the reductive elimination is intramolecular is proved beyond reasonable doubt however [52, 53]. Reductive elimination from a vinylgold(III) species occurs with retention of stereochemistry about the C=C bond [54]. The energies for all possible structures of the intermediates R_3Au have been calculated when R = Me, and it is found that a T-shaped molecule is most stable, followed by a Y-shaped molecule, and that a regular trigonal-planar molecule is least stable. For a d^8 complex a trigonal-planar molecule would be paramagnetic, and would be expected to undergo Jahn—Teller distortion to the diamagnetic T- or Y-shaped species. Interconversion between T-shaped species R_2AuR' occurs *via* Y-shaped species leading to *cis—trans* isomerisation, and reductive elimination occurs by coupling of the two closely spaced alkyl groups from the Y-shaped species. The situation is summarised in Fig. 9.4.

It is also shown that this concerted reductive elimination process is "orbitally-allowed" and an approximate activation energy to reductive elimination from $[Me_3Au]$ is calculated to be 0.8 eV or roughly 80 kJ mol^{-1}, while the activation energy to isomerisation is half this value. The activation energy to reductive elimination is only about half the estimated Au—C bond energy of *ca.* 150 kJ mol^{-1}, again due to the concerted nature of the reaction [53]. A very similar result is obtained for reductive elimination from $[PtIMe_3(Ph_2PCH_2CH_2\text{-}PPh_2)]$, where the experimental activation energy was only half the Pt—C bond energy [55].

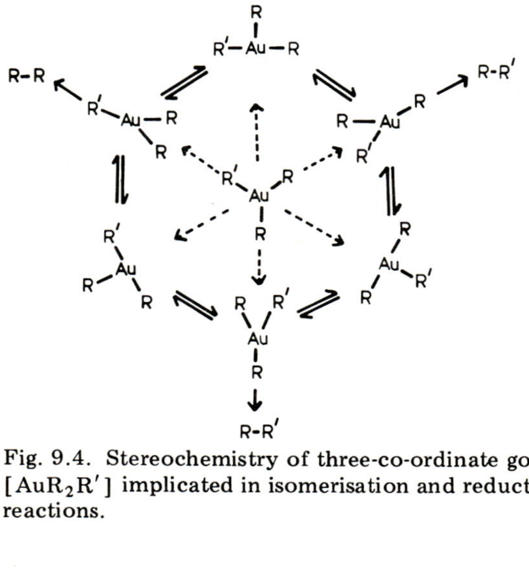

Fig. 9.4. Stereochemistry of three-co-ordinate gold(III) complexes $[AuR_2R']$ implicated in isomerisation and reductive elimination reactions.

The importance of three-co-ordinate intermediates in reductive elimination from dialkylgold(III) species has also been emphasised. For the complexes *cis*-$[AuMe_2L_2]^+$, reductive elimination of ethane is faster for bulkier ligands L, so that the rates follow the order L = $SbPh_3 > AsPh_3 > PPh_3 > PMePh_2 > PMe_2Ph > PMe_3$. The reactions are also retarded by excess of ligand L, suggesting that pre-dissociation of a ligand occurs before reductive elimination. The ease of elimination of alkyl groups from gold(III) follows the order Bu ~ Et > Me, as for the trialkylgold(III) complexes. Reaction is again intramolecular since $[AuMe(CD_3)(PPh_3)_2]^+$ gave very largely CH_3CD_3 with little C_2H_6 or C_2D_6 on decomposition. It is suggested that reductive elimination occurs from the species $[AuMe_2XL]$ where X^- is the initial counterion or from $[AuMe_2X]$ formed by dissociation of the second ligand L. The latter is favoured since reductive elimination from *cis*-$[AuMe_2XL]$ is also retarded in the presence of small amounts of L, insufficient to convert this to *cis*-$[AuMe_2L_2]^+X^-$. The rates of reductive elimination from both *cis*-$[AuMe_2L_2]^+X^-$ and *cis*-$[AuMe_2XL]$ are strongly dependent on the anion X, following the series X = CF_3SO_3, $ClO_4 > NO_3 \gg CF_3CO_2$, Cl. Again then, the experimental evidence strongly favours an intramolecular concerted elimination process probably from a three-co-ordinate gold(III) species [56, 57].

Perhaps more surprising is that a related mechanism is found for decomposition of alkylgold(I) complexes, RAuL, which again proceeds after dissociation of the tertiary phosphine ligand L. The following mechanism is suggested, although the precise nature of the intermediate is still debatable [58].

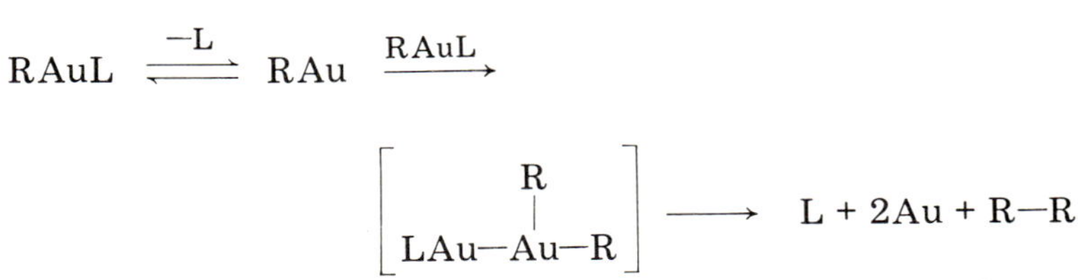

REFERENCES

1 L. Cattalini, *Prog. Inorg. Chem.*, 13 (1970) 263.
2 L. F. Warren and M. A. Bennett, *Inorg. Chem.*, 15 (1976) 3126.
3 R. J. Charlton, C. M. Harris, H. Patil and N. C. Stephenson, *Inorg. Nucl. Chem. Lett.*, 2 (1966) 409.
4 A. J. Hall and D. P. N. Satchell, *J. Chem. Soc., Chem. Commun.*, (1976) 163.
5 A. J. Hall and D. P. N. Satchell, *Chem. Ind.*, (1976) 373.
6 W. H. Baddley and F. Basolo, *Inorg. Chem.*, 3 (1964) 1087; B. I. Peschevitskii and G. I. Shamovskaya, *Zh. Neorg. Khim.*, 17 (1972) 2648.
7 R. L. Rich and H. Taube, *J. Phys. Chem.*, 58 (1954) 1.
8 L. Cattalini, A. Orio and M. L. Tobe, *J. Am. Chem. Soc.*, 89 (1967) 3130.
9 L. Cattalini, A. Orio and M. L. Tobe, *Inorg. Chem.*, 6 (1967) 75.
10 L. Cattalini, G. Marangoni and M. Martelli, *Inorg. Chem.*, 7 (1968) 1145.
11 L. Cattalini, M. Nicolini and A. Orio, *Inorg. Chem.*, 5 (1966) 1674.
12 L. Cattalini, A. Doni and M. Orio, *Inorg. Chem.*, 6 (1967) 280.
13 L. Cattalini and M. L. Tobe, *Inorg. Chem.*, 5 (1966) 1145.
14 C. F. Weick and F. Basolo, *Inorg. Chem.*, 5 (1966) 576.
15 D. L. Fant and C. F. Weick, *Inorg. Chem.*, 12 (1973) 1864.
16 L. Cattalini, V. Ricevuto, A. Orio and M. L. Tobe, *Inorg. Chem.*, 7 (1968) 51.
17 G. Bandoli, D. A. Clemente, G. Marangoni and L. Cattalini, *J. Chem. Soc., Dalton Trans.*, (1973) 886.
18 N. F. Borkett and M. I. Bruce, *Inorg. Chim. Acta*, 12 (1975) L33.
19 H. Schmidbaur and K. C. Dash, *J. Am. Chem. Soc.*, 95 (1973) 4855.
20 G. Annibale, L. Cattalini and G. Natile, *J. Chem. Soc., Dalton Trans.*, (1975) 188.
21 G. Annibale, L. Cattalini, A. A. El-Awady and G. Natile, *J. Chem. Soc., Dalton Trans.*, (1974) 802.
22 G. Annibale, G. Natile and L. Cattalini, *J. Chem. Soc., Dalton Trans.*, (1976) 285.
23 G. Annibale, G. Natile and L. Cattalini, *J. Chem. Soc., Dalton Trans.*, (1976) 1547.
24 G. E. Glass and R. S. Tobias, *J. Organomet. Chem.*, 15 (1968) 481.
25 A. Shiotani, H-F. Klein and H. Schmidbaur, *J. Am. Chem. Soc.*, 93 (1971) 1555; H. Schmidbaur, A. Shiotani and H-F. Klein, *Chem. Ber.*, 104 (1971) 2831.
26 A. J. Hall and D. P. N. Satchell, *J. Chem. Soc., Perkin Trans.*, 2 (1976) 1278.
27 D. S. Matteson, *Organometallic Reaction Mechanisms*, Academic, New York, 1974.
28 S. Komiya and J. K. Kochi, *J. Am. Chem. Soc.*, 98 (1976) 7599.
29 A. Tamaki and J. K. Kochi, *J. Chem. Soc., Dalton Trans.*, (1973) 2620.
30 A. Johnson and R. J. Puddephatt, *J. Chem. Soc., Dalton Trans.*, (1975) 115.
31 R. J. Puddephatt and P. J. Thompson, *J. Organomet. Chem.*, 117 (1976) 395.
32 B. J. Gregory and C. K. Ingold, *J. Chem. Soc. B*, (1969) 276.
33 R. J. Puddephatt and P. J. Thompson, *J. Chem. Soc., Dalton Trans.*, (1975) 1810.
34 G. W. Rice and R. S. Tobias, *J. Organomet. Chem.*, 86 (1975) C37.
35 A. Tamaki and J. K. Kochi, *J. Organomet. Chem.*, 64 (1974) 411.
36 A. Johnson and R. J. Puddephatt, *J. Organomet. Chem.*, 85 (1975) 115.
37 A. Johnson and R. J. Puddephatt, *J. Chem. Soc., Dalton Trans.*, (1976) 1360.
38 R. Kaptein, P. W. N. M. van Leeuwen and R. Huis, *J. Chem. Soc. Chem., Commun.*, (1975) 568.
39 P. W. N. M. van Leeuwen , R. Kaptein, R. Huis and C. F. Roobeek, *J. Organomet. Chem.*, 104 (1976) C44.
40 C. M. Mitchell and F. G. A. Stone, *J. Chem. Soc., Dalton Trans.*, (1972) 102.

41 M. H. Ford-Smith, J. J. Habeeb and J. H. Rawsthorne, *J. Chem. Soc., Dalton Trans.*, (1972) 2116.
42 B. T. Heaton and R. J. Kelsey, *Inorg. Nucl. Chem. Lett.*, 11 (1975) 363.
43 E. G. Perevalova, D. A. Lemenovskii, T. V. Baukova, E. I. Smyslova, K. I. Grandberg and A. N. Nesmeyanov, *Dokl. Akad. Nauk SSSR*, 206 (1972) 883.
44 A. Johnson and R. J. Puddephatt, *J. Chem. Soc., Dalton Trans.*, (1977) 1384.
45 R. L. Rich and H. Taube, *J. Phys. Chem.*, 58 (1954) 6.
46 R. Roulet, N. Quang Lan, W. R. Mason and G. P. Fenske, *Helv. Chim. Acta.* 56 (1973) 2405.
47 D. de Filippo, F. Devillanova and C. Preti, *Inorg. Chim. Acta*, 5 (1971) 103.
48 E. Bordignon, L. Cattalini, G. Natile and A. Scatturin, *J. Chem. Soc., Chem. Commun.*, (1973) 878; G. Natile, E. Bordignon and L. Cattalini, *Inorg. Chem.*, 15 (1976) 246.
49 (a) A. Peloso, *Coord. Chem. Rev.*, 16 (1975) 95.
(b) K. G. Moodley and M. J. Nicol, *J. Chem. Soc., Dalton Trans.*, (1977) 993.
50 G. Nord, L. H. Skibsted and A. S. Khalonin, *Acta Chem. Scand., Ser. A*, A29 (1975) 505.
51 V. M. Shulman, Z. A. Savel'eva and R. I. Novoselov, *Russ. J. Inorg. Chem.*, 18 (1973) 376.
52 A. Tamaki, S. A. Magennis and J. K. Kochi, *J. Am. Chem. Soc.*, 96 (1974) 6140.
53 S. Komiya, T. A. Albright, R. Hoffmann and J. K. Kochi, *J. Am. Chem. Soc.*, 98 (1976) 7255.
54 J. A. J. Jarvis, A. Johnson and R. J. Puddephatt, *J. Chem. Soc., Chem. Commun.*, (1973) 373.
55 M. P. Brown, R. J. Puddephatt and C. E. E. Upton, *J. Chem. Soc., Dalton Trans.*, (1974) 2457.
56 P. L. Kuch and R. S. Tobias, *J. Organomet. Chem.*, 122 (1976) 429.
57 S. Komiya and J. K. Kochi, *J. Am. Chem. Soc.*, 98 (1976) 7599.
58 A. Tamaki and J. K. Kochi, *J. Organomet. Chem.*, 61 (1973) 441.

CHAPTER 10

SPECTROSCOPIC STUDIES OF GOLD COMPLEXES

Most published papers on gold compounds now contain spectroscopic data, most commonly from the infrared and n.m.r. spectra. It would be inappropriate to attempt to give a comprehensive account of such spectra, and only enough data to show the kind of information about structure and bonding in gold complexes which can be obtained from spectroscopic studies will be given here.

VIBRATIONAL SPECTROSCOPY

Studies of gold—ligand stretching and other vibrations by infrared and Raman spectroscopy can yield useful information about the metal—ligand bond strength and the degree of covalency in the metal—ligand bond. The metal—ligand stretching frequency is determined by the masses of the metal and ligand, by the stretching force constant which is related to the strength of the metal—ligand bond, and by coupling of this vibration with other molecular vibrational modes belonging to the same symmetry species. Ideally therefore, when comparing metal—ligand bond strengths the stretching force constants should be calculated but this is difficult except for the most symmetrical molecules. We shall discuss such molecules first.

(a) Linear molecules $[AuX_2]^-$

There are three normal modes of vibration for these molecules, namely the symmetric stretching vibration (ν_1, symmetry Σ_g^+), the bending vibration (ν_2, symmetry Π_u) and the asymmetric stretching vibration (ν_3, symmetry Σ_u^+). ν_1 is Raman

TABLE 10.1

Infrared and Raman spectra of ions $[AuX_2]^-$

Complex	Band maxima/cm^{-1}			
	ν_1(R)[a]	ν_2(ir)[a]	ν_3(ir)[a]	Ref.
$[Bu_4N][AuCl_2]$	329	116	350	1
$[Bu_4N][AuBr_2]$	209	77	254	1
$[Bu_4N][AuI_2]$	158	63	210	1
$Li[AuMe_2]$	526	—	—	2
$Li[Au(CD_3)_2]$	477	—	—	2
$K[Au(CN)_2]$	445—452	ca. 100	427	3
$[Et_4N][Au\{Mn(CO)_5\}_2]$	—	—	185	4
$[Et_4N][Au\{Mo(CO)_3(C_5H_5)\}_2]$	—	—	170	4
$[Et_4N][Au\{Fe(CO)_2(C_5H_5)\}_2]$	—	—	193	4
$[Et_4N][Au\{Co(CO)_4\}_2]$	—	—	189	4

[a] Strictly, this nomenclature should only be used for triatomic MX_2 species (linear).

active only, while ν_2 and ν_3 are infrared active only. Some stretching frequencies for ions of this type are given in Table 10.1, in which data for some more complex linear molecules are also given.

The stretching force constants for all species $[AuX_2]^-$ are lower than for the isoelectronic HgX_2 species [1–5]. A com-

TABLE 10.2

Stretching force constants, $k/N\ m^{-1}$, for some linear molecules and ions

Species	X = Cl	X = Br	X = I
$[CuX_2]^-$	175	157	140
$[AgX_2]^-$	145	129	117
$[AuX_2]^-$	207[a], 204[b]	189[a], 191[b]	165[a], 171[b]
$[HgX_2]$	265	228	186

[a] Ref. 1.
[b] Ref. 5.

parison with similar derivatives of copper(I) and silver(I) is given in Table 10.2. It can be seen that in all cases the force constants decrease along the series X = Cl > Br > I, and Hg > Au > Cu > Ag. For the metal(I) complexes the series is the same as the series of first ionisation energies of the metals, and the opposite of the promotion energy $d^{10} \rightarrow d^9 s$ for the unipositive ions. Both these effects should lead to greater covalence in the M—X bond in the series Au > Cu > Ag as explained in Chapter 1, and it is probable that the trend in force constants reflects the degree of covalence in the M—X bonds. The mean amplitudes of vibration for the halide complexes $[AuX_2]^-$ are calculated to be higher than in the corresponding HgX_2 species, as expected from the difference in force constants [6].

(b) Square-planar complexes $[AuX_4]^-$

There are seven fundamental modes of vibration for square-planar compounds, of which three are Au—X stretching vibrations. Two of them (ν_1 and ν_3 of symmetry a_{1g} and b_{1g}, respectively) are Raman active and the other (ν_6 of e_u symmetry) is infrared active. Some data on complexes of this type are given in Table 10.3. It can be seen that the stretching force constants for $[AuCl_4]^-$ and $[AuBr_4]^-$ are almost the same as for $[AuCl_2]^-$ and $[AuBr_2]^-$, respectively. In general, increasing the oxidation state of a metal leads to an increase in the stretching force constant while an increase in co-ordination number leads to a decrease. These factors evidently cancel out here and the force constants are almost equal for the linear gold(I) and square-planar gold(III) complexes.

The vibrational frequencies of complexes $M[AuX_4]$ in the solid state are found to depend on the cation M^+, particularly for the bending vibrations. Also of interest is the observation that a resonance Raman effect is observed for solid $[Et_4N]$-$[AuBr_4]$ if the exciting line is close in energy to that of the intense charge transfer transition in the electronic spectrum of the complex. Thus, using the 457.9 nm excitation, an overtone progression in ν_1 (AuBr) up to $9\nu_1$ was observed [12].

TABLE 10.3

Stretching frequencies/cm^{-1} for some square-planar gold(III) complex ions

Ion	$\nu_1(a_{1g})$	$\nu_2(a_{2u})$	$\nu_3(b_{1g})$	$\nu_4(b_{2g})$	$\nu_6(e_u)$	$\nu_7(e_u)$	$k(Au-X)^a$	Ref.
$[AuF_4]^-$	588	—	561	237, 230	—	—	—	7
$[AuCl_4]^{-\,b}$	349	142	325	172	361	166	220	8
$[AuBr_4]^-$	214	99	196	106	252	106	185	8
$[AuI_4]^-$	148	—	—	—	—	—	—	9
$[Au(CN)_4]^-$	459	107	450	110	415	160	—	10
$[AuMe_4]^-$	530	—	522	—	484	—	—	2
$[Au(NH_3)_4]^{3+}$	570	—	545	—	556	—	—	11

[a] N m^{-1}.
[b] ν_5 is calculated to be at 112 cm^{-1}.

(c) The octahedral complex $[AuF_6]^-$

Only the Raman spectra of $Cs[AuF_6]$ and some other salts of $[AuF_6]^-$ have been reported [7, 13]. The Raman active stretching vibrations are ν_1 of a_{1g} symmetry and ν_2 of e_g symmetry, and the bending vibration ν_5 of t_{2g} symmetry is also active. These frequencies are observed at 595, 520 and 224 cm^{-1}, respectively, and the similarities between the fundamental frequencies of this ion and those of $[AuF_4]^-$ are again noteworthy [7].

(d) Gold—halogen stretching frequencies in complexes [LAuX]

There have been several studies of complexes [LAuX], aimed at correlating ν(AuX) with the *trans*-influence of the ligand L. It is, however, difficult to draw any positive conclusions from this work since there is strong coupling between the AuX and AuL stretching modes, and thus the AuX stretching frequencies may not follow the same sequence as the bond strengths. Some typical results are given in Table 10.4. The complexes are arranged in order of decreasing

TABLE 10.4

Gold—halogen stretching frequencies/cm^{-1} for some complexes [AuXL], X = Cl, Br or I

Complex	ν(AuCl)	ν(AuBr)	ν(AuI)	Ref.
$[Ph_3PSAuX]$	349	—	—	14
$[(PhO)_3PAuX]$	340	—	—	14
$[Ph_3PAuX]$	329—333	229—234	187	14—18
$[Ph_3AsAuX]$	328—338	232—235	190	14, 16—18
$[Ph_3SbAuX]$	333	250	—	16
$[(4\text{-}FC_6H_4)_3PAuX]$	327	226	—	18
$[(MeO)_3PAuX]$	326	—	—	19
$[Me_2SAuX]$	325	229—234	—	20, 21
$[Me_3AsAuX]$	317	210	155	17, 22
$[Me_3PAuX]$	311	205—206	164	17, 22

$\nu(AuX)$, and the series may correspond roughly with the series of increasing *trans*-influence of L. The *trans*-influence of Me_3P is certainly greater than those of the halides, since the gold—halogen stretching force constants for $[Me_3PAuX]$ are lower than for $Bu_4N[AuX_2]$ [23].

(e) Gold—halogen stretching frequencies in gold(III) complexes

In complexes $[AuX_3L]$, there are three gold—halogen stretching modes all of which are both infrared and Raman active. Two of these are due to the symmetric and asymmetric stretching vibrations of the *trans*-AuX_2 grouping (of a_1 and b_1 symmetry, respectively) and the other due to the AuX group *trans* to the ligand L (also of a_1 symmetry). All three bands are observed if there is a large difference between the *trans*-influence of X and L, but otherwise two or more of the bands may be accidentally degenerate [24]. Some data are given in Table 10.5. From the frequencies of $\nu(AuX)$ *trans* to L, the *trans*-influence series $Ph^- > Ph_3P \sim Ph_3As \sim Me_2S > C_5H_5N \sim Cl^-$ may be deduced, though, as for the gold(I)

TABLE 10.5

Gold—halogen stretching frequencies/cm^{-1} in some complexes $[AuX_3L]$

Species	$\nu(AuX_2)$		$\nu(AuX)$ *trans*-L	Ref.
	b_1	a_1	a_1	
$[PhAuCl_3]^-$	359	341, 337	289	25
	365	332	280	26
$[Ph_3PAuCl_3]$	366	340	312	27
$[Ph_3AsAuCl_3]$	374	—	302	14
$[Me_2SAuCl_3]$	361	340	313	20
$[C_5H_5NAuCl_3]$	362	—	333	24, 27
$[PhAuBr_3]^-$	253	204	182	25
$[Ph_3PAuBr_3]$	254	222	202	14
$[Ph_3AsAuBr_3]$	254	—	215	14
$[Me_2SAuBr_3]$	265	232	202	20
$[PhAuI_3]^-$	183	153	162	25

TABLE 10.6

Gold—halogen stretching frequencies for some gold(III) halide complexes

Species	ν(AuCl)	ν(AuBr)	ν(AuI)	Ref.
trans-$[Au(CN)_2X_2]^-$	362, 342	258	203	28
trans-$[Au(CN)_2X(py)]$	363	250	—	27
cis-$[AuMe_2X_2]^-$	281, 268	197, 179	—	29
cis-$[AuMe_2X(py)]$	273	—	150	30
cis-$[AuMe_2Cl(PPh_3)]$	281	—	—	31
cis-$[AuMe_2Cl(PMe_3)]$	284	—	—	32
cis-$[AuCl_2Ph(SPr_2)]$	328 (*trans* S)	—	—	26
	294 (*trans* C)	—	—	
cis-$[AuCl_2Ph(PMe_2Ph)]$	310 (*trans* P)	—	—	26
	286 (*trans* C)	—	—	

complexes [AuXL], problems arise due to coupling of ν(AuX) with ν(AuL). The stretching frequencies for the *trans*-AuX_2 group in $[AuX_3L]$ are almost independent of the ligand L, indicating that there is little *cis*-influence of L on these AuX bonds.

The relatively low *trans*-influence of pyridine is shown by the similarity of ν(AuX) in *trans*-$[Au(CN)_2X_2]^-$ and in *trans*-$[Au(CN)_2X(C_5H_5N)]$ where X = Cl or Br [27, 28] and the high *trans*-influences of alkyl and aryl groups are shown by numerous studies on organogold(III) halides. Some results are given in Table 10.6.

(f) Bridging gold—halogen stretching frequencies

The infrared and Raman spectra of the polymeric gold(I) halides show that the gold—halogen stretching frequencies and force constants decrease in the order AuCl > AuBr > AuI, and this is also the order of AuXAu bond angles in the compounds [33]. Comparison of the spectra for $[Au_2Cl_6]$ [34, 35] and for the dimethylgold(III) halides [36] shows that the bridging AuCl stretching frequencies are lower for $[(Me_2AuCl)_2]$ than for $[Au_2Cl_6]$, again indicating the high

TABLE 10.7

Bridging gold—halogen stretching frequencies/cm^{-1}

Species	Raman[a]		Infrared[a]		∠XAuX	Ref.
AuCl	341 (as)	289 (s)	355 (as)	320 (s)	99	33
AuBr	192 (as)	211 (s)	247 (as)	236 (s)	85	33
AuI	107 (as)	156 (s)	169 (as)	201 (s)	69	33
Au_2Cl_6	289 (as)	328 (s)	309	309	—	35
$(Me_2AuCl)_2$	273	273	273	245	—	36
$(Me_2AuBr)_2$	181	181	—	—	—	36

[a] as = asymmetric stretch; s = symmetric stretch.

trans-influence of the methyl group compared to chloride. Some data on bridging halogen complexes are given in Table 10.7.

In general, bridging gold—halogen stretching frequencies are lower than terminal stretching frequencies, though ν(AuX) for *cis*-$[AuMe_2Cl_2]^-$ and $[(Me_2AuCl)_2]$ are almost the same. Also the stretching force constants for $[AuCl_2]^-$ and AuCl, and for $[AuBr_2]^-$ and AuBr are similar.

To illustrate the effect of bridging and also of the *trans*-ligand on gold—halogen bond strengths, Table 10.8 contains the force constants which have been calculated for various complexes.

(g) Gold—oxygen and gold—sulphur stretching frequencies

Gold—oxygen and gold—sulphur stretching vibrations have been studied rather little. The complexes *cis*-$[AuMe_2(OH_2)_2]^+$, *cis*-$[AuMe_2(OH)_2]^-$ and $[AuMe_2(acac)]$ have ν(AuO) at 418, 441 and 445 cm^{-1}, respectively [37]. The Au—O stretching force constant in the hydroxo species *cis*-$[AuMe_2(OH)_2]^-$ (182 N m^{-1}) is rather higher than in the aquo complex *cis*-$[AuMe_2(OH_2)_2]^+$ (162 N m^{-1}) [29].

There has been some disagreement over the assignment of ν(AuS) in $[AuX(SMe_2)]$ where X = Cl or Br. The infrared and

TABLE 10.8

Gold—halogen stretching force constants/N m^{-1}

Complex	$[AuCl_4]^-$	*cis*-$[AuMe_2Cl_2]^-$	$[Au_2Cl_6]$	$[AuCl_2]^-$	AuCl	$[AuClPMe_3]$
k(AuX) terminal	220	138	222, 238	207		185
k(AuX) bridging			115, 174		209	
Ref.	8	29	35, 34	1	33	23
Complex	$[AuBr_4]^-$	*cis*-$[AuMe_2Br_2]^-$	$[AuBr_2]^-$	AuBr	$[AuBrPMe_3]$	
k(AuX) terminal	185	122	186		158	
k(AuX) bridging				192		
Ref.	8	29	1	33	23	
Complex	$[AuI_2]^-$	AuI				
k(AuX) terminal	165					
k(AuX) bridging		134				
Ref.	1	33				

Raman spectra contain bands in the regions 270—279 cm^{-1} and 330—345 cm^{-1}. Those in the latter region are probably assignable to the Au—S stretching vibration in each case and in the former to an SC_2 deformation mode [17, 21], but the opposite assignment has also been suggested [20]. It is likely that extensive mixing of these modes occurs. In *cis*-$[AuMe_2(SMe_2)_2]^+$ the Au—S stretching vibration is observed in the Raman spectrum at 293 cm^{-1} [29].

(h) Gold—nitrogen, —phosphorus and —arsenic stretching vibrations

The gold—nitrogen stretching vibrations occur between 545—570 cm^{-1} in $[Au(NH_3)_4]^{3+}$, but at 459 cm^{-1} in *cis*-$[AuMe_2(NH_3)_2]^+$ and at 490 cm^{-1} in $[AuMe_2(en)]^+$. These lower frequencies for the methylgold derivatives are due to the high *trans*-influence of the methyl groups. In pyridine complexes, ν(AuN) is expected to mix with ring modes of the pyridine ligand, but a decrease in ν(AuN) from 234—306 cm^{-1} in a variety of pyridine complexes of structure $[AuX_3$-(pyridine)], in which the ligand is *trans* to halogen, to 192—197 cm^{-1} in *cis*-$[AuMe_2X$(pyridine)], where it is *trans* to a methyl group [24, 30], is again probably due to the high *trans*-influence of the methyl group.

In trimethylphosphine complexes of gold, ν(AuP) is considered to occur in the region 335—381 cm^{-1} [17, 19, 22, 32], but it has been shown that significant mixing takes place between this vibrational mode and the PC_3 bending mode and that a band at ca. 200 cm^{-1} also has some Au—P stretching character [23]. Some data are given in Table 10.9. It is significant that ν(AuP) is lower in the tetrahedral $[Au(PMe_3)_4]^+$ than in the linear $[Au(PMe_3)_2]^+$ ion. Weakening of the AuP bonds with increasing co-ordination number is predicted as discussed in Chapter 3.

The gold—arsenic stretching vibration occurs in the region 254—273 cm^{-1} in trimethylarsinegold complexes, but considerably lower Au—P and Au—As stretching frequencies are found in triphenylphosphine and triphenylarsine complexes of gold as seen in Table 10.9.

TABLE 10.9

Gold—phosphorus and gold—arsenic stretching frequencies/cm^{-1} and force constants/N m^{-1}

Complex	ν(AuP) or ν(AuAs) /cm^{-1}	k/N m^{-1}	Ref.
$[AuCl(PMe_3)]$	381	230	17, 22, 23
$[AuBr(PMe_3)]$	378	222	17, 22, 23
$[AuI(PMe_3)]$	371	190	17, 22, 23
$[AuMe(PMe_3)]$	357	196	17, 23
$[AuMe_3(PMe_3)]$	351	200	23
cis-$[AuMe_2Cl(PMe_3)]$	346	—	32
$[Au(PMe_3)_2]^+$	353 or 351	—	19, 22
$[Au(PMe_3)_4]^+$	335	—	19
$[AuCl(AsMe_3)]$	267	—	22
$[AuMe_3(AsMe_3)]$	254	177	23
$[AuCl(PPh_3)]$	182	—	18
$[AuCl(AsPh_3)]$	171	—	18

Some AuP and AuAs stretching force constants are also included in Table 10.9. The higher values of the AuP force constants are expected in view of the greater affinity of gold for phosphine than for arsine ligands. The *trans*-influence series Cl > Br > Me, I is indicated by the AuP stretching force constants, but the overall range of force constants is not large and suggests that AuP bonds are not as susceptible to weakening by the *trans* ligand as are gold—halogen bonds. It is possible that a strong σ-donor in the *trans*-position weakens the AuP σ-bond, but enhances π-bonding between gold and phosphorus to counterbalance this effect.

(i) Gold—carbon stretching vibrations

The overall range of gold—carbon stretching frequencies in methylgold complexes is 457—591 cm^{-1}. As for other stretching vibrations, there is little difference between gold(I) and gold(III) derivatives, but the frequencies are primarily determined by the nature of the *trans* ligand. Some values are

TABLE 10.10

Gold—carbon stretching frequencies/cm^{-1} and force constants/$N\ m^{-1}$ in some methylgold complexes

Complex	$\nu(AuC)^a$			k	Ref.
	trans to X^-	*trans* to L	*trans* to Me		
$MeAuPMe_3$		539		230	17, 23
Me_3AuPMe_3		541		242	17, 23
			512, 498	225	
$Me_3AuAsMe_3$		546		255	23
			516, 499	231	
$(Me_3Au)_2en$		504	— , 477		17
cis-$[AuMe_2(OH_2)_2]^+$		590, 583		301	37
cis-$[AuMe_2(OH)_2]^-$	568, —			296	37
cis-$[AuMe_2(NH_3)_2]^+$		579, 572			30
cis-$[AuMe_2Cl_2]^-$	572, 563			281	29
cis-$[AuMe_2Br_2]^-$	558, 552			272	29
cis-$[AuMe_2(SMe_2)_2]^+$		555, —			29
cis-$[AuMe_2(AsPh_3)_2]^+$		543, 532			31
cis-$[AuMe_2(PMe_3)_2]^+$		516, 510			32
cis-$[AuMe_2Cl(PMe_3)]$	557	536			32
cis-$[AuMe_2Cl(AsPh_3)]$	560	545			31

[a] X^- = anionic ligand except Me^-, L = neutral ligand.

given in Table 10.10 to illustrate this effect. In particular the AuC stretching force constants indicate the *trans*-influence series $Me^- > PMe_3 > AsMe_3 > Br^- > Cl^- > OH^- > H_2O$. The greater *trans*-influence of the heavier halogens is illustrated by decreasing values of ν(AuC) in the dimeric $[(AuMe_2X)_2]$ in the order X = Cl > Br > I [36].

It is difficult to assign a gold—carbon stretching frequency in phenylgold complexes since there are several fundamental modes involving partial Au—C stretching and phenyl-ring deformation vibrations. One such vibration (Whiffen's y-mode) occurs in the region 471—478 cm^{-1}, and another which probably contains more Au—C stretching character (the t-mode) occurs either at ca. 285 cm^{-1} or, more likely, in the region 220—235 cm^{-1} [25, 26].

(j) Further aspects of vibrational spectroscopy

In the above sections we have concentrated on identifying the gold—ligand stretching frequencies in complexes of known structure. Inorganic chemists are often confronted however, with the problem of trying to deduce the structure of a complex from its vibrational spectra. This can often be achieved, although it is preferable to have both infrared and Raman spectra, and also to use other spectroscopic techniques to confirm the predictions. There are two useful ways to approach this, which are readily illustrated by the complex ion $[AuMe_2Cl_2]^-$. First, we can consider how the symmetry of the molecule will govern the number of stretching vibrations. For the *trans*-isomer, there will be two Au—C stretching vibrations, one of which will be infrared active only and the other Raman active only, while for the *cis* isomer the two Au—C stretching vibrations will be both infrared and Raman active. The Au—Cl stretching vibrations behave similarly. We can therefore deduce the stereochemistry from the number of bands in the Au—Me and Au—Cl stretching regions of the infrared and Raman spectra and by whether the infrared and Raman frequencies are coincident or not. There are problems in using only this criterion however, and it is better to consider also the values of the stretching frequencies. Thus, from data given in earlier sections, we predict ν(AuCl) to be 340—370 cm^{-1} in the *trans* isomer but 270—285 cm^{-1} in the *cis* isomer due to the different *trans*-influences of the methyl and chloro ligands. Using both these criteria, the experimental evidence (Tables 10.6 and 10.10) clearly indicates that the stereochemistry of $[AuMe_2Cl_2]^-$ is *cis*.

It is also possible to make deductions about the bonding in complexes by considering changes in ligand stretching frequencies on co-ordination. This has often been used for complexes of CN^-, CO, $-C{\equiv}CR^-$, RNC and $PR_2(C{\equiv}CR)$ in which π-bonding in the gold—ligand bond is expected to influence the strength of the C≡N, C≡O or C≡C bond, as measured by the corresponding stretching frequency. Studies of this kind have already been described in earlier chapters and will not be treated in detail here.

From a study of the infrared spectrum of $[(Me_2AuI)_2]$ in a caesium iodide disc from 90—298 K, the barrier to rotation about the Au—C bonds is calculated to be $270 \pm 70\ cm^{-1}$ [38].

NUCLEAR MAGNETIC RESONANCE SPECTROSCOPY

The isotope ^{197}Au has nuclear spin $I = \frac{3}{2}$ but the associated nuclear quadrupole moment leads to very short relaxation times. Hence, it has neither been possible to observe a ^{197}Au n.m.r. spectrum nor to observe coupling between ^{197}Au and other nuclei, even in spherically symmetrical ions such as $[AuF_6]^-$. Nevertheless n.m.r. spectroscopy is a very useful technique, particularly in organogold chemistry, for determining the stereochemistry of gold(III) complexes, for studying the rates and mechanisms of reactions, and for investigating the nature of bonding in complexes. We shall look at a few examples to illustrate these aspects.

To determine stereochemistry, it is often sufficient to count the number of peaks in the n.m.r. spectrum. For example, complexes of formula $[AuMe_2XL]$ should give one methylgold resonance in the ^1H n.m.r. spectrum for the *trans* isomer but two resonances for the *cis* isomer corresponding to the methyl groups *trans* to X and L, respectively. Thus $[AuMe_2$-Cl(pyridine)] gives two methylgold resonances showing that the *cis*-isomer is formed [30], and many other examples of this type are known [31, 32, 36, 39]. The complex $[Au(N_3)$-$(PPh_3)]$ gives three resonances in the ^{14}N n.m.r. spectrum showing that the azide ligand is σ-bonded to gold with each nitrogen in a different chemical environment [40].

Chemical shifts can also be useful in structure determination. Thus, in methylgold(III) complexes, it is found that a methyl group *trans* to another alkyl group resonates at higher field than a methyl group *trans* to halide or neutral ligands such as tertiary phosphines [2, 23, 41—43]. Thus, the *trans*-$AuMe_2$ group is easily recognised by examination of the ^1H n.m.r. spectrum. ^{31}P chemical shifts cover a wider range than ^1H shifts and the compounds present in complex mixtures can be identified by ^{31}P n.m.r. spectroscopy. Thus, the species [AuClL], $[AuL_2]^+$, $[AuL_3]^+$ and $[AuL_4]^+$ can be identified by their ^{31}P chemical shifts in a mixture of [AuClL] with excess of L, where L = $P(p\text{-tolyl})_3$ [44]. Similarly, separate peaks due to all isomers of all species $[AuCl_nBr_{3-n}(PEt_3)]$ can be observed in the ^{31}P n.m.r. spectrum of a solution of $[AuBr(PEt_3)]$ with chlorine added [45]. Integration of the spectrum allows the relative concentrations of the components to be estimated.

The magnitude and sign of coupling constants to ^{31}P in alkyl-(tertiary phosphine)gold(III) complexes are also useful in determining stereochemistry. Thus, in the two isomers of $[AuMe_2(CF_3)(PMe_3)]$ the magnitude of the coupling constant $^3J(PF)$ is found to be much higher in the *trans*-isomer (66 Hz) than in the *cis*-isomer (7 Hz). The complex $[AuI(Me)(CF_3)(PMe_3)]$ has $^3J(PF)$ 69 Hz, and hence contains the CF_3 and PMe_3 groups in mutually *trans* positions, thus defining the stereochemistry about gold(III). Unfortunately, the magnitude of the related coupling constant $^3J(PH)$ in methyl(tertiary phosphine)gold(III) complexes is not much different for *cis* or *trans* stereochemistry, being 5—10 Hz in both cases. However, the sign of the coupling constant is positive when the methyl group is *cis* to phosphine, and negative when it is *trans* to phosphine, so that if the sign of the coupling constant can be determined, then so can the stereochemistry [46]. In complexes such as *cis*-$[AuMe_2(PMe_3)_2]^+$ or $[Au_2Me_4(\mu\text{-}PPh_2)_2]$, the methyl protons and ^{31}P atoms give a second-order $A_3A_3^1XX^1$ spin system. Half the intensity in the methylgold resonance is present as a sharp 1 : 1 doublet with separation of the peaks given by $J(AX) + J(AX^1)$, which is the sum of the *cis* and *trans* coupling constants $^3J(PH)$. Since these are of similar magnitude but opposite sign, the doublet separation is very small [32, 48].

N.m.r. spectroscopy is often used to study the rates and

mechanisms of fast reactions, particularly those involving intramolecular isomerisation. For example, the resonance in the ^{1}H n.m.r. spectrum of the CH_2 protons in $[(PhCH_2)_2$-$SAuCl_3]$ appears as an AB quartet, since the two hydrogen atoms on each benzyl group are non-equivalent due to the lack of a plane of symmetry containing the S—C bond. In turn, this is due to the pyramidal structure at sulphur which has a stereochemically active lone pair of electrons. Now at temperatures above 62 °C the AB quartet collapses to a single peak, since at high temperatures inversion at the sulphur centre occurs rapidly on the n.m.r. time scale and the protons appear equivalent. Analysis of the spectra at different temperatures allows the activation parameters to be calculated. Inversion at sulphur in $[(PhCH_2)_2SAuCl]$ is much faster than in $[(PhCH_2)_2SAuCl_3]$ and only a singlet for the CH_2 protons of the former compound is observed even at low temperature [49]. Other examples of the use of n.m.r. spectroscopy to study fluxional molecules have been described in earlier chapters.

It is also possible to study intermolecular reactions by n.m.r. spectroscopy. For example, the exchange of phosphine ligands, when $[AuMe(PMe_3)]$ is treated with PMe_3, is indicated by loss of coupling of the methylgold protons to ^{31}P so that the initial 1 : 1 doublet resonance collapses to a singlet [50]. Other examples include the exchange of co-ordinated and free water molecules in an aqueous solution of *cis*-$[AuMe_2$-$(OH_2)_2]^+$ studied by ^{17}O n.m.r. spectroscopy [51], and the collapse of the two methylgold resonances in *cis*-$[AuMe_2Cl$-(pyridine)] to a single peak due to an exchange process in the presence of free pyridine [30]. Finally, n.m.r. spectroscopy can be used as a normal analytical method for studying the rates of reactions by integration of peaks due to either starting material or product. The advantage of the method is that the products can often be identified simultaneously by their n.m.r. spectra. Several applications are known in alkylgold chemistry [42, 43, 48].

There have been only a few studies of bonding in gold complexes using n.m.r. spectroscopy. The best example is based on the theory that the ^{19}F chemical shift is dominated by σ-inductive effects only in *m*-fluorophenyl compounds whereas both σ- and π-bonding will influence the ^{19}F chemical shift in *p*-fluorophenyl compounds. Thus, it is possible to separate the inductive and resonance effects. Examination of complexes [*p*-FC_6H_4AuL] and [*m*-FC_6H_4AuL] suggests that gold acts as both a σ-donor and π-donor to the aryl groups. If L is a tertiary phosphine ligand, then it acts as a π-acceptor ligand, with the strength of the back-bonding increasing in the series L = $Et_3P < Ph_3P < (C_6F_5)Ph_2P < (PhO)_3P$. On the other hand, if L is an aryl isocyanide ligand, then it acts as both a σ- and π-donor to gold, so that back-bonding is negligible in this case [52, 53].

MÖSSBAUER SPECTROSCOPY

Gold-197 is the only naturally occurring isotope of gold and possesses one of the best resonances for chemical applications of Mössbauer spectroscopy. The Mössbauer transition in ^{197}Au is from the 77.3 KeV level of spin $\frac{1}{2}$ to the ground state of spin $\frac{3}{2}$, and hence, in the presence of an asymmetric electric field, the transition is split into two lines of equal intensity as a result of the nuclear quadrupole interaction. The centre of the resonances is called the isomer shift (I.S.) and the separation between the two lines is called the quadrupole splitting (Q.S.).

Although the natural linewidth in ^{197}Au Mössbauer spectra of *ca.* 1.9 mm s^{-1} is much larger than for the commonest Mössbauer isotope ^{57}Fe, the range of isomer shifts (I.S.) and quadrupole splittings (Q.S.) is also much larger for gold, and thus the technique is chemically valuable. Nevertheless, only in recent years has ^{197}Au Mössbauer spectroscopy been applied to chemical problems and the technique is still rarely used. One reason for this is that the γ-ray source is ^{197}Pt, which decays with a half-life of only 18 hours to excited states of ^{197}Au. The ^{197}Pt sources are prepared by neutron bombardment of platinum metal and have a useful life of only a few days. Secondly, to obtain useful spectra, it is necessary to cool both source and absorber to liquid helium temperature.

In the following sections we shall consider trends in the isomer shifts and quadruople splittings for gold compounds as a function of the oxidation state of gold, the co-ordination number at gold and the ligands, and review some applications of ^{197}Au Mössbauer spectroscopy. Some typical values of Mössbauer parameters are given in Tables 10.11 and 10.12.

(a) Isomer shifts

The isomer shift is given by the expression:

$$\text{I.S.} = \text{K} \cdot \frac{\delta R}{R} \cdot \{[\Psi(\text{O})_s]_a{}^2 - [\Psi(\text{O})_s]_s{}^2\}$$

where K is a constant for a given isotope, $\delta R = R_{ex} - R_{gr}$ (the difference in radius of the excited and ground nuclear states, respectively) and $[\Psi(\text{O})_s]_a{}^2$ and $[\Psi(\text{O})_s]_s{}^2$ are the total s-electron densities at the absorber and source gold atoms, respectively. For gold, $\delta R/R$ is positive so that an increase in s-electron density at the ^{197}Au nucleus of the absorber will lead to an increase in isomer shift [54, 55].

Gold(I) can be considered to use sp hybrid orbitals in forming the usual linear complexes, and so stronger σ-donor ligands should give rise to higher s-electron density at the gold nucleus, and hence to higher isomer shifts. This prediction is confirmed by the data in Table 10.11. In addition π-acceptor ligands should decrease the 5d-electron density about gold and, since these d-electrons shield the 6s-electrons from the nuclear charge, this effect will also lead to higher 6s-electron density at the gold nucleus. Thus strong σ-donor, π-acceptor ligands should give the highest isomer shift values. These are the same properties which lead to a ligand being high in the spectrochemical series of ligands, and it is not surprising therefore that there is an excellent correlation between the effect of a ligand on the ^{197}Au isomer shift in gold(I) complexes and its position in the spectrochemical series [5].

TABLE 10.11

Mössbauer parameters for some gold(I) complexes

Compound	I.S./mm s^{-1} [a]	Q.S./mm s^{-1}	Ref.
AuCl	−1.4	4.5	54–56
AuBr	−1.4	4.2	54, 56
AuI	−1.3	4.4	54–56
$[AuCl_2]^-$	−1.4	4.1	54
$[Au(S_2O_3)_2]^{3-}$	0.72	7.08	54
$[AuCl(C_{10}H_{12})]$	+0.83	6.04	55
$[AuI(PPh_3)]$	+1.24	8.3	56
$[AuCl(SMe_2)]$	+1.26	6.42	56
$[Au(N_3)_2]^-$	+1.43	6.84	55
$[AuCl(C_5H_5N)]$	+1.7	6.4	56
$[AuCl(AsPh_3)]$	1.92	7.00	56
[AuCN]	2.3	8.0	55–58
$[Au(SCN)(PPh_3)]$	2.28	8.53	60
$[AuBr(PPh_3)]$	2.76	7.40	56
$[AuCl(PPh_3)]$	2.96	7.47	56
$[Au(N_3)(PPh_3)]$	3.3	8.4	56
$[Au(CN)_2]^-$	3.2	10.2	54, 55, 57, 58
$[Au(CN)(PPh_3)]$	3.9	10.5	56
$[AuMe(PMe_3)]$	4.9	10.2	59
$[AuMe(PPh_3)]$	4.93	10.35	56

[a] Relative to ^{197}Pt/Pt source.

Thus for the complexes $[AuX(PPh_3)]$ the isomer shift follows the series $X^- = Me^- > CN^- > N_3^- > Cl^- > Br^- > I^-$ and for [AuClL] the series is $L = PPh_3 > AsPh_3 > C_5H_5N > Me_2S > Cl^-$. In each case, this is the same as the order of the ligands in the spectrochemical series and, since the σ-donor properties of the ligand are dominant, it also resembles the series of decreasing *trans*-influence of the ligands.

Gold(III) can be considered to use dsp^2-hybrid orbitals in forming square-planar complexes. Thus again σ-donor ligands should increase the s-electron density at the ^{197}Au nucleus and cause an increase in the isomer shift. For the complex ions $[AuX_4]^-$ the isomer shifts decrease in the series $CN^- > N_3^- > SCN^- > Cl^- > Br^- > I^- > F^-$ as expected on this basis

TABLE 10.12

Mössbauer parameters for some gold(III) and gold(V) complexes

Compound	I.S./mm s^{-1} [a]	Q.S./mm s^{-1}	Ref.
Gold(III)			
AuF_3	−1.1	−2.74	54, 61
$Cs[AuF_4]$	+0.09	−0.75	55
$AuBr_3$	+0.2	+1.5	54—56
$K[AuI_4]$	+0.3	+1.28	55
$K[AuBr_4]$	+0.34	+1.13	54—56
$K[AuCl_4]$	+0.4	+0.5	54—56
$[AuCl_3(CNC_6H_4Me)]$	+0.75	+2.00	56
$[AuCl_3(SMe_2)]$	+1.26	+2.20	56
$Au_2O_3 \cdot H_2O$	+1.31	+1.69	55
$K[Au(SCN)_4]$	+1.63	+2.04	55
$Ph_4As[Au(N_3)_4]$	+1.66	+2.89	55
$[AuCl_3(PPh_3)]$	+2.06	+3.25	56
$[AuBr_3(PPh_3)]$	+2.11	+3.37	56
$K[Au(CN)_2Cl_2]$	+2.56	+5.26	54
$K[Au(CN)_2Br_2]$	+2.7	+5.4	54, 55
$K[Au(CN)_2I_2]$	+2.78	+5.87	54
$Et_4N[AuS_4C_4(CN)_4]$	+3.26	+1.97	54
$K[Au(CN)_4]$	+4.1	+6.9	54, 55
$[AuMe_3(PMe_2Ph)]$	+4.8	+9.0	59
Gold(V)			
$[Xe_2F_{11}][AuF_6]$	+2.28	0	61
$Cs[AuF_6]$	+2.39	0	61

[a] Relative to $^{197}Pt/Pt$ source.

(Table 10.12). The low isomer shift of the fluoride complex $[AuF_4]^-$ may be due to the π-donor ability of F^- which will lead to an increase in 5d-electron density, and hence to a decrease in the 6s-electron density at the gold nucleus by the extra shielding effect [61].

The isomer shift may be expected to increase with the oxidation state of gold along the series Au(I), Au(III), Au(V) as the number of 5d electrons formally decreases from 10 to 8 and 6, respectively, and the shielding of s-electrons from

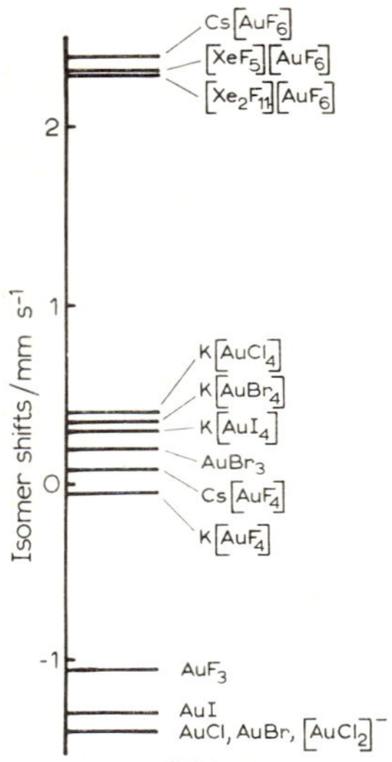

Fig. 10.1. Graphical representation of ^{197}Au isomer shifts for halides and halide complexes of gold(I), gold(III) and gold(V).

the nucleus decreases accordingly. The prediction is upheld for the gold halide complexes as illustrated in Fig. 10.1 [61]. The anomalous value of the isomer shift of AuF_3 is probably related to its unique structure, with tetragonal rather than square-planar co-ordination about the gold(III) centres.

Although it might be thought possible, from the above arguments, to distinguish between oxidation states of gold purely on the basis of isomer shift, examination of the data in Tables 10.11 and 10.12 clearly shows that this is not the case. Thus there is extensive overlap of the ranges of isomer shifts for all oxidation states.

Finally, it is interesting to note that for both gold(I) and gold(III) complexes an approximate law of average environment operates. Thus AuCN contains gold(I) linearly co-ordinated to one carbon and one nitrogen atom, and its isomer shift (2.3 mm s^{-1}) is approximately the mean of those for $[Au(N_3)_2]^-$ (1.43 mm s^{-1}) and $[Au(CN)_2]^-$ (3.2 mm s^{-1}).

Similarly, for gold(III) the complexes *trans*-$[Au(CN)_2X_2]^-$ have isomer shifts midway between those of $[Au(CN)_4]^-$ and $[AuX_4]^-$, where X = Cl, Br or I [54, 55].

(b) Quadrupole splittings

For ^{197}Au the quadrupole splitting is given by the expression:

$$\text{Q.S.} = \tfrac{1}{2}\, e^2qQ(1 + 1/3\eta^2)^{1/2}$$

where $eq = -V_{zz} = -$ the z component of the electric field gradient, Q is the quadrupole moment of the nucleus and $\eta = (q_{xx} - q_{yy})/q_{zz} =$ the asymmetry parameter. For ^{197}Au Q is positive, and hence the sign of the quadrupole splitting will be determined by the sign of V_{zz}. Generally the magnitude of the quadrupole splitting is dominated by the degree of asymmetry in occupation of the p-orbitals about the Mössbauer nucleus.

For linear gold(I) complexes, the z-axis is taken as the molecular axis and σ-bonding involves donation of electrons into the $6s6p_z$ hybrid orbitals of gold. Thus V_{zz} and hence the sign of the quadrupole splitting is expected to be negative, and this has been confirmed experimentally for $K[Au(CN)_2]$ [57]. Furthermore, stronger σ-donor ligands are expected to give higher occupation of both 6s and $6p_z$ orbitals, leading to higher values of both isomer shift and quadrupole splitting. Indeed, an approximate relationship between isomer shift and quadrupole splitting for gold(I) complexes is observed as illustrated in Fig. 10.2 and is given by the equation:

$$\text{Q.S.} = -\text{I.S.} - 5.6 \text{ mm s}^{-1}$$

For gold(III) the z-axis is the axis perpendicular to the plane defined by the AuL_4 group. For an ionic Au^{3+} complex a negative quadrupole splitting would be expected, since there would be a concentration of electron density along the z-axis due to the doubly occupied $5d_{z^2}$ orbital and vacant $5d_{x^2-y^2}$ orbital. However, in covalent σ-bonding, gold(III) uses the $5d_{x^2-y^2}$, 6s, $6p_x$ and $6p_y$ orbitals, and occupation of these p- and d-orbitals will produce a

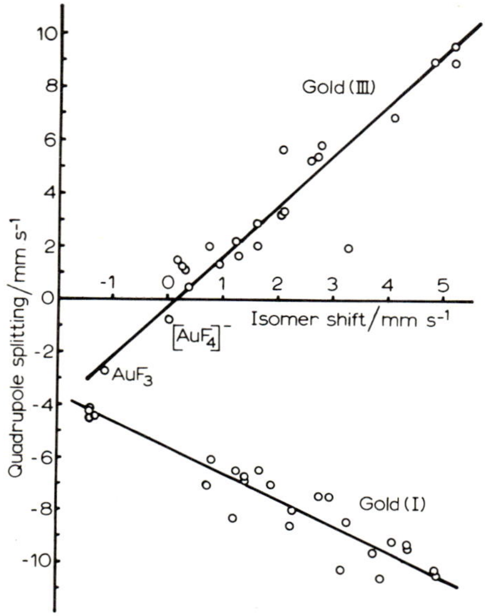

Fig. 10.2. Correlation of isomer shifts and quadrupole splittings for gold(I) and gold(III) complexes. The signs of quadrupole splittings for gold(III) are assumed (see text).

positive contribution to the field gradient. Calculations indicate that covalent bonding dominates and that the quadrupole splittings are positive, except for the ionic fluoride complexes, although the predictions have not yet been confirmed experimentally. Strong indirect evidence is obtained from the observation that the quadrupole splitting increases along with the isomer shift as the σ-donor properties of the ligands increase. Again, there is a direct correlation illustrated in Fig. 10.2 and given by the equation:

$$Q.S. = 2.0\ I.S. - 0.5\ mm\ s^{-1}$$

It can be seen from Fig. 10.2 that there is a considerable spread of points about the best correlation for both gold(I) and gold(III) complexes, but if both the isomer shift and quadrupole splitting for an unknown complex are determined, it should be possible to distinguish between oxidation states (I) and (III) for gold in most cases. It should also be possible to distinguish between gold(II) and mixed valence gold(I)—gold(III) complexes.

(c) Applications to some chemical problems

In recent years ^{197}Au Mössbauer spectroscopy has been useful in aiding the characterisation of gold complexes. For example, reaction of but-2-yne with [Au_2Cl_6] gives a complex $Au_2Cl_4 \cdot$ 2MeCCMe. The ^{197}Au Mössbauer spectrum of this complex contains two doublets with I.S. 1.46, Q.S. 6.74 and I.S. 0.97, Q.S. 1.37 mm s^{-1}, respectively [62]. The former peaks are due to a gold(I) complex and the latter to a gold(III) complex as determined by the relationship between the I.S. and Q.S. values. Further, the gold(III) complex can be identified as $[AuCl_4]^-$ by the fingerprint technique by comparison with [Bu_4N][$AuCl_4$] which has I.S. 1.02, Q.S. 1.31 mm s^{-1}. Hence, the unknown complex is formulated as $[Au(MeC{\equiv}CMe)_2]^+[AuCl_4]^-$. A similar method is useful for studying other alkene and alkyne complexes of gold [63, 64].

In another example, the reaction of [$AuMe(PMe_3)$] with hexafluoro-but-2-yne was found to give a complex of empirical formula [$\{AuMe(PMe_3)\}_2C_4F_6$]. Initially, a symmetrical structure was proposed, but the ^{197}Au Mössbauer spectrum showed the presence of both gold(I) and gold(III) centres and the correct structure [$Me_3PAuC(CF_3){=}C(CF_3)AuMe_2$-($PMe_3$)] was then confirmed [59].

Recently, the Mössbauer spectrum of a gold(II) complex has been obtained and the I.S. and Q.S. found to be intermediate between those for similar gold(I) and gold(III) complexes. Thus further applications of Mössbauer spectroscopy to characterisation of unusual oxidation states of gold may be predicted [65].

Finally, it is suggested that both isomer shifts and quadrupole splittings are lower for three-co-ordinate gold(I) complexes such as [$AuCl(PPh_3)_2$] than for similar linear gold(I) complexes such as $[AuCl(PPh_3)]$. Hence, it may be possible to determine the co-ordination number of a gold(I) complex from its ^{197}Au Mössbauer spectrum [66].

NUCLEAR QUADRUPOLE RESONANCE SPECTROSCOPY

The pure quadrupole resonance spectrum of the ^{197}Au nucleus has been observed only from AuCl and AuI. The quadrupolar nuclear coupling for AuCl is about 514 MHz or 4.1 mm s^{-1} in reasonable agreement with the values obtained from the Mössbauer spectrum. This high value is interpreted in terms of gold(I) using $6s6p$ hybrid orbitals and with the bonding being largely covalent [66, 67]. It is also suggested that ligand-to-metal π-bonding is more important for AuI than for AuCl.

From the ^{35}Cl and ^{37}Cl n.q.r. spectra of $[AuCl_2]^-$ and $[AuCl_4]^-$ the extent of ionic character in the AuCl bonds is estimated as 68 and 41%, respectively, leading to effective charges on gold of +0.36 and +0.64, respectively [68, 69]. There is rather poor agreement with the calculated charges on gold of +0.11 and +0.20 for $[AuCl_2]^-$ and $[AuCl_4]^-$, respectively, from m.o. calculations [70].

For the complexes [LAuCl], a good σ-donor ligand L increases the isomer shift in the ^{197}Au Mössbauer spectrum, but causes an increase in ionic character of the Au—Cl bond, thus decreasing the value of ν(Cl) in the chlorine n.q.r. spectrum. Thus a reasonable correlation is found between these two parameters as L is varied [60].

PHOTOELECTRON SPECTROSCOPY

Very little work has been done on either the X-ray photoelectron spectra (ESCA) or the ultraviolet photoelectron spectra (UPE) of gold complexes, though both techniques have considerable potential interest.

Thus the ESCA spectra of related gold(I), gold(II) and gold(III) complexes derived from phosphine ylid ligands show that the core Au($4f_{7/2}$) binding energies increase with oxidation state of gold. Thus it should be possible to determine oxidation states of gold in related complexes using the ESCA technique [65].

Ultraviolet photoelectron spectroscopy is more limited in scope, since spectra can only be obtained for volatile compounds. Thus complexes [AuMeL] and [$AuMe_3L$], but not [AuClL] or [$AuCl_3L$], are found to give satisfactory spectra. Some results with tentative assignments are given in Table 10.13. It is difficult to make specific assignments

TABLE 10.13

Valence ionisation energies of methylgold complexes/eV

Assignment	$MeAuPMe_3$	Me_3AuPMe_3	$Me_3AuPMePh_2$
σ(AuC)		7.80	7.64
	8.27	8.63	8.34
π(Ph)			9.48
5*d*	9.34		
	9.95	10.18	9.98
	10.63	10.70	10.56
	11.45	11.06	10.98

for the 5d ionisations because ligand-field effects and spin—orbit coupling have similar magnitudes in gold complexes. Note, however, that the ionisation energy is less for the trimethylgold(III) complexes than for the methylgold(I) complex. This is probably due to the destabilising effect of two mutually *trans* methyl groups in the gold(III) complexes causing the σ(AuC) ionisation to be lowered. There is also little difference between the mean 5d levels for the gold(I) and gold(III) complexes. Clearly the effect of increased oxidation state is compensated by the extra σ-donor methyl ligand in the gold(III) complexes [71].

Greatest interest in the electronic spectra of gold(III) complexes has been in the identification and assignment of the d—d bands with the aim of determining the relative energies of the 5d-orbitals in square-planar complexes.

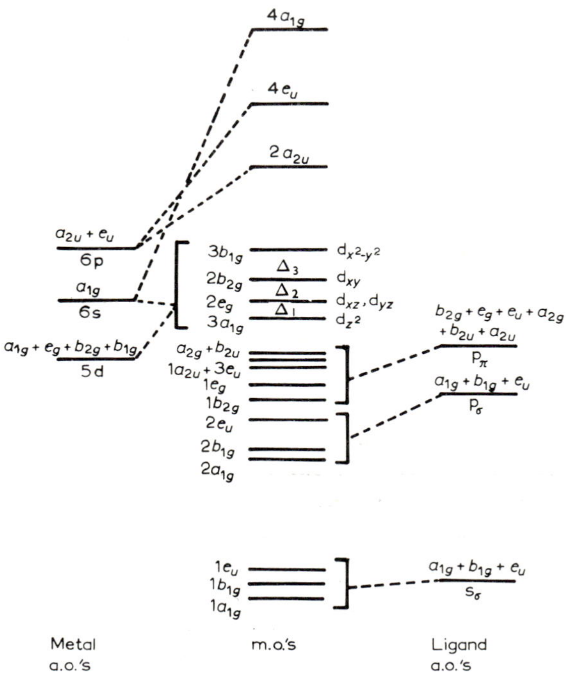

Fig. 10.3. Qualitative molecular-orbital energy-level diagram for a square-planar gold(III) complex.

A qualitative molecular orbital energy level diagram for a square-planar halide complex $[MX_4]^{n-}$ is shown in Fig. 10.3 [72, 73]. The order of d-orbital energies is $d_{x^2-y^2} > d_{xy} > d_{xz}, d_{yz} > d_{z^2}$ and is confirmed by two further molecular orbital calculations on $[AuX_4]^-$ [74, 75]. Of these 5d-orbitals, all but $d_{x^2-y^2}$ are doubly occupied in the ground state. There are three ligand-field parameters illustrated in Fig. 10.3, the Δ_1, Δ_2 and Δ_3, defining d-orbital splitting in the complexes. Calculations on the $[AuCl_4]^-$ ion indicate that the $5d_{z^2}$ orbital is actually lower in energy than the largely ligand p_π-orbitals and also the p_σ-orbital, $2e_u$ (Fig. 10.3) [70]. In general then, the lowest energy transitions in the electronic spectrum of $[AuX_4]^-$ are expected to be (using the orbital approximation for simplicity) $5d_{xy} \rightarrow 5d_{x^2-y^2}$, $5d_{xz,yz} \rightarrow 5d_{x^2-y^2}$, ligand $p_\pi \rightarrow 5d_{x^2-y^2}$ and ligand $p_\sigma \rightarrow 5d_{x^2-y^2}$, in order of increasing energy.

The electronic spectra of complexes $[AuX_4]^-$ do indeed contain one or two weak bands at low energy as expected for orbitally forbidden d—d transitions, followed at higher energy by two strong ligand-to-metal charge-transfer bands. Some data together with the assignments are given in Table 10.14, which also includes the calculated transition energies [70, 74]. The assignments given in Table 10.14 are those due to Mason and Gray [73, 76], and are in reasonable agreement with the predictions according to molecular orbital theory [70, 74]. However, a later study of the electronic spectrum of single crystals of $Na[AuCl_4] \cdot 2H_2O$ indicates strongly that the assignment of the band at 26,300 cm^{-1} to the $5d_{xz}$, $d_{yz} \rightarrow 5d_{x^2-y^2}$ transition is not correct. Instead, this was thought due to the $5d_{xy} \rightarrow 5d_{x^2-y^2}$ transition, with the lowest energy band due to the $5d_{xy} \rightarrow 6s$ transition [75]. Other authors have suggested that all the bands are charge transfer in origin and that no d—d bands are observed [77]. No doubt, arguments will continue for some time, and the difficulty of making positive assignments may be complicated yet further by the effects of spin—orbit coupling.

What is clear is that the spectra change markedly as the ligand X in $[AuX_4]^-$ is varied. As expected the ligand-to-metal charge-transfer bands move to lower energy along the series X=Cl > Br > I as it becomes progressively easier to oxidise the halide, and these bands are also at lower energy than in the corresponding $[PtX_4]^{2-}$ ions as a result of the higher oxidation state of gold(III) over platinum(II). It is more difficult to comment on trends in the ligand-field parameters owing to the

TABLE 10.14

Observed and calculated electronic spectra for some $[AuX_4]^-$ ions (band maxima/1000 cm^{-1})

Assignment[a]	X = Cl[b]		X = Br[c]		X = I[d]		X = SCN[e]	
	obs.	calc.	obs.	calc.	obs.	calc.	obs.	calc.
$2b_{2g}(d_{xy}) \rightarrow 2b_{1g}(d_{x^2-y^2})$ / $^1A_{1g} \rightarrow {}^1A_{2g}$	21.7	22.3	18.5	17.9	10.8	6.8	20.0	20.1
$2e_g(d_{xz}, d_{yz}) \rightarrow 3b_{1g}(d_{x^2-y^2})$ / $^1A_{1g} \rightarrow {}^1E_g$	26.3	26.4	21.2	21.4	10.8	7.5	25.2	22.8
$p\pi \rightarrow 3b_{1g}(d_{x^2-y^2})$ / $^1A_{1g} \rightarrow {}^1A_{2u}, {}^1E_u(1)$	30.4	30.9—32.3	24.8	25.0—26.2	12.7?	8.2—9.0	31.2	25.6—26.6
$2e_u(p\sigma) \rightarrow 3b_{1g}(d_{x^2-y^2})$ / $^1A_{1g} \rightarrow {}^1E_u(2)$	44.5	46.5	39.7	40.7	17.0, 23.2	21.3	—	—

[a] The alternative orbital and state transition nomenclatures are given.
[b] Refs. 7, 73, 75—80.
[c] Refs. 70, 73, 76—80.
[d] Refs. 70, 77.
[e] Refs. 74, 76.

uncertainty of the assignments, but some trends become apparent. Thus, there is a remarkable change by a factor of two in the energies of the d—d bands between $[AuCl_4]^-$ and $[AuI_4]^-$, and the bands are all at lower energy than for the corresponding $[PtX_4]^{2-}$ ions. This latter observation is exceptional to the general rule that increase in oxidation state leads to higher ligand-field splittings. It is suggested, that the σ^*-orbital $5d_{x^2-y^2}$ has more pure metal character for complexes of Au^{3+} than for Pt^{2+}, but that the $5d_{xy}$ level is more strongly destabilised by interaction with the ligand π-orbitals for Au^{3+} than for Pt^{2+}. The result is that the separation $5d_{xy} \rightarrow 5d_{x^2-y^2}$ is less for gold(III) complexes than for platinum(II) complexes [73, 76]. The very low separation in $[AuI_4]^-$ is due to very strong ligand-to-metal π-bonding in this complex. Another way of looking at this is to say that the orbital $2b_{2g}$ $(5d_{xy})$ has considerable ligand character, and hence that the transition $2b_{2g}(5d_{xy}) \rightarrow 3b_{1g}(5d_{x^2-y^2})$ has considerable ligand-to-metal charge-transfer character and its energy takes on the character of such a transition, decreasing sharply for the larger halide ions.

As well as the gold(III) complexes discussed above, electronic spectra have been discussed for $[Au(N_3)_4]^-$ [81], $[Au(NH_3)_4]^{3+}$ [73], $[Au(CN)_4]^-$ [73] and for several other amine complexes of gold(III) [79, 80, 82]. It has also been noted that the ligand-to-metal charge-transfer transitions in organogold halides $(R_2AuBr)_2$ and $[R_2AuBr(NH_3)]$ are at much higher energies than in $[AuBr_4]^-$, and the result is ascribed to the strong destabilisation of the σ^*-orbital $5d_{x^2-y^2}$ on gold by the strong σ-donor alkyl groups R [83].

Although the u.v. spectrum of $[Au(CN)_4]^-$ is clear up to 50,000 cm^{-1}, the spectrum of $[Au(CN)_2]^-$ contains many bands below this energy. These bands are due to metal-to-ligand charge-transfer transitions, which are, as expected, at lower energy for the gold(I) than for gold(III). A simplified molecular orbital energy level diagram for $[Au(CN)_2]^-$ is shown in Fig. 10.4, and detailed analysis of the u.v. spectrum of $[Au(CN)_2]^-$ confirms that the order of d-orbitals is $5d_{z^2} \geqslant 5d_{xy}, 5d_{x^2-y^2} > 5d_{xz}, 5d_{yz}$, taking the z-axis as the molecular

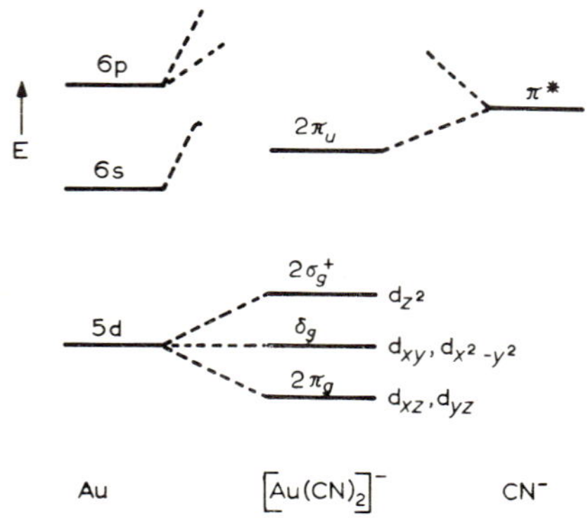

Fig. 10.4. Simplified molecular-orbital energy-level diagram for $[Au(CN)_2]^-$.

axis. Since the $5d_{xy}$ and $5d_{x^2-y^2}$ are non-bonding, it is clear that $5d_{z^2}$ is slightly destabilised by σ-bonding, and that $5d_{xz}$ and $5d_{yz}$ are slightly stabilised by π-bonding to vacant π^*-orbitals of the cyanide ligands. However the total splitting of the 5d levels is estimated to be no more than 5000 cm^{-1}, indicating that the ten 5d-electrons are essentially core electrons and are involved only weakly in bonding. There is reasonable agreement between the observed and calculated transition energies, lending support to these conclusions [74, 84]. The u.v. spectra of $[AuX_2]^-$, X = halide, have not been reported, but are expected to contain only very high energy bands, since d—d transitions are not possible for gold(I), and ligand-to-metal charge-transfer transitions are expected to be at high energy for this low oxidation state of gold.

REFERENCES

1 P. Braunstein and R. J. H. Clark, *J. Chem. Soc., Dalton Trans.*, (1973) 1845.
2 G. W. Rice and R. S. Tobias, *Inorg. Chem.*, 14 (1975) 2402; 15 (1976) 489.
3 L. H. Jones, *J. Chem. Phys.*, 43 (1965) 594.
4 P. Braunstein and J. Dehand, *J. Organomet. Chem.*, 88 (1975) C24.
5 G. A. Bowmaker and R. Whiting, *Aust. J. Chem.*, 29 (1976) 1407.
6 E. J. Baran, *Spectrosc. Lett.*, 8 (1975) 151.
7 K. Leary and N. Bartlett, *J. Chem. Soc., Chem. Commun.*, (1972) 903.

8 P. L. Goggin and J. Mink, *J. Chem. Soc., Dalton Trans.*, (1974) 1479.
9 R. S. Tobiàs, in A. Anderson (Ed.), *The Raman Effect*, Vol. 2, Marcel Dekker, New York, 1973, p. 457.
10 L. H. Jones and J. M. Smith, *J. Chem. Phys.*, 41 (1964) 2507.
11 M. Weishaupt and J. Straehle, *Z. Naturforsch. B*, 31B (1976) 554.
12 Y. M. Bosworth and R. J. H. Clark, *Chem. Phys. Lett.*, 28 (1974) 611.
13 J. H. Holloway and G. J. Schrobilgen, *J. Chem. Soc., Chem. Commun.*, (1975) 623.
14 D. R. Williamson and M. C. Baird, *J. Inorg. Nucl. Chem.*, 34 (1972) 3393.
15 J. M. Meyer and A. L. Allred, *J. Inorg. Nucl. Chem.*, 30 (1968) 1328.
16 A. D. Westland, *Can. J. Chem.*, 47 (1969) 4135.
17 G. E. Coates and C. Parkin, *J. Chem. Soc.*, (1963) 421.
18 A. G. Jones and D. B. Powell, *Spectrochim. Acta*, 30A (1974) 563; 30A (1974) 1001.
19 H. Schmidbaur and R. Franke, *Chem. Ber.*, 105 (1972) 2985.
20 E. A. Allen and W. Wilkinson, *Spectrochim. Acta*, 28A (1972) 2257.
21 P. L. Goggin, R. J. Goodfellow, S. R. Haddock, F. J. S. Reed, J. G. Smith and K. M. Thomas, *J. Chem. Soc., Dalton Trans.*, (1972) 1904.
22 D. A. Duddell, P. L. Goggin, R. J. Goodfellow, M. G. Norton and J. G. Smith, *J. Chem. Soc. A*, (1970) 545.
23 C. F. Shaw and R. S. Tobias, *Inorg. Chem.*, 12 (1973) 965.
24 L. Cattalini, R. J. H. Clark, A. Orio and C. K. Poon, *Inorg. Chim. Acta*, 2 (1968) 62.
25 P. Braunstein and R. J. H. Clark, *Inorg. Chem.*, 13 (1974) 2224.
26 K. S. Liddle and C. Parkin, *J. Chem. Soc., Chem. Commun.*, (1972) 26.
27. T. Boschi, B. Crociani, L. Cattalini and G. Marangoni, *J. Chem. Soc. A*, (1970) 2408.
28 L. H. Jones, *Inorg. Chem.*, 3 (1964) 1581.
29 W. M. Scovell and R. S. Tobias, *Inorg. Chem.*, 9 (1970) 945.
30 H. Hagnauer, G. C. Stocco and R. S. Tobias, *J. Organomet. Chem.*, 46 (1972) 179.
31 S. W. Krauhs, G. C. Stocco and R. S. Tobias, *Inorg. Chem.*, 10 (1971) 1365; G. C. Stocco and R. S. Tobias, *J. Am. Chem. Soc.*, 93 (1971) 5057.
32 A. Shiotani and H. Schmidbaur, *Chem. Ber.*, 104 (1971) 2838.
33 D. Breitinger and H. Leuchtenstein, *Z. Naturforsch. B*, 29B (1974) 806.
34 D. M. Adams and R. G. Churchill, *J. Chem. Soc. A*, (1968) 2141. I. R. Beattie, T. R. Gilson and G. A. Ozin, *J. Chem. Soc. A*, (1968) 2765.
35 R. Fornieris, J. Hiraishi, F. A. Miller and M. Uehara, *Spectrochim. Acta, A*, 26A (1970) 581.
36 W. M. Scovell, G. C. Stocco and R. S. Tobias, *Inorg. Chem.*, 9 (1970) 2682.
37 M. G. Miles, G. E. Glass and R. S. Tobias, *J. Am. Chem. Soc.*, 88 (1966) 5738.
38 R. C. Leech, D. B. Powell and N. Sheppard, *Spectrochim. Acta*, 22 (1966) 1931.
39 F. Stocco, G. C. Stocco, W. M. Scovell and R. S. Tobias, *Inorg. Chem.*, 10 (1971) 2639.
40 W. Beck, W. Becker, K. F. Chew, W. Derbyshire, N. Logan, D. M. Revitt and D. B. Sowerby, *J. Chem. Soc., Dalton Trans.*, (1972) 245.
41 H. Schmidbaur and A. Shiotani, *Chem. Ber.*, 104 (1971) 2821.
42 A. Tamaki and J. K. Kochi, *J. Organomet. Chem.*, 64 (1974) 411.
43 A. Johnson and R. J. Puddephatt, *J. Organomet. Chem.*, 85 (1975) 115.
44 E. L. Muetterties and C. W. Allegranti, *J. Am. Chem. Soc.*, 92 (1970) 4114.
45 B. T. Heaton and R. J. Kelsey, *Inorg. Nucl. Chem. Lett.*, 11 (1975) 363.
46 J. D. Kennedy, W. McFarlane and R. J. Puddephatt, *J. Chem. Soc., Dalton Trans.*, (1976) 745.
47 A. Johnson and R. J. Puddephatt, *J. Chem. Soc., Dalton Trans.*, (1976) 1360.
48 R. J. Puddephatt and P. J. Thompson, *J. Organomet. Chem.*, 117 (1976) 395.

49 F. Coletta, R. Ettore and A. Gambaro, *Inorg. Nucl. Chem. Lett.*, 8 (1972) 667.
50 H. Schmidbaur, H. F. Klein and A. Shiotani, *Chem. Ber.*, 104 (1971) 2831.
51 G. E. Glass, W. B. Schwabacher and R. S. Tobias, *Inorg. Chem.*, 7 (1968) 2471.
52 D. I. Nichols, *J. Chem. Soc. A*, (1970) 1216.
53 L. G. Vaughan and W. A. Sheppard, *J. Am. Chem. Soc.*, 91 (1969) 6151; *J. Organomet. Chem.*, 22 (1970) 739.
54 M. O. Faltens and D. A. Shirley, *J. Chem. Phys.*, 53 (1970) 4249.
55 H. D. Bartunik, W. Potzel, R. L. Mössbauer and G. Kaindl, *Z. Phys.*, 240 (1970) 1.
56 J. S. Charlton and D. I. Nichols, *J. Chem. Soc. A*, (1970) 1484.
57 H. Prosser, G. Wortman, K. Syassen and W. B. Holzapfel, *Z. Phys. B*, 24 (1976) 7.
58 L. Pfeiffer, R. S. Raghavan, C. P. Lichtenwalner and K. W. West, *Phys. Rev. Lett.*, 30 (1973) 635.
59 J. A. J. Jarvis, A. Johnson and R. J. Puddephatt, *J. Chem. Soc., Chem. Commun.*, (1973) 373; A. Johnson, *Ph.D. Thesis*, University of Liverpool, 1975.
60 P. G. Jones, A. G. Maddock, M. J. Mays and A. F. Williams, Abstract of *16th Meeting, Mössbauer Discussion Group, University of Liverpool, 1976.*
61 G. Kaindl, K. Leary and N. Bartlett, *J. Chem. Phys.*, 59 (1972) 5050.
62 R. Hüttel and H. Forkl, *Chem. Ber.*, 105 (1972) 1664.
63 P. Tauchner and R. Hüttel, *Chem. Ber.*, 107 (1974) 3761.
64 R. Hüttel, P. Tauchner and H. Forkl, *Chem. Ber.*, 105 (1972) 1.
65. H. Schmidbaur, J. R. Mandl, F. E. Wagner, D. F. van der Vondel and G. P. van der Kelen, *J. Chem. Soc., Chem. Commun.*, (1976) 170.
66 P. Machmer, M. Read and P. Cornil, *C. R. Acad. Sci. Ser. A,B*, 262B (1966) 650; *Inorg. Nucl. Chem. Lett.*, 3 (1967) 215.
67 P. Machmer, *Z. Naturforsch.*, 21 (1966) 1025; *J. Inorg. Nucl. Chem.*, 30 (1968) 2627.
68 G. A. Bowmaker and R. Whiting, *Aust. J. Chem.*, 29 (1976) 1407.
69 A. Sasane, T. Matuo, D. Nakamura and M. Kubo, *J. Magn. Reson.*, 4 (1971) 257.
70 J. Reinhold, H. Zwanziger, E. Hoyer and C. Zwanziger, *Z. Chem.*, 14 (1974) 314; H. Zwanziger, J. Reinhold and E. Hoyer, *Z. Chem.*, 14 (1974) 489.
71 J. Behan, R. A. W. Johnstone and R. J. Puddephatt, unpublished results.
72 H. Basch and H. B. Gray, *Inorg. Chem.*, 6 (1967) 365.
73 W. R. Mason and H. B. Gray, *J. Am. Chem. Soc.*, 90 (1968) 5721.
74 H. Zwanziger, J. Reinhold and E. Hoyer, *Z. Chem.*, 15 (1975) 69.
75 D. H. Brown and W. E. Smith, *J. Chem. Soc., Dalton Trans.*, (1976) 848.
76 W. R. Mason and H. B. Gray, *Inorg. Chem.*, 7 (1968) 55.
77 J. L. Ryan, *Inorg. Chem.*, 8 (1969) 2058.
78 A. J. McCaffery, P. N. Schatz and P. J. Stephens, *J. Am. Chem. Soc.*, 90 (1968) 5730.
79 H. Ito, J. Fujita and K. Saito, *Bull. Chem. Soc. Jpn.*, 40 (1967) 2584.
80 A. K. Gangopadhyay and A. Chakravorty, *J. Chem. Phys.*, 35 (1961) 2206.
81 H.-H. Schmidtke and D. Garthoff, *J. Am. Chem. Soc.*, 89 (1967) 1317; W. Beck, W. P. Fehlhammer, P. Pöllmann, E. Schuirer and K. Feldl, *Chem. Ber.*, 100 (1967) 2335.
82 W. H. Baddley, F. Basolo, H. B. Gray, C. Nolting and A. J. Poë, *Inorg. Chem.*, 2 (1963) 921.
83 A. Chakravorty, *Naturwissenschaften*, 48 (1961) 643.
84 W. R. Mason, *J. Am. Chem. Soc.*, 95 (1973) 3573; 98 (1976) 5182.

CHAPTER 11

ANALYSIS AND APPLICATIONS OF GOLD COMPLEXES

ANALYTICAL METHODS FOR GOLD [1—3]

Many analytical procedures for gold determination rely on gold being present as $[AuCl_4]^-$. Gold as the pure metal, as a constituent of an ore, or in many complexes can readily be converted to the required tetrachloroaurate(III) by dissolution in aqua regia. Nitric acid and oxides of nitrogen are then removed by evaporation, addition of more concentrated hydrochloric acid and then further evaporation. This procedure is repeated twice to ensure complete removal of oxides of nitrogen [1]. Alternatively, dissolution of gold by chlorine or hydrogen peroxide in hydrochloric acid is occasionally used [4]. It is, of course, possible to analyse gold in the form of other complexes than $[AuCl_4]^-$ and examples will be found in later sections of this chapter.

Another general technique, which is widely used in analysis for gold, is solvent extraction. This may be used to concentrate a dilute aqueous solution of a gold complex or to separate gold from interfering impurities. Thus, $H[AuCl_4]$ may be extracted from solution in hydrochloric acid in very high yield by such solvents as diethyl ether, ethyl acetate, methyl iso-butyl ketone, *iso*-amyl alcohol and dichloromethane—tetrahydrofuran [5, 6]. The organic phase may then be estimated directly by neutron activation, atomic absorption or spectrophotometric methods [2, 3, 5].

Gold can also be concentrated prior to analysis by chromat graphy using an ion-exchange resin or, in a recent developme a polymeric resin containing pendant thiol groups which selectively complex to gold [7—9]. For example, a solution containing gold(III) can be chromatographed on a hexylth glycollate resin. Interfering ions such as Ag(I), Hg(II) and can then be removed by elution with hydrochloric acid, and finally the gold is eluted with thiorea solution as the complex $[Au\{S{=}C(NH_2)_2\}_2]^+$ and estimated by spectrophotometry. Enrichment factors of 10^8–10^{11} have been obtained by chromatographic methods [8].

In the following sections a brief mention will be made of the methods used for analysis of gold in chemical laboratories. It would not be appropriate to give detailed recipes here, and only the outlines of the methods will be given. The traditional fire assay method of analysis of gold in ores is used only in specialist laboratories and has been extensively reviewed elsewhere [10]. It will therefore be omitted here.

(a) Gravimetric methods [11]

Most gravimetric methods for gold are based on reduction of a gold-containing compound to the metal and weighing the gold in this form. If gold is present as $[AuCl_4]^-$, reduction with oxalic acid, sulphur dioxide, hydroquinone or sodium nitrite is most common, the preferred method depending on the interfering ions present. Formaldehyde, hydrazine salts and hypophosphorous acid are used less commonly. The disadvantage of this general method is that the gold may be formed in a finely divided state which adheres strongly to the vessel in which it is precipitated. Thus filtration and transfer may be difficult.

It is also possible to precipitate gold quantitatively as insoluble complexes. These are often of uncertain composition as when citarin, dimethylglyoxime or 8-quinolinol are used as precipitants. The compounds are then filtered off and ignited to gold metal. Other complexes are precipitated in pure form and may be weighed directly. Thus stable salts of $[AuCl_4]^-$ or $[AuBr_4]^-$ are precipitated by bulky cations such as N-(N-bromo-C-tetradecylbetainyl)-C-tetra-decylbetaine, bis(trimethyl)hexamethylenediammonium or 2,4,6-triphenylpyrilium [1, 11, 12]. Derivatives of sulphur–donor ligands may also be suitable. Thus thioglycollic acid, thiophenol, di-2-thienylketoxime and *m*-phenylenebis(1-tetrazoline-5-thione) all form derivatives which may be weighed directly [11, 13, 14]. The derivative of Reinecke's salt $Au[Cr(NH_3)_2(SCN)_4]$ has also been used [11].

(b) Titrimetric methods [1, 2]

Most titrimetric methods for gold depend on reduction of gold(III) to metallic gold. For example, hydroquinone reduces gold(III) to gold rapidly and quantitatively. The titration may be carried out using indicators such as O-anisidine or benzidine to determine the endpoint or by potentiometric titration. Alternatively, excess of hydroquinone may be added, and the excess back-titrated with cerium(IV) [15]. Methods based on reduction of gold(III) to gold by iron(II), titanium(III), oxalate or ascorbic acid have also been described [1].

Reduction of gold(III) to AuI by excess of iodide, with titrimetric estimation of the released iodine, is the basis of another method, while estimation of the cyanide used in oxidation of gold to $[Au(CN)_2]^-$ by air has also been used to determine gold.

(c) Spectrophotometric methods [2, 5]

Spectrophotometric methods are particularly useful when insufficient sample is present for the gravimetric or titrimetric methods. Although interference by other metal ions is often a serious problem, suitable systems have been developed for analysis of gold in the presence of most of the likely impurities.

The simplest method is to convert the gold complex to $[AuCl_4]^-$ or $[AuBr_4]^-$ in aqueous solution and to measure the absorbance due to these ions. Many impurities interfere, however, and it is usual to extract the $[AuCl_4]^-$ or $[AuBr_4]^-$ into an organic solvent before measuring the absorbance. In dichloromethane—tetrahydrofuran the species present is thought to be $[thf\text{-}H]^+[AuCl_4]^-$, while in chloroform containing tetraphenylarsonium chloride the ion pair $[Ph_4As]^+[AuCl_4]^-$ is formed [5, 6].

In other methods, various reagents can be added to form coloured complexes of gold and these are then estimated spectrophotometrically. One popular reagent is rhodanine which forms an intensely red complex with $[AuCl_4]^-$. In aqueous solution the complex is formed as a colloidal suspension and the colour is unstable, but this problem can be overcome by solvent extraction [5], or by addition of pyridine, since the complex is soluble in aqueous pyridine [16]. The number of other reagents used is too great to allow detailed discussion

TABLE 11.1

Colorimetric reagents for gold

Aqueous phase	Reagent/organic solvent	Ref.
$[AuCl_4]^-$, HCl	Phenyl-α-pyridyl ketoxime in chloroform	17
$[AuCl_4]^-$, HCl	4-(2-Pyridylazo)resorcinol in xylene	18
$[AuCl_4]^-$, HNO_3	4-(2-Thiazolylazo)resorcinol in xylene-butanol	19
$[AuCl_4]^-$, HCl	Chrompyrazole I in toluene	20
$[AuCl_4]^-$, HCl	*o*-Phenylenediamine	21
$[AuCl_4]^-$, HCl	Malachite green in benzene	22
$[AuCl_4]^-$, HCl	Victoria pure blue dye	23
Au(III), alkaline	Phthalophenone	24
$[AuCl_4]^-$, HCl	Rhodamine-B in di-*iso*propyl ether	25

[1—3, 5], but some are included in Table 11.1. Most methods rely on extraction of the coloured complexes into an organic solvent, a procedure which often separates gold from interfering ions and which can concentrate the gold-containing species and thus make possible the analysis of very dilute solutions of $[AuCl_4]^-$.

A related method is based on oxidation of luminol by $[AuCl_4]^-$ and observation of the resultant chemiluminescence. The luminescence is however suppressed by ions such as thiocyanate [26].

(d) Electrochemical methods

These methods are not often used, but ion-selective electrodes with high specificity for $[AuCl_4]^-$ or $[Au(CN)_2]^-$ have recently been developed and the method may become more popular as a result [27]. The liquid ion-exchange electrode recommended for $[AuCl_4]^-$ is based on bis(4-ethyl-3,5-dipropyl-1-pyrazolyl)-methane [27]. Several ions including Zn^{2+} and Ag^+ interfere.

(e) Atomic absorption and emission spectrography

Atomic absorption spectrophotometry is becoming a particularly popular technique for determination of gold in trace quantities [28, 29]. Thus, using a carbon rod atomiser, concentrations of gold in aqueous solution of about 1 ppm can be estimated accurately [30, 31], and considerably lower concentrations can be analysed if atomic absorption is used in conjunction with concentration of gold by solvent extraction or chromatography [2]. Using chromatographic concentration and emission spectroscopy, concentrations of gold in sea-water in the region of 10^{-12} ppm were determined to an accuracy of about ±10% [8].

(f) Neutron activation

This is the most sensitive analytical method for gold and is the preferred technique when gold is present in very low concentrations as in sea-water or meteorites [4, 7]. The great sensitivity arises since ^{197}Au is the only naturally occurring isotope of gold and has a high neutron capture cross section of 98 barns. The product ^{198}Au decays by β- and γ-emission with a half-life of 2.7 days. Again, the gold is normally concentrated and separated from impurities by solvent extraction or chromatography prior to analysis [2, 4, 7, 32, 33].

BIOLOGICAL APPLICATIONS OF GOLD [34]

Although gold has been administered since the earliest times for the cure of all diseases, with the *aurum potabile* of the thirteenth century perhaps being the most famous cure-all, it is only recently that a sound basis for gold chemotherapy has emerged. The chief uses are of various gold(I) thiol compounds in the treatment of rheumatoid arthritis and of radioactive ^{198}Au in radiotherapy.

(a) Gold in the treatment of arthritis [35, 36]

The gold compounds which are presently used in the treatment of rheumatoid arthritis are shown in Table 11.2, and their use is generally known as chrysotherapy. Other compounds such as $Na_3[Au(S_2O_3)_2]$, known as Sanocrysin, have formerly been used and these drugs were also used in the period 1925–1935 in the treatment of tuberculosis.

The gold(I) compounds are administered intramuscularly and appear to be as effective as any other known drug for treatment of arthritis, though toxic side-effects are often a problem [37].

In the evaluation of new gold drugs, it is customary to test if the drug suppresses adjuvant arthritis in rats, since there is a

TABLE 11.2

Commonly used gold(I) drugs

Chemical formula[a]	Chemical name	Trade name
$Au{-}S{-}CH(CO_2^-\ Na^+)CH_2CO_2^-\ Na^+$	Disodium gold(I) thiomalate	Myocrisin
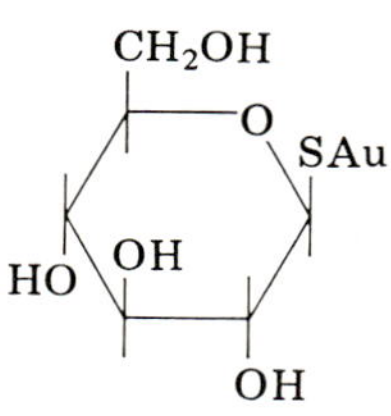	Gold(I) thioglucose	Solganol

[a] The detailed structures are not known. The compounds may be polymeric like other complexes $(RSAu)_n$.

strong similarity between this disease and rheumatoid arthritis in humans. Although there is some disagreement [38], controlled tests show that sodium gold(I) thiomalate does suppress both the primary and secondary lesions of adjuvant arthritis in rats, although it has little effect when given therapeutically after arthritis has begun [39, 40].

A disadvantage of the present chrysotherapy drugs is that they must be administered parenterally in doses of about 50 mg per week, and it is difficult to control toxic side-effects which may continue for a considerable time after chrysotherapy is discontinued. An orally administered drug has considerable advantages in this respect, and a number of trialkylphosphine-

TABLE 11.3

Some tertiary phosphine complexes of gold(I) with anti-arthritis properties

$Et_3P{-}Au{-}Cl$ (ref. 41)

$Et_3P{-}Au{-}SCN$ (ref. 42)

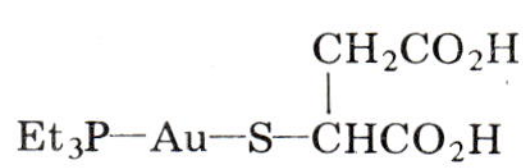

(ref. 45)

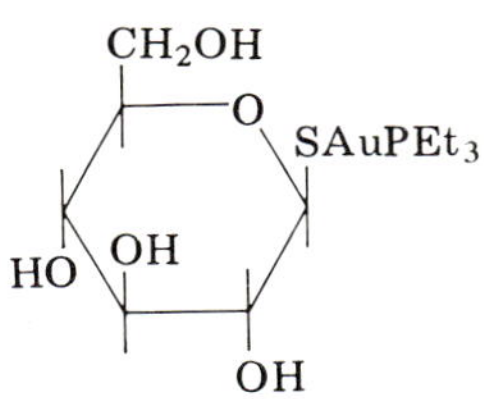

(ref. 43)

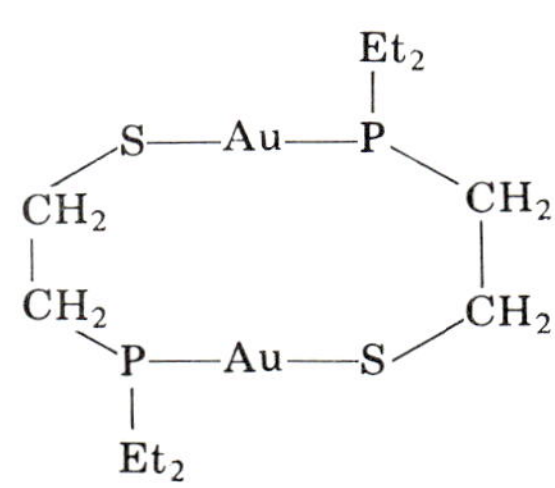

(ref. 46)

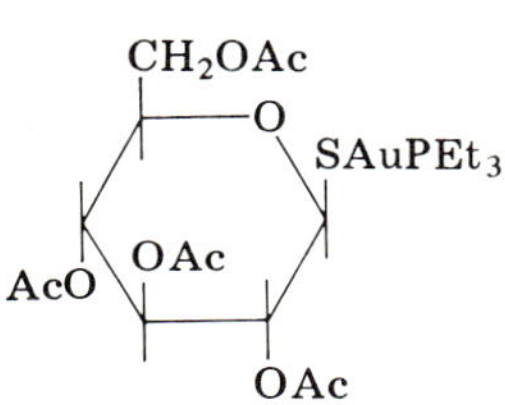

(ref. 44)

gold(I) complexes have recently been shown to suppress adjuvant-induced arthritis in rats and to give a high serum gold level. Some of the more effective reagents are shown in Table 11.3. The effectiveness depends primarily on the phosphine ligand, with triethylphosphine being most effective, but also on the anionic group. The role of the tertiary phosphine is probably to enhance the lipid solubility of the gold drug and so to facilitate its absorption and distribution in the body. Also as a result of the different solubility properties, the distribution of gold throughout the body is different from that resulting from administration of the conventional water-soluble gold drugs. In particular, concentration of gold in the kidneys with resultant harmful effects is considerably less for the phosphine complexes, and it is likely that the new phosphine—gold drugs will be used therapeutically in the not too distant future.

(b) The mechanism of action of gold drugs

Studies using radioactively labelled gold(I) thiomalate show that about 20% of the gold is excreted in the first two weeks, but that after this time further excretion is very slow. Gold is concentrated in the inflamed joints and probably acts by inhibiting enzymes involved in the inflammatory process [47, 48]. Certainly, there is considerable evidence that gold(I) thiomalate inhibits such enzymes as acid phosphatase, β-glucuronidase, cathepsin and glucosamine-6-phosphate synthetase and, it is suggested, that there may be a general effect on SH-containing enzymes arising through co-ordination of the thio group to gold [49—54]. Thus the inhibition may not be observed in the presence of SH protecting groups or by addition of other SH-containing compounds such as cysteine or DL-penicillamine [49, 53], which can compete successfully for the gold centre.

For this kind of behaviour to be observed, it is necessary that gold(I) thiomalate should be able to react with more thiol or that exchange between co-ordinated RS^- and uncomplexed RSH should occur readily. Exchange is demonstrated by the observation that after administration of gold drugs labelled at both the gold and sulphur atoms, the distribution of radioactive gold and sulphur in the body is not identical [55], showing that cleavage of the gold—sulphur bond occurs in vivo. Both exchange and further co-ordination of thiolate to gold have been demonstrated by ^{13}C n.m.r. spectroscopy, and evidence for formation of the cluster $[Au_4(thiomalate)_7]^{3-}$ in the presence of excess of thiomalate has been obtained [56, 57].

Further evidence for the interaction of gold with thiol groups is obtained by X-ray diffraction studies of crystalline proteins and enzymes containing a gold complex. The heavy gold atom is used to enable solution of the phase problem associated with determination of structures by X-ray methods. The complex ions $[Au(CN)_2]^-$, $[AuCl_4]^-$ and $[AuI_4]^-$ are frequently used, but it is possible that substitution of a halide ion or reduction to gold(I) may occur in the presence of the protein or enzyme. Some examples of structures investigated by this method are given in Table 11.4, and it will be clear that gold is often associated with the amino-acid cysteine which contains an SH group.

In conclusion, it is shown that gold(I) thiol derivatives can undergo exchange with free thiols, that gold has a high tend-

TABLE 11.4

Gold derivatives of enzymes and proteins [34]

Gold labelling agent	Protein or enzyme	Gold binding site	Ref.
$[Au(CN)_2]^-$	Flavodoxin	Cysteine	58
	Liver alcohol dehydrogenase	Cysteine	59
$[AuCl_4]^-$	Cytochrome c_2	Histidine	60
	Lysozyme	Arginine	61
	Myoglobin	Histidine	62
	Haemoglobin	Cysteine	63
	Lactate dehydrogenase	Cysteine	64
$[AuI_4]^-$	Myoglobin	Near haem	62
	Chymotrypsin	?	65

ency to bind to sulphur-containing groups in enzymes and proteins, and that gold(I) thiol drugs inhibit enzymes in which a free SH group is necessary for activity to be observed. Thus, the circumstantial evidence that gold therapy in arthritis treatment is successful due to inhibition of thiol-containing enzymes which cause the inflammation is strong, and it is likely that direct evidence on this matter will be sought in the near future.

(c) Other biological effects of gold

Gold therapy may have harmful or beneficial side-effects. An example of the latter is given by the anti-histamine properties of gold(I) thiomalate, which is present in the therapeutic concentration range during normal chrysotherapy treatment [66]. Gold drugs also have beneficial effects against Mycoplasma and Leishmania [66].

Also potentially interesting is the observation that $[AuCl_4]^-$ forms a complex with DNA probably by co-ordination of nitrogen atoms of amino-acids [67]. This could interfere with cell division and indicates that gold(III) complexes may be found to have anti-cancer properties. Thus the complex *cis*-$[PtCl_2(NH_3)_2]$ is thought to have anti-tumour properties due to co-ordination of DNA to platinum. A preliminary report suggests that $[AuCl_2(\text{5-diazouracil})_2]^+Cl^-$ does indeed have anti-tumour properties [68]. It is also known that gold(I) thiomalate inhibits DNA synthesis in sheep lymphocytes [69]. Further exciting progress in this area is obviously possible.

(d) Toxicity of gold [54, 70]

Labelling studies have shown that gold(I) drugs are concentrated in the kidneys and, to a lesser extent, in the thymus, liver, spleen and hypothalamus [71, 72]. The precise distribution is dependent to some extent on the particular compounds administered. The concentration in the kidney can be particularly serious and may result in dermatitis, stomatitis, thrombocytopenia and renal damage [73].

Gold toxicity may be treated by injection of chelating agents which hasten the excretion of gold from the body. Recommended for this purpose are DL-penicillamine and British anti-lewisite (BAL), both of which contain SH groups which are particularly effective towards gold(I) [72, 74]. Also recommended is treatment with prednisone and corticosteroids generally [74, 75].

Although the overall toxicity of gold(I) thioglucose is no greater than that of the thiomalate, the former leads to considerably greater concentration of gold in the hypothalamus of the brain [76, 77]. In animal studies, this is found to result in obesity and hyperphagia in mice and guinea-pigs and to brain damage in rats [76—80].

(e) Applications of radioactive gold in medicine

Radioactive ^{198}Au finds use in medicine in radiotherapy. Gold may be in the form of grains which are placed where required or sometimes in colloidal form. Colloidal ^{198}Au is used in radiotherapy of pleural or peritoneal effusions and bladder cancer where an insoluble radiopharmaceutical capable of uniform radiation of irregular surfaces is required [81].

Colloidal ^{198}Au is also used for various diagnostic purposes, for example in bone marrow scanning or in delineating the lungs and liver. The colloidal gold is taken up by the organ which can then be observed by scintigraphy.

CATALYTIC PROPERTIES OF GOLD

Gold is generally regarded as being the least useful of the noble metals for catalytic purposes. The low catalytic activity of gold is due to the d-band being filled, with the result that gold is unable to chemisorb small molecules. Thus hydrogen is not adsorbed by gold at temperatures below 473 K (above this temperature, thermal excitation of electrons from the d-band to the 6s-band occurs) and ethylene is adsorbed only weakly.

If gold is to be used as a hydrogenation catalyst, it is therefore necessary to supply hydrogen atoms for reduction at the gold surface. This can be effected either electrolytically or by the gold-catalysed decomposition of formic acid, but neither method is attractive. Better results have been obtained by using very finely divided gold particles on alumina or silica supports. Such small metal crystallites become electron-deficient, and their properties then resemble those of the preceding element in the Periodic Table, so that very finely divided gold particles take on some of the properties of platinum metal [82]. In accord with this theory, the supported gold catalysts, which are usually prepared by heating $H[AuCl_4]$ adsorbed on the support material, are able to catalyse the hydrogenation of alkenes, alkynes and dienes under conditions where massive gold is inactive. These catalysts also effect exchange between deuterium gas and hydrogen atoms present in the support material. Thus, in the catalytic hydrogenation of butadiene, it is suggested that butadiene is adsorbed onto the gold and is then reduced by hydrogen atoms which migrate from the support to the metal surface [83, 84].

Gold films prepared under ultra-high vacuum have been shown to catalyse exchange between hydrogen and deuterium groups of $MeSiH_3$ and $MeSiD_3$, apparently by reversible loss of H_2 or D_2 from the chemisorbed molecules with formation of chemisorbed MeSiH or MeSiD species [85]. Gold is also a catalyst for isomerisation and hydrocracking of hydrocarbons, but is less effective than other noble metals. One interesting example is the catalysis of the isomerisation of the complex strained-ring molecule diademane to snoutene and then to basketene by gold [86]. Some reactions of oxygen are also catalysed by gold, for example the oxidation of propene to propene oxide and of methanol to formaldehyde [87].

Gold is also frequently used to modify the catalytic properties of another metal [88]. Often, gold decreases the reactivity but increases the specificity of the catalytic reaction For example, addition of gold to platinum or iridium surfaces gives increased selectivity to the isomerisation of *iso*-butane rather than to its hydrogenolysis [89], and a gold—palladium catalyst has greater specificity for isomerisation of α-pinene to β-pinene and for oxidation of ethylene to acetaldehyde than has palladium alone. Interestingly, the isomerisation of 2-methylpentane to 3-methylpentane occurs predominantly by a different mechanism with a Pt catalyst than with the Pt—Au mixed catalyst, as demonstrated by labelling studies [90].

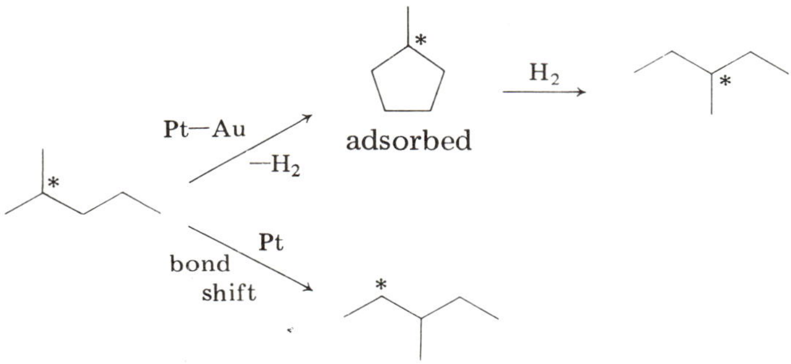

Exceptionally, the addition of gold to palladium has little effect on the selectivity towards the catalytic isomerisation or cracking of hydrocarbons though the reactivity is decreased [90]. Addition of gold to rhenium enhances the catalytic properties, for example in the conversion of C_6 hydrocarbons to benzene [91].

Gold compounds may also have catalytic properties, either alone or in conjunction with a second transition metal compound. For example, the manufacture of vinyl acetate from ethylene, oxygen and acetic acid is catalysed by palladium acetate on silica or alumina promoted by $H[AuCl_4]$ or, preferably $BaAu_2(OAc)_8$ [92].

Some catalytic reactions are based on the oxidising ability of gold(III) complexes. For example, acetylene is oxidised to glyoxal in water—acetonitrile by $[AuCl_3(PPh_3)]$.

$$HCCH + 2[AuCl_3(PPh_3)] + 2H_2O \rightarrow CHO \cdot CHO + 2[AuCl(PPh_3)] + 4HCl$$

Since $[AuCl(PPh_3)]$ can be oxidised back to $[AuCl_3(PPh_3)]$ by chlorine or tetrachloro-*o*-quinone, the system can be made catalytic and yields of 250% based on gold have been obtained [93]. Primary amines are also oxidised to ketones or aldehydes by gold(III) complexes [94], and alkenes can be converted to epoxides by $[AuCl(PPh_3)]$ and O_2 apparently by a free-radical mechanism [95, 96], but none of these systems is yet attractive commercially. Gold(III) complexes can also be used stoichiometrically for the oxidation of alkenes, alkynes and cyclopropanes [96b]. Cyclopropanes are first isomerised to alkenes which then give 1,2-adducts. Other catalytic reactions are the hydrogenation of alkenes with Au(III)—Sb bonded complexes [97], the dimerisation of acrylonitrile to adiponitrile using $[AuBr(PMe_3)]$ and sodium amalgam [98], the metathesis of alkenes using $[AuCl(PPh_3)]$ and organoaluminium compounds [99] and the decomposition of diazoalkanes by $[AuCl_3(pyridine)]$ [100]. The catalysis of conversion of isocyanides to formamidines was discussed in Chapter 7.

In conclusion, neither gold nor gold complexes have been used extensively as catalysts in the past but recent research suggests that their use is likely to increase in the future.

METHODS OF GOLD COATING

One of the most rapidly expanding uses of gold is in the electronics industry. Gold is used in switching devices and in micro-electronic circuits. For these purposes, gold is usually plated onto another material electrolytically. The nature of the gold deposited is critically dependent on the nature of the electrolyte solution used. Most such solutions are based on $K[Au(CN)_2]$ which may be used in acid, neutral or alkaline conditions, but solutions based on thiosulphate or sulphite complexes of gold are also used [101]. It was originally thought that gold was deposited by direct reduction of adsorbed $[Au(CN)_2]^-$ but it now seems that the reducible species is AuCN which is present in very low concentration in equilibrium with $[Au(CN)_2]^-$. Only at high overpotentials does reduction of $[Au(CN)_2]^-$ occur [102]. Gold is also used to connect components in micro-electronic circuits. For this use, various gold pastes may be used, or gold may be deposited by the metal sputtering technique from the vapour phase. The technology involved in the various uses of gold in electronics is now extremely sophisticated, with the need to use the minimum quantities of gold being a particularly important factor.

Gold is also used to decorate glass and ceramics. Here, another technique for producing thin, strongly adhering gold coatings is to adsorb a solution of a gold compound onto the surface and then to pyrolyse the compound to gold. Recommended for this purpose are such compounds as gold(I) thiolates and phosphine complexes of gold(I) [103, 104]. Where particular designs are required the "liquid gold" solutions can be painted onto the pottery or glassware.

OTHER APPLICATIONS OF GOLD

Applications of gold which do not primarily involve its chemical properties will be reviewed only very briefly. The chief uses are as a monetary standard, in jewellery and in dentistry. The infrared reflectivity of gold leads to its use in the aeronautics and space industries, and its low chemical reactivity leads to use in lining specialised chemical plant and, alloyed with platinum, in the production of viscose rayon. It is encouraging to the chemist that the fastest growing uses of gold are in industrial rather than decorative applications [105].

REFERENCES

1 E. M. Wise (Ed.) in *Gold. Recovery, Properties and Applications*, Van Nostrand, New York, 1964.

2 F. E. Beamish and J. C. van Loon, *Miner. Sci. Eng.*, 4 (1972) 3; *Recent Advances in the Analytical Chemistry of Noble Metals*, Pergamon, Oxford, 1972.

3 A. I. Busev and V. M. Ivanov, *Analytical Chemistry of Gold*, Nauka, Moscow, 1973.

4 L. L. Sundberg, *Anal. Chem.*, 47 (1975) 2037.

5 N. R. Das and S. N. Bhattacharya, *Talanta*, 23 (1976) 535.

6 M. Ziegler and H. Winkler, *Mikrochim. Acta*, 1 (1976) 457.
7 J. P. Riley, in J. P. Riley and G. Skirrow (Eds.), *Chemical Oceanography*, Vol. 3, Academic, London, 1975, 2nd ed., Ch. 19.
8 R. Buijle and F. J. Haftka, *Bol. Geol. Min.*, 85 (1974) 421.
9 E. M. Moyers and J. S. Fritz, *Anal. Chem.*, 48 (1976) 117.
10 G. L. Putnam, *J. Chem. Educ.*, 30 (1953) 579.
11 F. E. Beamish, *Talanta*, 13 (1966) 773.
12 T. C. Chadwick, *Anal. Chem.*, 46 (1974) 1326.
13 W. J. Holland and J. Gerard, *Anal. Chim. Acta*, 41 (1968) 327.
14 G. S. Johar, *Labdev, Part A*, 7 (1969) 85.
15 S. C. S. Rajan and N. A. Raju, *Talanta*, 22 (1975) 185.
16 I. E. Lichtenstein, *Anal. Chem.*, 47 (1975) 465.
17 B. Sen, *Anal. Chim. Acta*, 21 (1959) 35; C. K. Bhaskare and S. G. Kawarkar, *J. Indian Chem. Soc.*, 52 (1975) 520.
18 S. G. Nagarkar and M. C. Eshwar, *Anal. Chim. Acta*, 71 (1974) 461.
19 B. Subrahmanyam and M. C. Eshwar, *Anal. Chim. Acta*, 82 (1976) 435.
20 A. I. Busev, N. N. Gorbunova and V. M. Ivanov, *Zavod. Lab.*, 37 (1971) 26.
21 M. Furnica, *Chem. Abstr.*, 70 (1969) 63887.
22 Z.-T. Hu and L.-C. Tsou, *K'o Hsueh T'ung Pao*, 20 (1975) 216.
23 G. C. Constantinescu, *Rev. Chim. (Bucharest)*, 26 (1975) 255.
24 S. A. Shahine, M. F. El-Shahat and R. M. Mahmoud, *Indian J. Chem.*, 13 (1975) 626.
25 B. J. McNulty and L. D. Woollard, *Anal. Chim. Acta*, 13 (1955) 154.
26 N. M. Lukovskaya, A. V. Terletskaya and T. A. Bogoslovskaya, *Zh. Anal. Khim.*, 29 (1974) 2268; N. M. Lukovskaya and T. A. Bogoslovskaya, *Ukr. Khim. Zh. (Russ. Ed.)*, 41 (1975) 529.
27 V. N. Golubev, N. K. Evseeva, I. N. Kremenskaya and S. K. Timofeeva, *Elektrokhimiya*, 12 (1976) 263; A. S. Bychkov, O. M. Petrukhin, V. A. Zarinskii, Yu. A. Zolotov, L. V. Bakhtinova and G. G. Shanina, *Zh. Anal. Khim.*, 31 (1976) 2114.
28 J. G. S. Gupta, *Miner. Sci. Eng.*, 5 (1973) 207.
29 N. L. Fishkova, *Zh. Anal. Khim.*, 29 (1974) 2121.
30 F. J. M. J. Maessen, F. D. Posma and J. Balke, *Anal. Chem.*, 46 (1974) 1445.
31 K. Dittrich and W. Mothes, *Talanta*, 22 (1975) 318.
32 R. W. Hummel, *Analyst*, 82 (1957) 483.
33 D. F. Schutz and K. J. Turekian, *Geochim. Cosmochim. Acta*, 29 (1965) 259.
34 P. J. Sadler, *Gold Bull.*, 9 (1976) 110.
35 D. T. Walz, M. J. Di Martino and B. M. Sutton, *Med. Chem., Ser. Monogr.*, 13 (1974) 209; T. Y. Shen, *ibid.*, p. 179.
36 M. J. Lefevre, *Rev. Belge Homoeopath.*, 26 (1974) 75.
37 T. J. Constable, A. P. Crockson, R. A. Crockson and B. McConkey, *The Lancet* (1975) 1176; R. H. Freyberg, in J. L. Hollander (Ed.), *Arthritis and Allied Conditions*, Lea and Febiger, Philadelphia, 1966.
38 J. D. Jessop and H. L. F. Currey, *Ann. Rheum. Dis.*, 27 (1968) 577.
39 D. T. Walz, M. J. Di Martino and A. Misher, *Ann. Rheum. Dis.*, 30 (1971) 303.
40 R. D. Sofia and J. F. Douglas, *Agents Actions*, 3 (1973) 335.
41 D. T. Walz, M. J. Di Martino, B. Sutton and A. Misher, *J. Pharmacol. Exp. Ther.*, 181 (1972) 292.
42 B. M. Sutton, D. T. Walz and J. Weinstock, Ger. Pat., 2,434,920, Feb. 1975; *Chem. Abstr.*, 83 (1975) 84857.
43 B. M. Sutton, E. McGusty, D. T. Walz and M. J. di Martino, *J. Med. Chem.*, 15 (1972) 1095.
44 D. T. Walz, M. J. di Martino, W. L. Chakrin, B. M. Sutton and A. Misher, *J. Pharmacol. Exp. Ther.*, 197 (1976) 145.
45 E. R. McGusty and B. M. Sutton, *U.S. Pat.*, 3,718,679 and 3,718,680; *Chem. Abstr.*, 78 (1973) 135672 and 135673.

46 J. Weinstock, B. M. Sutton, G. Y. Kuo, D. T. Walz and M. J. di Martino, *J. Med. Chem.*, 17 (1974) 139.
47 J. S. Lawrence, *Ann. Rheum. Dis.*, 20 (1961) 341.
48 F. Scheiffarth, H. W. Baenkler and G. Schoerg, *Z. Rheumaforsch.*, 29 (1970) 42.
49 R. S. Ennis, J. L. Granda and A. S. Posner, *Arthritis Rheum.*, 11 (1968) 756.
50 S. Paltemaa, *Acta Rheumatol. Scand.*, 14 (1968) 161.
51 M. Davies, J. B. Lloyd and F. Beck, *Biochem. J.*, 121 (1971) 21.
52 E. Fujihara, N. Tsubota and M. Nakazawa, *Chem. Pharm. Bull.*, 19 (1971) 190.
53 W. J. Westwick, J. Allsop and R. W. E. Watts, *Biochem. Pharmacol.*, 23 (1974) 153.
54 R. J. Martin and P. Lamprey, *Life Sci.*, 14 (1974) 1121.
55 H. A. Swartz, J. E. Christian and F. N. Andrews, *Am. J. Physiol.*, 202 (1962) 743.
56 P. J. Sadler, *Struct. Bonding (Berlin)*, 29 (1976) 171.
57 A. A. Isab and P. J. Sadler, *J. Chem. Soc., Chem. Commun.*, (1976) 1051.
58 R. M. Burnett, G. D. Darling, D. S. Kendall, M. E. LeQuesne, S. G. Mayhew, W. W. Smith and M. L. Ludwig, *J. Biol. Chem.*, 249 (1974) 4383.
59 H. Eklund, B. Nordström, E. Zeppezauer, G. Söderlund, I. Ohlsson, T. Boieve, B.-O. Söderburg, O. Tapia, C.-I. Bränden and A. Akeson, *J. Mol. Biol.*, 102 (1976) 25.
60 F. R. Salemme, S. T. Freer, N. H. Xuong, R. A. Alden and J. Kraut, *J. Biol. Chem.*, 248 (1973) 3910.
61 C. C. F. Blake, *Adv. Protein Chem.*, 28 (1968) 39.
62 H. Scouloudi, *Proc. Roy. Soc. London*, A258 (1960) 181.
63 E. A. Padlan and W. E. Love, *J. Biol. Chem.*, 249 (1974) 4067.
64 M. J. Adams, D. J. Haas, B. A. Jeffery, A. McPherson, H. L. Mermall, M. G. Rossmann, R. W. Schevitz and A. J. Wonacott, *J. Mol. Biol.*, 41 (1969) 159.
65 A. Tulinsky, N. V. Mani, C. N. Morimoto and R. L. Vandeen, *Acta Crystallogr.*, A29 (1973) 1309.
66 E. R. Trethewie, *Med. J. Aust.*, 2 (1970) 1136; C. E. Berkoff, P. Actor, B. M. Sutton, D. T. Walz and N. Klinikowski, *Arzneim.-Forsch.*, 24 (1974) 1988.
67 C. K. S. Pillai and U. S. Nandi, *Biopolymers*, 11 (1973) 1431; J. Kollman, C. Schrocter and E. Hoyer, *J. Prakt. Chem.*, 317 (1975) 515.
68 C. Dragulescu, J. Heller, A. Maurer, S. Policec, V. Topcui, M. Csalci, S. Kirchner, S. Kravitz and R. Moraski, *Proc. of XVI Int. Conf. Co-ord. Chem., Dublin, 1974*, Paper 1.9.
69 R. N. P. Cahill, *Experimentia*, 27 (1971) 913.
70 H. C. Hodge, L. J. Leach, F. A. Smith, W. H. Strain and D. R. Taves, *Drill's Pharmacol. Med.*, McGraw-Hill, New York, 4th ed., 1971, p. 1120.
71 H. L. Atkins, R. M. Lambrecht, A. P. Wolf, A. N. Ansari and M. Guillaume, *Radiology*, 115 (1975) 723.
72 M. Rubin, A. Sliwinski, M. Photias, M. Feldman and N. Zvaifler, *Proc. Soc. Exp. Biol. Med.*, 124 (1967) 290.
73 N. O. Rothermich, V. K. Philips, W. Bergen and M. H. Thomas, *Arthritis Rheum.*, 19 (1976) 1321.
74 D. F. Hill, *Med. Clin. North Am.*, 52 (1968) 733.
75 B. Deren, R. Masi, M. Weksler and R. L. Nachman, *Arch. Intern. Med.*, 134 (1974) 1012.
76 A. E. Debons, L. Silver, E. P. Cronkite, H. A. Johnson, G. Brecher, D. Tenzer and I. L. Schwartz, *Am. J. Physiol.*, 202 (1962) 743.
77 A. E. Debons, I. Krimsky, A. From and R. J. Cloutier, *Am. J. Physiol.*, 219 (1970) 1403.
78 R. J. Chang and R. H. Persellin, *Proc. Soc. Exp. Biol. Med.*, 129 (1968) 568.
79 P. de Laey, C. Dent, A. C. Teny and E. H. Quinn, *Arch. Int. Pharmacodyn. Ther.*, 211 (1974) 341.
80 B. B. Sandrew and J. Mayer, *Physiol. Behav.*, 10 (1973) 1061.

81 E. E. Rogoff, R. Romano and E. W. Hahn, *Radiology*, 114 (1975) 225.
82 G. C. Bond and P. A. Sermon, *Gold Bull.*, 6 (1973) 102. G. C. Bond, *Gold Bull.*, 5 (1972) 11.
83 G. C. Bond, P. A. Sermon, G. Webb, D. A. Buchanan and P. B. Wells, *J. Chem. Soc., Chem. Commun.*, (1973) 444.
84 D. A. Buchanan and G. Webb, *J. Chem. Soc., Faraday Trans. I*, 71 (1975) 134.
85 D. I. Bradshaw, R.B. Moyes and P. B. Wells, *J. Chem. Soc., Chem. Commun.*, (1975) 137.
86 L. U. Meyer and A. de Meijere, *Tetrahedron Lett.* (1976) 497.
87 C. N. Hodges and L. C. Roselaar, *J. Appl. Chem. Biotechnol.*, 25 (1975) 609.
88 E. G. Allison and G. C. Bond, *Catal. Rev.*, 7 (1972) 233.
89 D. I. Hagen and G. A. Somorjai, *J. Catal.*, 41 (1976) 466.
90 A. O'Cinneide and F. G. Gault, *J. Catal.*, 37 (1975) 311.
91 J. K. A. Clarke and J. F. Taylor, *J. Chem. Soc., Faraday Trans. I*, 71 (1975) 2063.
92 H. Fernholz, F. Wunder and H. J. Schmidt, *Ger. Pat.*, 2,057,087, 1972; *Chem. Abstr.*, 77 (1972) 87897; S. Schott, D. W. Lum and I. L. Mador, *Ger. Pat.*, 2,420,374, 1974; *Chem. Abstr.*, 82 (1975) 90659.
93 R. A. C. Rennie, *Brit. Pat.*, 1,071,902, 1967; *Chem. Abstr.*, 67 (1967) 53685.
94 M. E. Kuehne and T. C. Hall, *J. Org. Chem.*, 41 (1976) 2742.
95 A. Fusi, R. Ugo, F. Fox, A. Pasini and S. Cenini, *J. Organomet. Chem.*, 26 (1971) 417.
96 K. Masada and A. Yasui, *Japan Pat.*, 71 09,691, 1971; *Chem. Abstr.*, 75 (1971) 19786.
(b) R. O. C. Norman, W. J. E. Parr and C. B. Thomas, *J. Chem. Soc., Perkin Trans. I*, (1976) 1983.
97 M. C. Muller, *Gold Bull.*, 7 (1974) 39.
98 K. Yamamoto and M. Kawamata, *Japan Pat.*, 71 16,730, 1971; *Chem. Abstr.*, 75 (1971) 64570.
99 D. H. Kubicek and E. A. Zuech, *U.S. Pat.*, 3,558,520, 1971; *Chem. Abstr.*, 74 (1971) 99441.
100 M. G. Krakovyak and S. S. Skorokhodov, *Vysokomol. Soedin., Ser. A.*, 11 (1969) 794.
101 R. T. Page, *Met. Finish. J.*, 20 (1974) 87 and 122.
102 J. A. Harrison and J. Thompson, *Electrochim. Acta.*, 18 (1973) 829.
103 H. M. Fitch, *Belg. Pat.*, 621, 886, 1973.
104 E. M. Potrafke, *U.S. Pat.*, 3,625,755, 1971; *Chem. Abstr.*, 76 (1972) 62691.
105 D. Lloyd-Jacob, *Gold Bull.*, 4 (1971) 25.

APPENDIX

GOLD—ELEMENT BOND LENGTHS[a]

A. GOLD—HALOGEN BOND LENGTHS

Compound	r(AuX)/pm	Comments	Ref.
AuF_3	191	terminal AuF distance	1
	204	bridging AuF distance	
$K[AuF_4]$	195		2
AuCl	236	bridging AuCl	3
$[Au_4Cl_8]$	230	bridging Au(I)—Cl	4
	233	bridging Au(III)—Cl	
	224	terminal Au(III)—Cl	
$[Au_2Cl_6]$	233	bridging Au(III)—Cl	5
	224	terminal Au(III)—Cl	
$Cs_2[AuCl_2][AuCl_4]$	228.1	in $[AuCl_2]^-$	6
	229.5	in $[AuCl_4]^-$	
$[AuCl(PPh_3)]$	227.9		7
$[AuCl(PCl_3)]$	233		8
[AuCl(piperidine)]	225.6		8b
$[(AuCl)_2(PhSCH_2CH_2SPh)]$	232.9, 229.3		9
$[(AuCl)_2(Ph_2PCH_2PPh_2)]$	228.8		9b
$[AuCl(PPh_3)_2]$	250.0	trigonal planar, $\angle ClAuP = 118.7°$	10
[AuCl(P—P)]	281.8	three-co-ord. gold(I) $\angle ClAuP = 90.4, 93.4°$	11
$[Ph_4As][AuCl_4]$	227.1		12
$[C_5H_5N_4O][AuCl_4] \cdot 2H_2O$	227.5—229.2		13
$[AuCl_3(PPh_3)]$	228	*trans* to Cl	14
	234.7	*trans* to PPh_3	
cis-$[AuCl(C_6F_5)_2(PPh_3)]$	237	*trans* to C_6F_5	15
cis-$[AuCl_2Ph(SPr_2)]$	227	*trans* to SPr_2	16
	238	*trans* to Ph	
$[AuCl_3(biquinolyl)]$	235, 238	distorted square pyramid, *trans* to Cl	17
	219	*trans* to N	

[a] For abbreviations, see p. 266.

APPENDIX A—*continued*

Compound	r(AuX)/pm	Comments	Ref.
$[AuCl_3(2,9\text{-dimethylphen})]$	227.5, 228.5	distorted square pyramid, *trans* to Cl	18
	226.7	*trans* to N	
$[AuCl(Ph_4porphyrin)]$	301	distorted square pyramid, axial Au—Cl	19
$[Au_2Cl_2\{\mu\text{-}(CH_2)_2PEt_2\}_2]$	236	gold(II) complex	20
$[Au(S_2CNBu_2)_2][AuBr_2]$	234.9	r(AuBr)	21
$[AuBr(AsPh_3)]$	237.7		22
AuBrSe	246		45
$[Br_2Au(\mu\text{-Br})_2AuMe_2]$	240.2	terminal AuBr distance	71
	245.3	bridging AuBr *trans* to Br	
	260.7	bridging AuBr *trans* to Me	
cis-$[\overline{AuBr_2(Ph_2PC_6H_4C}HCH_2Br)]$	249	*trans* to P and C	23
cis-$[\overline{AuBr_2(Ph_2PC_6H_4CH_2C}HCH_2Br)]$	249	*trans* to P	23
	251	*trans* to C	
$[AuBr_3(2,9\text{-dimethylphen})]$	241.9, 241.0	distorted square pyramid, *trans* to Br	18
	239.8	*trans* to N	
AuI	262	bridging iodide	24
trans-$[AuI_2(diars)_2]I$	335	tetragonal gold(III)	25
$[Au_{11}I_3\{P(C_6H_4F)_3\}_7]$	260.0	cluster complex	26
$[Au_2I_2\{\mu\text{-}CH_2)_2PMe_2\}_2]$	269.9	gold(II) complex	27
trans-$[AuI_2\{C(NHC_6H_4Me)_2\}_2]^+$	260.1, 260.6	carbene complex	28

B. GOLD—GROUP VIB ELEMENT BOND LENGTHS

Compound	r(AuX)/pm	Comments	Ref.
$K[Au(NO_3)_4]$	202	monodentate nitrate	29
$[(Me_2AuOH)_4]$	215	bridging AuO	30
$[\{\overline{Au(CPh{=}CPhCPh{=}C}Ph)OH\}_2]$	223, 226	bridging AuO	31

APPENDIX B—*continued*

Compound	r(AuX)/pm	Comments	Ref.
$Na_3[Au(S_2O_3)_2]\cdot 2H_2O$	228	r(AuS)	32
$[Au\{S{=}C(NH)_2C_2H_4\}_2]Cl\cdot H_2O$	227.8, 227.9		33
$[enH_2][Au(SO_3)_2(en)]$	230.1, 230.8	S-bonded sulphite	34
$[(AuCl)_2(PhSCH_2CH_2SPh)]$	225.8, 226.0		9
$[WS_4(AuPMePh_2)_2]$	253	trigonal gold(I)	35
cis-$[AuCl_2Ph(SPr_2)]$	231		16
$[Au(S_2CNEt_2)(PPh_3)]$	234, 301	asymmetrically bonded dithiocarbamate	36
$[\{Au(S_2CNPr_2)\}_2]$	228		37
$[\{AuS_2P(O^iPr)_2\}_n]$	228		38
$[Au(S_2CNBu_2)_2]Br$	230.9, 233.4		39
$[Au(S_2CNBu_2)_2][AgBr_2]$	235.7, 232.4		40
$[Au(S_2CNBu_2)_2][AuBr_2]$	233.2, 233.3		21
$[Au(S_2CNEt_2)_3]$	234.3	bidentate dithiocarbamate	41
	234.1	monodentate dithiocarbamate	
$[Au(S_2CNBu_2)\{S_2C_2(CN)_2\}]$	232.9, 232.4	AuS(dithiocarbamate)	42
	228.4, 230.3	AuS(mnt)	
$[Au(S_2CNBu_2)_2][Au\{S_2C_2(CN)_2\}_2]$	233.2, 233.7	AuS(dithiocarbamate)	43
	231.2, 230.6	AuS(mnt)	
$[PClPh_3][Au\{S_2C_2(CF_3)_2\}_2]$	229.1		44
AuBrSe	249, 245	bridging AuSe	45
$AuTe_2$	267	r(AuTe) calaverite	46
$AuTe_2$	256	r(AuTe) in krennerite	46

C. GOLD—GROUP VB ELEMENT BOND LENGTHS

Compound	r(AuX)/pm	Comments	Ref.
$[Au(CCPh)(^iPrNH_2)]$	203		47
[AuCl(piperidine)]	206.8		8b
$[Au(bipy)(PPh_3)][PF_6]$	216.6, 221.2	distorted-trigonal gold(I)	48
$[Au(NH_3)_4][NO_3]_3$	202		49
$[enH_2]^{2+}[Au(SO_3)_2(en)]^{2-}$	211.9, 213.3		34
$[AuCl(Ph_4porph)]$	200 av.	square-pyramidal gold(III)	19
$[AuCl_3(2,9\text{-dimethylphen})]$	209	*trans* to Cl	18

APPENDIX C—*continued*

Compound	r(AuX)/pm	Comments	Ref
	258	axial position of square pyramid	
[AuBr$_3$(2,9-dimethylphen)]	208	*trans* to Br	18
	261	axial nitrogen	
[AuCl$_3$(biquinolyl)]	228	*trans* to Cl	17
	240	axial nitrogen	
[AuCl(PPh$_3$)]	224.3		7
[AuCN(PPh$_3$)]	227		50
[AuMe(PPh$_3$)]	227.9		51
[Au(C$_6$F$_5$)(PPh$_3$)]	227		52
[Ph$_3$PAuC(CF$_3$)=C(CF$_3$)AuPPh$_3$]	228		53
[Au(PPh$_3$)$_2$] [TCNQ]	228.6		10
[Au(PMePh$_2$)$_2$] [PF$_6$]	231.6		54
[(AuCl)$_2$(Ph$_2$PCH$_2$PPh$_2$)]	223.8		9b
[AuCl(PCl$_3$)]	219		8
[Au{Co(CO)$_4$}(PPh$_3$)]	223		55
[Au{Mn(CO)$_4$P(OPh)$_3$}(PPh$_3$)]	233		56
[Au{W(CO)$_3$(C$_5$H$_5$)}(PPh$_3$)]	225		57
[Au(S$_2$CNEt$_2$)(PPh$_3$)]	225		38
[Au(PPh$_3$)$_3$] [B$_9$H$_{12}$S]	238.2		58
[Au(bipy)(PPh$_3$)] [PF$_6$]	221.2		48
[AuCl(PPh$_3$)$_2$]	232.3		10
[(C$_5$H$_5$)Fe{C$_5$H$_4$(AuPPh$_3$)$_2$}] [BF$_4$]	227.5, 228.3		59
[WS$_4$(AuPMePh$_2$)$_2$]	227		35
[AuCl(P—P)]	230.7, 231.0		11
[Me$_3$PAuC(CF$_3$)=C(CF$_3$)AuMe$_2$PMe$_3$]	229	Au(I)—P	60
	230	Au(III)—P	
[Au$_{11}$I$_3${P(C$_6$H$_4$F)$_3$}$_7$]	221—229	cluster complex	26
[Au$_9${P(C$_6$H$_4$Me)$_3$}$_8$] [PF$_6$]$_3$	230	cluster complex	61
[Au$_6${P(C$_6$H$_4$Me)$_3$}$_6$] [BPh$_4$]$_2$	227.4—231.1	cluster complex	62
[AuCl$_3$(PPh$_3$)]	233.5		14
[AuBr$_3$(PPh$_3$)]	250		63
[AuMe$_3$(PPh$_3$)]	235		27
cis-[AuCl(C$_6$F$_5$)$_2$(PPh$_3$)]	237		15
cis-[AuBr$_2$(Ph$_2$PC$_6$H$_4$CHCH$_2$Br)]	228		23
cis-[AuBr$_2$(Ph$_2$PC$_6$H$_4$CH$_2$CHCH$_2$Br)]	229		23
[AuBr(AsPh$_3$)]	234.2		22
trans-[AuI$_2$(diars)$_2$]I	243, 246		25

D. GOLD—CARBON BOND LENGTHS

Compound	r(AuX)/pm	Comments	Ref.
$[Au(CN)_2]^-$	212		64
$[Au(CN)(PPh_3)]$	185		50
$[Au(CN)(MeNC)]$	201	r(AuCN)	50b
	198	r(AuCNMe)	
$[AuMe(PPh_3)]$	212.4		51
$[Au(C_6F_5)(PPh_3)]$	207		52
$[Ph_3PAuC(CF_3){=}C(CF_3)AuPPh_3]$	205		53
$[Au(CCPh)(^iPrNH_2)]$	194		47
$[(C_5H_5)Fe\{C_5H_4(AuPPh_3)_2\}][BF_4]$	213, 227	bridging carbon	59
$[Au_2\{\mu\text{-}(CH_2)_2PMe_2\}_2]$	216.5, 223.3		27
$[AuMe_3(CH_2SOMe_2)]$	211.5	r(AuMe) *trans* ylid	65
	214.3, 215.9	r(AuMe) *trans* Me	
	214.4	$r(AuCH_2)$	
$[AuMe_3(PPh_3)]$	198	*trans* to P	27
	211	*trans* to Me	
cis-$[AuCl(C_6F_5)_2(PPh_3)]$	212	*trans* to Cl	15
	218	*trans* to P	
$[(Me_2AuOH)_4]$	205		30
$[\{Au(CPh{=}CPhCPh{=}CPh)OH\}_2]$	209, 213		31
cis-$[AuCl_2Ph(SPr_2)]$	200		16
cis-$[AuBr_2(Ph_2PC_6H_4CHCH_2Br)]$	210		23
cis-$[AuBr_2(Ph_2PC_6H_4CH_2CHCH_2Br)]$	213		23
$[Me_3PAuC(CF_3){=}C(CF_3)AuMe_2(PMe_3)]$	210	Au(I)—C	60
	203	Au(III)—$C(CF_3)$	
	211, 214	Au(III)—Me	
trans-$[AuI_2\{C(NHC_6H_4Me\}_2]$	207, 209	carbene complex	28
$[Au_2Cl_2\{(\mu\text{-}CH_2)_2PEt_2\}_2]$	194.2	gold(II) complex	20
$[Au_2I_2\{(\mu\text{-}CH_2)_2PMe_2\}_2]$	209.8, 223.1	gold(II) complex	27
$[Au(S_2CNEt_2)_2][Au(1,2\text{-}C_2B_9H_{11})_2]$	277	mean value	66
$[Au(S_2CNEt_2)(1,2\text{-}C_2B_9H_{11})]$	276	mean value	66
$H[Au(CN)_4]2H_2O$	197		69
$Ph_4As[Au(CN_4{}^iPr)_4]$	198	tetrazolato complex	70
$[Br_2Au(\mu\text{-}Br)_2AuMe_2]$	212		71

E. GOLD—METAL BOND LENGTHS

Compound	r(AuX)/pm	Comments	Ref.
Au	288.3	cubic close packed	
Au_2	250	gas phase only	
$[\{Au(S_2CNPr_2)\}_2]$	276	r(AuAu) intra-molecular	37
	340	r(AuAu) inter-molecular	
$[\{AuS_2P(O^iPr)_2\}_n]$	304	r(AuAu)	38
$[(C_5H_5)Fe\{C_5H_4(AuPPh_3)_2\}][BF_4]$	276.8	r(AuAu)	59
	282	r(AuFe)	
$[Au_6\{P(C_6H_4Me)_3\}_6][BPh_4]_2$	293.4—309.3		62
$[Au_9\{P(C_6H_4Me)_3\}_8][PF_6]_3$	268.9—286.8		61
$[Au_{11}I_3\{P(C_6H_4F)_3\}_7]$	260.0—318.7		26
$[Au_{11}I_3\{P(C_6H_4Cl)_3\}_7]$	261.2—315.5		67
$[Au_{11}(SCN)_3(PPh_3)_7]$	267	Mean centre to periphery	68
$[Au(S_2CNEt_2)(1,2\text{-}C_2B_9H_{11})]$	222	Mean r(AuB)	66
$[Au(S_2CNEt_2)_2][Au(1,2\text{-}C_2B_9H_{11})]$	226	Mean r(AuB)	66
$[WS_4(AuPMePh_2)_2]$	284	r(AuW)	35
$[Ph_3PAuMn(CO)_4\{P(OPh)_3\}]$	257	r(AuMn)	56
$[Ph_3PAuW(CO)_3(C_5H_5)]$	269.8	r(AuW)	57
$[Ph_3PAuCo(CO)_4]$	250	r(AuCo)	55

ABBREVIATIONS

Bipy	= 2,2′-bipyridine
Bu	= butyl
Diars	= 1,2-bis(dimethylarsino)benzene
2,9-Dimethylphen	= 2,9-dimethyl-1,10-phenanthroline
En	= 1,2-diaminoethane
Et	= ethyl
Me	= methyl
P-P	= 2,11-bis(diphenylphosphinomethyl)benzo-[*c*]-phenanthrene
Ph	= phenyl
Ph_4porph	= $\alpha,\beta,\gamma,\delta$-tetraphenylporphinate
Pr	= propyl
TCNQ	= 7,7,8,8-tetracyanoquinodimethane

REFERENCES

1 F. W. B. Einstein, P. R. Rao, J. Trotter and N. Bartlett, *J. Chem. Soc. A*, (1967) 478.
2 A. J. Edwards and G. R. Jones, *J. Chem. Soc. A* (1969) 1936.
3 E. M. W. Janssen, J. C. W. Folmer and G. A. Wiegers, *J. Less-Common Met.*, 38 (1974) 71.
4 D. B. Dell'Amico, F. Calderazzo, F. Marchetti, S. Merlino and G. Perego, *J. Chem. Soc., Chem. Commun.*, (1977) 31.
5 E. S. Clark, D. H. Templeton and C. M. McGillavry, *Acta Crystallogr.*, 11 (1958) 284.
6 J. C. M. T. Eijndhoven and G. C. Verschoor, *Mater. Res. Bull.*, 9 (1974) 1667.
7 N. C. Baenziger, W. E. Bennett and D. M. Soboroff, *Acta Crystallogr.*, B32 (1976) 962.
8 G. J. Arai, *Rec. Trav. Chim. Pays-Bas*, 81 (1962) 307.
(b) J. J. Guy, P. G. Jones, M. J. Mays and G. M. Sheldrick, *J. Chem. Soc., Dalton Trans.*, (1977) 8.
9 (a) M. G. B. Drew and M. J. Riedl, *J. Chem. Soc., Dalton Trans.*, (1973) 52.
(b) H. Schmidbaur, A. Wohlleben, F. Wagner, O. Orama and G. Huttner, *Chem. Ber.*, 110 (1977) 1748.
10 N. C. Baenziger, K. M. Dittemore and J. R. Doyle, *Inorg. Chem.*, 13 (1974) 805.
11 M. Barrow, H. B. Buergi, D. K. Johnson and L. M. Venanzi, *J. Am. Chem. Soc.*, 98 (1976) 2356.
12 P. G. Jones, J. J. Guy and G. M. Sheldrick, *Acta Crystallogr.*, B31 (1975) 2687.
13 M. R. Caira, L. R. Nassimbeni and A. L. Rodgers, *Acta Crystallogr.*, B31 (1975) 1112.
14 G. Bandoli, D. A. Clemente, G. Marangoni and L. Cattalini, *J. Chem. Soc., Dalton Trans.*, (1973) 886.
15 R. W. Baker and P. J. Pauling, *J. Chem. Soc., Chem. Commun.*, (1969) 745.
16 M. McPartlin and A. J. Markwell, *J. Organomet. Chem.*, 57 (1973) C25.
17 R. J. Charlton, C. M. Harris, H. Patil and N. C. Stephenson, *Inorg. Nucl. Chem. Lett.*, 2 (1966) 409.
18 W. T. Robinson and E. Sinn, *J. Chem. Soc., Dalton Trans.*, (1975) 726.
19 R. Timkovich and A. Tulinsky, *Inorg. Chem.*, 16 (1977) 962.
20 H. Schmidbaur, J. R. Mandl, A. Frank and G. Huttner, *Chem. Ber.*, 109 (1976) 466.
21 P. T. Beurskens, H. J. A. Blaauw, J. A. Cras and J. J. Steggerda, *Inorg. Chem.*, 7 (1968) 805.
22 F. W. B. Einstein and R. Restivo, *Acta Crystallogr.*, B31 (1975) 624.
23 M. A. Bennett, K. Hoskins, W. R. Kneen, R. S. Nyholm, P. B. Hitchcock, R. Mason, G. B. Robertson and A. D. C. Towl, *J. Am. Chem. Soc.*, 93 (1971) 4591.
24 A. Weiss and A. Weiss, *Z. Naturforsch.*, 11b (1956) 604; H. Jagodzinski, *Z. Krist.*, 112 (1959) 80.
25 V. F. Duckworth and N. C. Stephenson, *Inorg. Chem.*, 8 (1969) 1661.
26 P. Bellon, M. Manassero and M. Sansoni, *J. Chem. Soc., Dalton Trans.*, (1972) 1481.
27 J. P. Fackler, Jr., personal communication.
28 Lj. Manojlovic-Muir, *J. Organomet. Chem.*, 73 (1974) C45.
29 C. D. Garner and S. C. Wallwork, *J. Chem. Soc. A*, (1970) 3092.
30 G. E. Glass, J. H. Konnert, M. G. Miles, D. Britton and R. S. Tobias, *J. Am. Chem. Soc.*, 90 (1968) 1131.

31 M. Peteau-Boisdenghien, J. Meunier-Piret and M. van Meersche, *Cryst. Struct. Commun.*, 4 (1975) 375.
32 H. Ruben, A. Zalkin, M. O. Faltens and D. H. Templeton, *Inorg. Chem.*, 13 (1974) 1836.
33 P. G. Jones, J. J. Guy and G. M. Sheldrick, *Acta Crystallogr.*, B32 (1976) 3321.
34 A. Dunand and R. Gerdil, *Acta Crystallogr.*, B31 (1975) 370.
35 J. C. Huffman, R. S. Roth and A. R. Siedle, *J. Am. Chem. Soc.*, 98 (1976) 4340.
36 J. G. Wijnhoven, W. P. J. H. Bosman and P. T. Beurskens, *J. Cryst. Mol. Struct.*, 2 (1972) 7.
37 R. Hesse and P. Jennische, *Acta. Chem. Scand.*, 26 (1972) 3855.
38 S. L. Lawton, W. J. Rohrbaugh and G. T. Kokotailo, *Inorg. Chem.*, 11 (1972) 2227.
39 P. T. Beurskens, J. A. Cras and J. G. M. van der Linden, *Inorg. Chem.*, 9 (1970) 475.
40 J. A. Cras, J. H. Noordik, P. T. Beurskens and A. M. Verhoeven, *J. Cryst. Mol. Struct.*, 1 (1971) 155.
41 J. H. Noordik, *Cryst. Struct. Commun.*, 2 (1973) 81.
42 J. H. Noordik, T. W. Hummelink and J. G. M. van der Linden, *J. Coord. Chem.*, 2 (1973) 185.
43 J. H. Noordik and P. T. Beurskens, *J. Cryst. Mol. Struct.*, 1 (1971) 339.
44 J. H. Enemark and J. A. Ibers, *Inorg. Chem.*, 7 (1968) 2636.
45 D. Mootz, A. Rabeneau, H. Wunderlich and G. Rosenstein, *J. Solid State Chem.*, 6 (1973) 583.
46 R. W. G. Wyckoff, *Crystal Structures*, Vol. 1, Interscience, New York, 2nd ed., 1963, p. 337.
47 P. W. R. Corfield and H. M. M. Shearer, *Acta Crystallogr.*, 23 (1967) 156.
48 W. Clegg, *Acta Crystallogr.*, B32 (1976) 2712.
49 M. Weishaupt and J. Straehle, *Z. Naturforsch.*, B31 (1976) 554.
50 P. L. Bellon, M. Manassero and M. Sansoni, *Ric. Sci.*, 39 (1969) 173.
(b) S. Esperas, *Acta. Chem. Scand.*, A30 (1976) 527.
51 P. D. Gavens, J. J. Guy, M. J. Mays and G. M. Sheldrick, *Acta Crystallogr.*, B33 (1977) 137.
52 R. W. Baker and P. J. Pauling, *J. Chem. Soc., Dalton Trans.*, (1972) 2264.
53 C. J. Gilmore and P. Woodward, *J. Chem. Soc., Chem. Commun.*, (1971) 1233.
54 J. J. Guy, P. G. Jones and G. M. Sheldrick, *Acta Crystallogr.*, B32 (1976) 1973
55 T. L. Blundell and H. M. Powell, *J. Chem. Soc. A.* (1971) 1685.
56 K. A. I. F. Mannan, *Acta Crystallogr.*, 23 (1967) 649.
57 J. B. Wilford and H. M. Powell, *J. Chem. Soc. A*, (1969) 8.
58 L. J. Guggenberger, *J. Organomet. Chem.*, 81 (1974) 271.
59 V. G. Andrianov, Yu. T. Struchkov and E. R. Rossinskaya, *J. Chem. Soc., Chem. Commun.*, (1973) 338.
60 J. A. J. Jarvis, A. Johnson and R. J. Puddephatt, *J. Chem. Soc., Chem. Commun.*, (1973) 373.
61 P. L. Bellon, F. Cariati, M. Manassero, L. Naldini and M. Sansoni, *J. Chem. Soc., Chem. Commun.*, (1971) 1423.
62 P. L. Bellon, M. Manassero and M. Sansoni, *J. Chem. Soc., Dalton Trans.*, (1973) 2423.
63 M. F. Perutz and O. Weisz, *J. Chem. Soc.*, (1946) 438.
64 A. Rosenzweig and D. T. Cromer, *Acta Crystallogr.*, 12 (1959) 709.
65 J. P. Fackler, Jr., and C. Paparizos, *J. Am. Chem. Soc.*, 99 (1977) 2363.
66 H. M. Colqhoun, T. J. Greenhough and M. G. H. Wallbridge, *J. Chem. Soc., Chem. Commun.*, (1976) 1019.
67 V. G. Albano, P. L. Bellon, M. Manassero and M. Sansoni, *J. Chem. Soc., Chem. Commun.*, (1970) 1210.

68 M. McPartlin, R. Mason and L. Malatesta, *J. Chem. Soc., Chem. Commun.*, (1969) 334.
69 R. A. Penneman and R. R. Ryan, *Acta Crystallogr.*, B28 (1972) 1629.
70 W. P. Fehlhammer and L. F. Dahl, *J. Am. Chem. Soc.*, 94 (1972) 3370.
71 S. Komiya, J. C. Huffman and J. K. Kochi, *Inorg. Chem.*, 16 (1977) 1253.

SUBJECT INDEX